STUDENT'S STUD

ALGEBRA FOR COLLEGE STUDENTS

SECOND EDITION

LIAL • MILLER • HORNSBY

Prepared with the assistance of

JANE BRANDSMA
Suffolk County Community College

HarperCollinsPublishers

Cover photo: Comstock/Georg Gerster

"Thy belly is like a heap of wheat set about with lilies" – Solomon's song of his beloved might also have celebrated the sensual curves of the grain–giving earth in California, where wheat and barley are grown on the crowns of rolling hills and in long narrow valleys at the northern approach to the Carrizo Plain. In May the pale gold winter wheat has just been harvested; these fields will be fallowed for eighteen months before being resown.

Student's Study Guide for ALGEBRA FOR COLLEGE STUDENTS, second edition.

Copyright © 1992 by HarperCollins Publishers, Inc.

ISBN 0–673–46471–7

91 92 93 94 9 8 7 6 5 4 3 2 1

PREFACE

This book is designed to be used along with <u>Algebra</u> <u>for</u> <u>College</u> <u>Students</u>, second edition, by Margaret L. Lial, Charles D. Miller, and E. John Hornsby, Jr.

In this book each of the objectives in the textbook is illustrated by a solved example or explained using a modified form of programmed instruction. This book should be used in addition to the instruction provided by your instructor. If you are having difficulty with an objective, find the objective in the <u>Student's</u> <u>Study</u> <u>Guide</u> and carefully read and complete the appropriate section.

Also, in this book you will find a short list of suggestions that may help you to become more successful in your study of mathematics. A careful reading should prove to be a valuable experience.

The following people have made valuable contributions to the production of this <u>Student's</u> <u>Study</u> <u>Guide</u>: Jane Haasch and Marjorie Seachrist, editors; Judy Martinez and Sheri Minkner, typists; Therese Brown and Charles Sullivan, artists; Carmen Eldersveld, proofreader.

We also want to thank Tommy Thompson of Seminole Community College for his suggestions for the essay, "To the Student: Success in Algebra" that follows this preface.

TO THE STUDENT: SUCCESS IN ALGEBRA

The main reason students have difficulty with mathematics is that they don't know how to study it. Studying mathematics *is* different from studying subjects like English or history. The key to success is regular practice.

This should not be surprising. After all, can you learn to play the piano or to ski well without a lot of regular practice? The same thing is true for learning mathematics. Working problems nearly every day is the key to becoming successful. Here is a list of things you can do to help you succeed in studying algebra.

1. *Attend class regularly.* Pay attention in class to what your instructor says and does, and make careful notes. In particular, note the problems the instructor works on the board and copy the complete solutions. Keep these notes separate from your homework to avoid confusion when you read them over later.

2. Don't hesitate to ask questions in class. It is not a sign of weakness, but of strength. There are always other students with the same question who are too shy to ask.

3. *Read your text carefully.* Many students read only enough to get by, usually only the examples. Reading the complete section will help you to be successful with the homework problems. Most exercises are keyed to specific examples or objectives that will explain the procedures for working them.

4. Before you start on your homework assignment, rework the problems the instructor worked in class. This will reinforce what you have learned. Many students say, "I understand it perfectly when you do it, but I get stuck when I try to work the problem myself."

5. Do your homework assignment only *after* reading the text and reviewing your notes from class. Check your work with the answers in the back of the book. If you get a problem wrong and are unable to see why, mark that problem and ask your instructor about it. Then practice working additional problems of the same type to reinforce what you have learned.

6. Work as neatly as you can. Write your symbols clearly, and make sure the problems are clearly separated from each other. Working neatly will help you to think clearly and also make it easier to review the homework before a test.

7. After you have completed a homework assignment, look over the text again. Try to decide what the main ideas are in the lesson. Often they are clearly highlighted or boxed in the text.

8. Use the chapter test at the end of each chapter as a practice test. Work through the problems under test conditions, without referring to the text or the answers until you are finished. You may want to time yourself to see how long it takes you. When you have finished, check your answers against those in the back of the book and study those problems that you missed. Answers are referenced to the appropriate sections of the text.

9. Keep any quizzes and tests that are returned to you and use them when you study for future tests and the final exam. These quizzes and tests indicate what your instructor considers most important. Be sure to correct any problems on these tests that you missed, so you will have the corrected work to study.

10. Don't worry if you do not understand a new topic right away. As you read more about it and work through the problems, you will gain understanding. Each time you look back at a topic you will understand it a little better. No one understands each topic completely right from the start.

CONTENTS

13 FURTHER TOPICS IN ALGEBRA

CHAPTER 1   THE REAL NUMBERS

1.1 Basic Terms

[1] Write sets. (Frames 1-9)

[2] Decide if one set is a subset of another. (Frames 10-14)

[3] Use number lines. (Frames 15-26)

[4] Find additive inverses. (Frames 27-32)

[5] Use absolute value. (Frames 33-44)

[6] Know the common sets of numbers. (Frames 45-56)

---

1. A set is a collection of objects or _____.   elements

2. In a set of chairs, the elements are _____.   chairs

3. The set containing the elements a, b, and c is written as _____.   {a, b, c}

4. The set {9, 11, 12, 13, 10} has a countable or _____ number of elements.   finite

5. Three dots are used following a list of elements to show that the list continues in the same pattern. The set {3, 4, 5, 6, ...} has an uncountable or _____ number of elements.   infinite

6. A set containing no elements is called the _____ set or _____ set, written _____ or _____.   empty null; ∅; { }

7. The set of all numbers in the alphabet is _____.   ∅ (or { })

8. Letters are sometimes used to represent numbers. A letter used to represent an element from a set of numbers is called a _____.   variable

9. Variables can be used to help define sets of numbers. For example,

    $\{x | x \text{ is a number between 0 and 1}\}$

    represents the _____ of all _____ between   set; numbers
    0 and 1.

10. Set A is a _____ of set B, or A _____ B, if   subset; $\subseteq$
    every element of A is an element of B.

Answer <u>true</u> or <u>false</u> for Frames 11–14.

11. $\{3, 5, 9\} \subseteq \{1, 3, 5, \ldots\}$ _____   True

12. $\{1/2, 3/2\} \not\subseteq \{x | x \text{ is a number between 0 and 2}\}$

                                        _____   False

13. Every set is a subset of itself.   True

14. The empty set is a subset of every set.   True

15. In studying sets of numbers, it is convenient if
    we can visualize them. This is done by using a
    _____ line.   number

16. Each number can be associated with one _____   point
    on the number line.

17. To draw a number line, choose any point on the
    line and call it ___.   0

18. Then choose a point to the right of 0 and call
    it ___. This gives a scale that can be used   1
    to identify points representing other numbers.

19. The numbers to the right of 0 are the _____ numbers, while the numbers to the left of 0 are the _____ numbers.

positive

negative

20. A number is called the _____ of the point on the line to which it corresponds, while the point is called the _____ of the number.

coordinate

graph

21. To graph the set {-3, -2, 0, 1}, place a _____ on the number line corresponding to each _____ of the set.

point

element
(or number)

22. This gives the following graph.

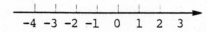

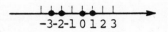

Graph the sets in Frames 23—26.

23. $\left\{6, 7\frac{1}{2}, 9, 10\frac{1}{2}\right\}$

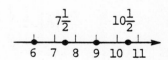

24. $\left\{\frac{3}{4}, \frac{3}{2}, 3, 5\right\}$

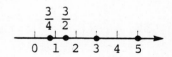

25. $\left\{-4, -3\frac{1}{2}, -\frac{5}{4}, -2\frac{3}{4}\right\}$

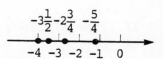

26. {0, 1, -3, -4, 2, -1}

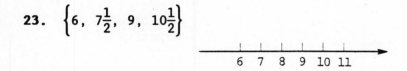

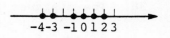

27. Pairs of numbers which lie the same distance from ___ on the number line but on opposite sides of zero are called additive _____.

0

inverses

28. The numbers 7 and ____ are additive inverses
    of each other, as are −9 and ___, and −(−6)
    and ____.

    −7
    9
    −6

29. Find the additive inverse of each number.

    −9 ____   12 ____ $\frac{1}{4}$ ____

    9; −12; −$\frac{1}{4}$

30. The number ____ is its own additive inverse.

    0

31. The additive inverse of a positive number is
    _____ and the additive inverse of a neg-
    ative number is _____.

    negative
    positive

32. If a represents a number, then ____ represents
    the additive inverse of a.

    −a

33. In general, the absolute value of a is defined as

    $|a| = \begin{cases} a \text{ if } a \text{ is } \text{_____ or \_\_} \\ -a \text{ if } a \text{ is } \text{_____}. \end{cases}$

    positive; 0
    negative

34. $|-3| = $ ___, $|18| = $ ___, and $|0| = $ ___.

    3; 18; 0

35. $|-7| = $ ___; therefore, $-|-7| = -(\text{___}) = $ ___.

    7; 7; −7

36. $-|-12| = $ ____ and $-|15| = $ ____.

    −12; −15

37. The distance on the number line from 0 to a
    is the _____ value of a, and is written
    as _____.

    absolute
    $|a|$

38. The distance from 0 to 3 is _____, and thus
    is ___.

    $|3|$
    3

Give the additive inverse and then the absolute value for the numbers given in Frames 39—44.

	Number	Additive inverse	Absolute value	
39.	6	_____	_____	−6; 6
40.	−4	_____	_____	4; 4
41.	−11	_____	_____	11; 11
42.	−(−4)	_____	_____	−4; 4
43.	−|−4|	_____	_____	4; 4
44.	0	_____	_____	0; 0

45. There are certain sets that we encounter frequently when studying sets of numbers. The set of _____ numbers is
    $\{x|x$ is a coordinate of a point on a number line$\}$.

    real

46. The set of _____ numbers or _____ numbers is $\{1, 2, 3, 4, ...\}$.

    natural; counting

47. Including zero in the set of natural numbers gives the set of _____ numbers,
    $\{\_\_, 1, 2, 3, 4, ...\}$.

    whole

    0

48. The set of integers is $\{..., \_\_\_, \_\_\_, \_\_\_, 0, 1, 2, 3, ...\}$.

    −3; −2; −1

49. The set of _____ numbers is $\left\{\frac{p}{q}|p$ and q are _____, $q \neq \_\_\right\}$ or all _____ or _____ decimals.

    rational

    integers; 0;
    terminating;
    repeating

50. The set of _____ numbers is $\{x|x$ is a _____ number that is not rational$\}$ or all _____ or _____ decimals.

    irrational

    real

    non-terminating;
    non-repeating

51. A _____ over the series of numerals that repeat | bar
    is used to indicate a repeating decimal.

Answer _true_ or _false_ for Frames 52–56.

52. -6 is an integer.   _____ | true

53. -6 is a rational number.   _____ | true

54. -6π is a rational number.   _____ | false

55. 4/2 is a natural number.   _____ | true

56. 0 is a rational number.   _____ | true

1.2   Equality and Inequality

1. Use the properties of equality.   (Frames 1–10)

2. Use inequality symbols.   (Frames 11–29)

3. Graph sets of real numbers.   (Frames 30–40)

4. Use the properties of inequality.   (Frames 41–44)

1 Water will boil when heated to 100°C. This fact | property
   is a _____ of water.

2. The symbol =, called the _____ symbol, is | equality
   used between two expressions to show that they
   both name the same number.

**Complete the four properties of equality for any real
numbers a, b, and c, in Frames 3–6.**

3. a = ___.   _____ property | a; reflexive

4. If $a = b$, then ___ = ___. _____ property | b; a; symmetric

5. If $a = b$, and $b =$ ___, | c
    then ___ = ___. _____ property | a; c; transitive

6. If $a = b$, then a may replace ___, or | b
    b _____ a, in any sentence with- | replace
    out changing the truth or falsity
    of the sentence. _____ property | substitution

**Name the property which justifies the statements in
Frames 7–10.**

7. If $3x = 5$, then $5 = 3x$. _____ property | symmetric

8. If $3x = 5$, and $5 = 4z$,
    then $3x = 4z$. _____ property | transitive or substitution

9. $9 = 9$. _____ property | reflexive

10. If $r = s$, and if $5s + 7k + 3z = 12$,
    then $5r + 7k + 3z = 12$. _____ property | substitution

11. A mathematical statement which says that two ex-
    pressions are not equal is called an _____. | inequality

12. To write "3 is less than x," use the _____ | inequality
    symbol and write _____. | $3 < x$

13. To write "is less than or equal to" use the sym-
    bol ___, while ___ is used for "is greater than," | $\leq$; $>$
    and ___ is used for "is greater than or equal to." | $\geq$

14. The symbol $\neq$ means "(*is/is not*) equal to." | is not

Write the statements of Frames 15–19 using inequality symbols.

15. 6 is less than or equal to 12.   _____ $6 \leq 12$

16. 11 is greater than k.   _____ $11 > k$

17. −7 is greater than or equal to r.   _____ $-7 \geq r$

18. 2 is not equal to 3.   _____ $2 \neq 3$

19. 6 is not greater than 7.   _____ $6 \not> 7$

20. To tell which of two numbers is smaller, we can use a _____ line. On the number line, the smaller of two numbers is always to the _____.   number
left

Answer _true_ or _false_ for Frames 21–27.

21. $-5 \leq 5$ _____   true

22. $7 \geq 7$ _____   true

23. $-4 \geq -2$ _____   false

24. $0 < -2$ _____   false

25. $-|-2| \geq 0$ _____   false

26. $6 \leq 2 + 4$ _____   true

27. $0 \leq -|-2|$ _____   false

28. $-|-9| \geq -|-2|$ _____   false

29. We can use inequality symbols and a variable to describe sets of numbers. For example, the set {x|x > -1} includes all numbers _____ than ___ .

greater
-1

30. The graph of this set consists of a _____ at -1 to show that ___ is not part of the graph.

parenthesis
-1

31. Complete the graph with a heavy line to the _____ of -1, as follows.

right

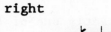

32. The interval of all numbers greater than -1 is written _____ .

(-1, ∞)

33. The graph of {x|x ≤ 6} consists of a _____ at ____, and a heavy line to the _____ of 6, as follows.

square bracket
6; left

34. The interval is written _____ .

(-∞, 6]

Write each set in Frames 35-40 in interval notation and graph the interval.

35. {x|x ≤ -1}   _____

(-∞, -1]

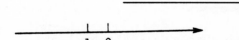

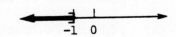

36. {x|x > 0}

(0, ∞)

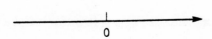

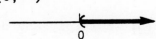

37. {x|2 ≤ x ≤ 5}

[2, 5]

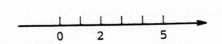

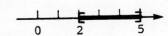

38. {x|-3 < x ≤ 4}

(-3, 4]

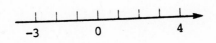

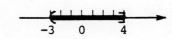

39. $\{x\,|\,-4 < x \le 2\}$

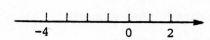

$(-4, 2]$

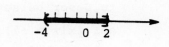

40. $\{x\,|\,0 > x\}$

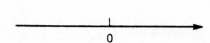

$(-\infty, 0)$

Complete the two properties of inequality in Frames 41–42 for any real numbers a, b, and c.

41. Exactly one of the following is true: $a = b$,
a ___ b, or a ___ b.   _____ property

$>; <;$ trichotomy

42. If $a < b$, and b ___ c, then
_____.   _____ property

$<$

$a < c;$ transitive

Name the property which justifies the expressions in Frames 43–44.

43. If k and r are real numbers, then either $k = r$,
$k < r$, or $k > r$.   _____ property

trichotomy

44. If $k < 6$, and $6 < m$,
then $k < m$.   _____ property

transitive

1.3   Operations on Real Numbers

[1] Add and subtract real numbers. (Frames 1–55)

[2] Find the distance between two points. (Frames 56–59)

[3] Multiply and divide real numbers. (Frames 60–98)

[4] Use exponents and roots. (Frames 99–112)

[5] Learn the order of operations. (Frames 113–128)

[6] Evaluate expressions for given values of variables. (Frames 129–133)

---

1. To add two negative numbers, use the rule that
    says, to add two numbers with the same sign, add
    their _____ values. The sign of the answer   absolute
    is the _____ as the sign of the two numbers.   same

2. Therefore, if two negative numbers are added,
    the result will be a _____ number.   negative

Add the negative numbers in Frames 3–8.

3. $(-2) + (-11) = -($_____$) = $_____   2 + 11; −13

4. $(-1) + (-3) = -($_____$) = $_____   1 + 3; −4

5. $(-12) + (-9) = -($_____$) = $_____   12 + 9; −21

6. $(-8) + (-7) = -($_____$) = $_____   8 + 7; −15

7. $(-15) + (-13) = -($_____$) = $_____   15 + 13; −28

8. $(-5) + (-2) = -($_____$) = $_____   5 + 2; −7

9. To add a positive and a negative number, first
    find the _____ value of each number.   absolute

10. Before adding −13 and 10, find $|-13| = $_____   13
    and $|10| = $____.   10

11. Then subtract the smaller absolute value from
    the _____.   larger

12. This gives 13 − 10 = _____.   3

13. If the number with the larger absolute value is
    positive, leave the answer _____.   positive

14. If the number with the larger _____ value   absolute
    is negative, make the answer _____.   negative

15. Since -13 has a larger absolute value than 10,
    make the answer _____: -13 + 10 = ____.   negative; -3

Add the positive and negative numbers in Frames 16—20.

16. -7 + 11 = ____   4

17. 7 + (-11) = ____   -4

18. -11 + 7 = ____   -4

19. 2 + (-14) = ____   -12

20. 9 + (-6) = ____   3

21. To add a series of numbers, start at the ____.   left

22. 4 + 7 + 2 + (-5) = _____ + 2 + (-5)   11
            = _____ + (-5) = _____   13; 8

Add the series of numbers in Frames 23—27.

23. 3 + (-5) + (-2) = ____   -4

24. -2 + (-4) + 7 = ____   1

25. 5 + (-8) + (-7) + 10 = ____   0

26. -8 + (-7) + 2 = ____   -13

27. 7 + (-9) + (-4) + 3 = ____   -3

28. To subtract two numbers, change the sign of the second number and _____ it to the first.   In symbols, a – b = a + (____).

add

–b

29. To subtract 11 from 6, written 6 – _____, change the sign of _____ and add.

11

11

30. 6 – 11 = 6 + (____) = _____

–11; –5

31. To subtract –5 from 7, written 7 – (____), change the sign of ____ and _____.

–5

–5; add

32. 7 – (–5) = 7 + ____ = ____

5; 12

Perform the indicated subtraction in Frames 33–45.

33. –4 – (–7) = –4 + _____ = _____

7; 3

34. –4 – 7 = –4 + _____ = _____

(–7); –11

35. –9 – (–5) = –9 + _____ = _____

5; –4

36. –9 – 5 = –9 + _____ = _____

(–5); –14

37. 9 – 5 = 9 + _____ = _____

(–5); 4

38. 9 – (–5) = 9 + _____ = _____

5; 14

39. –6 – (–4) = _____

–2

40. –8 – (–3) = _____

–5

41. –3 – (–15) = _____

12

42. 4 – (–5) = _____

9

43. 7 - (-12) = ____   19

44. 2 - 5 = ____   -3

45. 7 - 13 = ____   -6

46. When working a problem involving a series of
additions and subtractions, perform the addi-
tions and subtractions in order from the ____.   left

47. 9 + (-4) + (-2) - (-5) = ____ + (-2) - (-5)   5
                                = ____ - (-5) = ____   3; 8

48. -4 + (-6) - (-4) + (-7) = ____ - (-4) + (-7)   -10
                                = ____ + (-7) = ____   -6; -13

**Perform the additions and subtractions indicated in
Frames 49–55.**

49. -11 - (-8) + (-6) = ____   -9

50. -8 + 15 - (-6) = ____   13

51. -8 - |-12| + |-7| = ____   -13

52. -9 - |-6| + 5 = ____   -10

53. -2 - (-7) + |-3| = ____   8

54. |-8| - |-9| + |-3| + |4| = ____   6

55. 4 - (-2) + |-7| - 6 = ____   7

56. The distance between two points on a number line
is given by the _____ value of the   absolute
_____ of the numbers.   difference

Find the distance between the pairs of numbers in Frames 57–59.

57. 11, −4

   distance − |11 − (___)| − |___| − ____   −4; 15; 15

58. −6, −9

   distance − |−6 − (___)| − ____   −9; 3

59. −8, 10   ____   18

60. If two positive numbers are multiplied, the product will be a _____ number.   positive

61. In symbols, write this as "if a > 0 and b > 0, then ab ___ 0."   >

62. To multiply a positive and a negative number, use the rule that says the product of two numbers with different signs is _____.   negative

63. Therefore, the product of a positive number and a negative number is a _____ number.   negative

Multiply the positive and negative numbers in Frames 64–69.

64. (−9)7 − −(9 · 7) − ____   −63

65. 9(−7) − −(_____) − ____   9 · 7; −63

66. 8(−5) − −(_____) − ____   8 · 5; −40

67. −2(4) − ____   −8

68. $(4)\left(-\frac{1}{4}\right)$ − ____   −1

69. $-12\left(\frac{1}{2}\right) = $ _____ | -6

70. To multiply two negative numbers, use the rule that says the product of two numbers with the same sign is _____. | positive

71. Therefore, the product of two negative numbers is a _____ number. | positive

Multiply the negative numbers in Frames 72–76.

72. $(-6)(-7) = (6)(___) = $ _____ | 7; 42

73. $(-3)(-8) = (___)(8) = $ _____ | 3; 24

74. $(-8)\left(-\frac{1}{4}\right) = $ _____ | 2

75. $(-1)(-3) = $ _____ | 3

76. $(-9)(-10) = $ _____ | 90

77. To multiply a series of numbers, start at the _____. | left

78. $(-2)(-3)(-4)(2) = ___\ (-4)(2) = ____\ (2) = $ _____ | 6; -24; -48

Multiply the series of numbers in Frames 79–84.

79. $(-8)(3)(-1) = $ _____ | 24

80. $6(-7)(2)(-1) = $ _____ | 84

81. $(3)(-4)\left(\frac{1}{2}\right) = $ _____ | -6

82. $(-7)(-2)(-3) = $ _____ -42

83. $(-7)(-2)(-3)(-1) = $ _____ 42

84. $(-1)(-7)(-2)(-3)(-1) = $ _____ -42

85. To divide a by b (b $\neq$ 0), find the real number
q such that a = _____. bq

86. In symbols, _____ only if a = bq. $\frac{a}{b} = q$

87. Suppose for some nonzero real number a, $\frac{a}{0} = q$.
Then by the definition, a = _____. But, for $0 \cdot q$
every number q, $0 \cdot q = $ _____. So, no single 0
quotient q is _____. On the other hand,   possible
if a = 0, by the definition $\frac{0}{0} = q$ or 0 = _____. $0 \cdot q$
Again, this is true for all q, so there is no
_____ quotient $\frac{0}{0}$. Therefore, division by 0   single
is _____.   undefined

88. To divide 6 by -2, written $\frac{6}{-2}$, find the number
such that the _____ of it and -2 is 6.   product

89. $6 = -2 \cdot ($_____$)$ -3

90. The quotient $\frac{6}{-2}$ is ____. -3

Divide the numbers in Frames 91-95.

91. $\frac{-15}{-3} = $ _____ since $-15 = -3 \cdot ($____$)$ $5; \ 5$

92. $\frac{15}{-3} = $ _____ since $15 = -3 \cdot ($____$)$ $-5; \ -5$

93. $\dfrac{-15}{3}$ = _____ since $-15 = 3 \cdot$ (____) | $-5; -5$

94. $\dfrac{-18}{2}$ = _____ | -9

95. $\dfrac{-12}{4}$ = _____ | -3

96. From the definition of division, it can be proved that to divide a number a by a number b, b ≠ 0, multiply a by the reciprocal of b. In symbols,

$$\frac{a}{b} = a \cdot \text{\_\_\_\_\_}, \quad b \neq 0.$$

$\dfrac{1}{b}$

Use this rule to divide fractions.

97. The reciprocal of $-\dfrac{5}{8}$ is _____. | $-\dfrac{8}{5}$

98. $-\dfrac{1}{2} \div \left(-\dfrac{5}{8}\right) = -\dfrac{1}{2} \cdot$ _____ = _____ | $-\dfrac{8}{5}; \dfrac{4}{5}$

99. Exponents are used to write repeated _____. | products
For example, 3^4 means

_____, | $3 \cdot 3 \cdot 3 \cdot 3$

which equals _____. | 81

100. In 3^4, the base is _____, and the exponent | 3
is _____. | 4

Evaluate each expression in Frames 101–107.

101. $2^5 =$ _____   32

102. $(-5)^3 =$ _____   −125

103. $\left(-\dfrac{3}{4}\right)^2 =$ _____ $\dfrac{9}{16}$

104. $-4^3 =$ _____   −64

105. Since $7^2 = 49$, the number 7 is a _____   square root
      of 49.   Another square root of 49 is _____.   −7

106. The positive square root of a number is written
      with the symbol _____. $\sqrt{}$

107. The cube root of 8 is written _____. $\sqrt[3]{8}$
      Find the cube root of 8. _____   2

Find each root in Frames 108–112.

108. $\sqrt{100} =$ _____   10

109. $\sqrt{169} =$ _____   13

110. $\sqrt{\dfrac{121}{144}} =$ _____ $\dfrac{11}{12}$

111. $\sqrt[3]{216} =$ _____   6

112. $\sqrt[4]{256} =$ _____   4

When more than one operation is included in a problem, be careful in deciding which operation to perform
first. Complete the order of operations in Frames
113-118.

113.   If parentheses or square _____ are present,   brackets

    (a)   work separately above and below any

          _____ bars.   fraction

    (b)   Use the rules below within each set of

          _____ or square brackets.   Start   parentheses

          with the _____ and work outward.   innermost

114.   If no parentheses are present,

    (a)   simplify all _____ and _____,   powers; roots

          working from _____ to right.   left

    (b)   Do any _____ or divisions, in   multiplications

          the _____ in which they occur, working   order

          from _____ to right.   left

    (c)   Do any _____ or subtractions in   additions

          the order in which they occur, working

          from left to _____.   right

115.   To evaluate $(3 \cdot 5) + 6 \cdot 3$, first find _____ = $(3 \cdot 5)$

    15, since this product is inside the _____.   parentheses

116.   Then multiply _____ = 18. $6 \cdot 3$

117.   Now $(3 \cdot 5) + 6 \cdot 3$ has been simplified to

    _____ + _____.   15; 18

118.   Finally, add these two numbers:

    _____ + _____ = _____.   15; 18; 33

119.   To evaluate $5 \cdot 2 + 3 - 7 + 2[5 + (-8)]$, first

    find _____ = _____. $5 + (-8)$; -3

120. Then find _____ = 10.

$5 \cdot 2$

121. This gives _____ + 3 - 7 + 2 · (____).

10; -3

122. Now evaluate 2 · _____ - _____.

(-3); -6

123. Finally, add and subtract starting at the left:
$10 + 3 - 7 + (-6) = $ _____.

0

Simplify (solve) the problems in Frames 124-128.

124. $-6 + (-2)(-3) + 2[4 - (-2)] = $ _____

12

125. $\dfrac{3(-5 - 3)}{4} = $ _____

-6

126. $\dfrac{-2[3 - (-4)]}{2} = $ _____

-7

127. $\dfrac{3(-4) - (-2)(-3)}{-3} = $ _____

6

128. $10\sqrt{36} - 2^2 \cdot 5 = 10 \cdot $ _____ - _____ · 5

= _____ - _____

= _____

6; 4
60; 20
40

129. Sometimes an expression must be evaluated for
given values of the variables. For example,
suppose r = -2 and s = 3. Then

$-r - s = -($ _____ $) - ($ ____ $)$

= _____ .

-2; 3
-1

130. Evaluate m - n + q if m = 4, n = -9, and q = -4.

m - n + q = ____ - (____) + (____)

= _____

4; -9; -4
9

131. Evaluate $y + z - x$ if $y = -11$, $x = -8$, and $z = -6$. _____ | −9

132. Evaluate the expression

$$-2q^2 - 4q$$

when $q = -5$.

Replace q with _____. | −5

$$-2q^2 - 4q = -2(\underline{\quad})^2 - 4(\underline{\quad})$$ | −5; −5

$$= -2(\underline{\quad}) + \underline{\quad}$$ | 25; 20

$$= \underline{\quad} + 20$$ | −50

$$= \underline{\quad}$$ | −30

133. Evaluate $-y^2 + 8y - 1$ when $y = -3$.

_____ | −34

1.4 Properties of the Real Numbers

☐1 Use the distributive property. (Frames 1–4)

☐2 Use the inverse properties. (Frames 5–9)

☐3 Use the identity properties. (Frames 10–13)

☐4 Use the commutative and associative properties. (Frames 14–21)

☐5 Use the multiplication property of zero. (Frames 22–23)

☐6 Use the addition and multiplication properties of equality. (Frames 24–27)

1. For any real numbers a, b, and c, $a(b + c) = ab +$ _____. This is called the _____ property. | ac; distributive

Use the distributive property to rewrite each expression in Frames 2–4.

2. $-11(a + b) = \underline{\quad} a + \underline{\quad} b = \underline{\quad} + \underline{\quad} = \underline{\quad}$ | −11; −11; −11a; −11b; −11a − 11b

3. q(p + r) = _____ + _____

    qp + qr

4. 7c + 2c = (_____)c = _____

    7 + 2; 9c

5. For each real number a, there is a unique real
    number ____ such that a + (−a) = ____ and

    −a; 0

    −a + a = ___.

    0

    For each non−zero real number a, there is a

    unique real number ____ such that a · $\left(\frac{1}{a}\right)$ = ____

    $\frac{1}{a}$; 1

    and $\left(\frac{1}{a}\right)$ · a = ____.

    1

    These are called the _____ properties.

    inverse

Use the inverse properties to complete Frames 6−9.

6. 2 + ____ = 0

    −2

7. $\left(-\frac{1}{3}\right)$ · ____ = 1

    −3

8. −10 + ____ = 0

    10

9. $\left(\frac{4}{7}\right)$ ____ = 1

    $\frac{7}{4}$

10. There is a unique real number ____, called the

    0

    additive _____, such that a + 0 = a and

    identity

    0 + a = a.

    There is a unique real number ___, called the

    1

    multiplicative _____, such that a · 1 = a

    identity

    and 1 · a = a.

    These are called the _____ properties.

    identity

Use the identity properties to complete Frames 11−13.

11. 0 + ___ = 7

    7

12. 3 · ___ = 3

    1

13. $-8 +$ ___ $= -8$ 0

14. For any real numbers a and b, a + b = b + a and
     ab = ____. These are called the _____   ba; commutative
     properties of addition and multiplication. They
     are used to change the _____ of the terms in   order
     an expression.

15. For any real numbers a, b and c, (a + b) + c =
     a + (b + c) and a(bc) = _____. These are   (ab)c
     called the _____ properties of addition   associative
     and multiplication. They are used to _____   regroup
     the terms of an expression.

**Use the commutative and associative properties to
complete Frames 16–19.**

16. $7 + (-3) = -3 +$ ___   7

17. $\left(\frac{1}{3} + \frac{2}{5}\right) + \frac{3}{4} = \frac{1}{3} +$ (___ $+$ ___) $\frac{2}{5}$; $\frac{3}{4}$

18. $9 \cdot b = b \cdot$ ___   9

19. $q(rt) =$ (___ $)t$ qr

**Use the commutative and associative properties to
simplify the expressions in Frames 20–21.**

20. $9b + 2 - 6b + 5b - 7 =$ _____   8b – 5

21. $(8r)(-2s)(3t) =$ _____   –48rst

Complete the following in Frames 22–27.

22. If a is any real number, then $a \cdot 0 =$ ___. This   0
     is called the _____ property of zero.   multiplication

23. Using the _____ property of zero,
     9 • 0 = ___ , 0 • 11 = ___ , and (6 + 4 + 3) • 0 =
     ___ .

multiplication

0; 0

0

24. If a, b, and c are real numbers and if a = b,
     then a + c = b + ___ . This is called the
     _____ property of equality.

c

addition

25. If a, b, and c are real numbers and if a = b,
     then ac = ____ . This is called the _____
     property of _____ .

bc; multiplication

equality

26. Using the addition property of equality if a = b,
     then a + 7 = _____ ; and if r + 2 = 11, then
     r + 2 + (−2) = 11 + ____ .

b + 7

(−2)

27. Using the multiplication property of equality,
     if k = s, then 7k = ____ ; and if 3m = 9, then
     (1/3)(3m) = (1/3)(___) .

7s

9

**Name the property which justifies the expression in
Frames 28–37.**

28. 11 + (−11) = 0   _____ property

inverse

29. pq = qp   _____ property

commutative

30. m + (n + 3) = (m + n) + 3 _____ property

associative

31. 3p + (−3p) = 0   _____ property

inverse

32. 7 + 4 = 4 + 7   _____ property

commutative

33. (−9) • 1 = 1 • (−9)   _____ property

commutative

34. (5 • 0)7 = 5(0 • 7)   _____ property

associative

35. $x(2 + 4) = x \cdot 2 + x \cdot 4$ _____ property | distributive

36. $12 + 0 = 12$ _____ property | identity

37. $5(m + k) = 5m + 5k$ _____ property | distributive

Chapter 1 Test

The answers for these questions are at the back of this Study Guide.

Evaluate each of the following.

1. $-4 + 3 + 2$

1. _____

2. $6 - (-5)$

2. _____

3. $(-2) + (-5) - (-3)$

3. _____

4. $(-4)(-2)$

4. _____

5. $\dfrac{-8 - \sqrt[3]{64}}{-2}$

5. _____

6. $\dfrac{2^3 - 2[3 - (-7)]}{-4}$

6. _____

7. $|-2| - |-2| + 2$

7. _____

8. $\dfrac{7(3 - 2)6}{-14}$

8. _____

9. $\dfrac{-4[-3 - (-7 + 3^2)]}{\sqrt{144}}$

9. _____

10. $\dfrac{-7 - [-8 - (5 - 6 + 2)]}{3}$

10. _____

Let $A = \{-11, .\overline{7}, -9, \pi, 4, -\sqrt{5}, 3/8, .121122..., -2/3, 1.4, 0/5\}$. List the elements of set A that are:

11. counting numbers

11. _____

12. integers

12. _____

13. rational numbers

13. _____

14. irrational numbers   14. _____

15. real numbers   15. _____

Decide whether each statement is true or false.

16. $\{4, 5, 7] \subseteq \{2, 3, 4, 5, 6, 7, 8\}$ 16. _____

17. $\{0, 2, 4, 6\} \subseteq \{x|x \text{ is a natural number}\}$ 17. _____

18. $\{ \ \} \subseteq \{-3, 0, 3, 6, 9\}$ 18. _____

19. $\{-10, 6, 2, -5\} \subseteq \{-10, -5, 2, 6\}$ 19. _____

Evaluate each expression when $x = -2$, $y = 3$ and $z = -1$.

20. $3x + 2y$ 20. _____

21. $-2x^2 + z$ 21. _____

22. $\dfrac{y - z}{2xy}$ 22. _____

Name the property used in each of the following statements.

23. $8 + (-4) = (-4) + 8$ 23. _____

24. $(3 + 4) + 7 = 3 + (4 + 7)$ 24. _____

25. $-7 + 7 = 0$ 25. _____

26. $-4 + 0 = -4$ 26. _____

27. $\dfrac{2}{3} + \left(-\dfrac{2}{3}\right) = 0$ 27. _____

28. $a(3 + 2) = a \cdot 3 + a \cdot 2$ 28. _____

29. If x = 5, then 5 = x.

29. _____

30. If x = 6, and 6 = k, then x = k.

30. _____

31. If a = 9, and a + b + c = 11, then
     9 + b + c = 11.

31. _____

32. 3 · k = k · 3

32. _____

33. 27c · 0 = 0

33. _____

34. If 3 + k = 2, then (3 + k) + (−3) = 2 + (−3).

34. _____

35. Either a = b, a < b, or a > b.

35. _____

36. If 6 < m, and m < a, then 6 < a.

36. _____

37. 7 = 7

37. _____

Write each set in Frames 31–33 in interval notation and graph the interval.

38. $\{x \mid x \geq -7\}$ _____

38. ⟶

39. $\{x \mid x < -1\}$ _____

39. ⟶

40. $\{x \mid 0 < x \leq 3\}$ _____

40. ⟶

CHAPTER 2   LINEAR EQUATIONS AND INEQUALITIES

2.1   Linear Equations in One Variable

[1]   Define linear equations.   (Frames 1–6)

[2]   Solve linear equations using the addition and multiplication properties of equality.   (Frames 7–25)

[3]   Solve linear equations using the distributive property.   (Frames 26–47)

[4]   Solve linear equations with fractions and decimals.   (Frames 48–70)

[5]   Identify conditional equations, contradictions, and identities. (Frames 71–76)

---

1. A mathematical statement which says that two expressions are equal is called an _____.	equation
2. $x + 5 = 9$ is an example of an equation. If $x = 4$, the equation is [(*true/false*)]	true
3. Therefore, 4 is called the _____ of the equation.	solution
4. An equation of the form $ax + b = c$ is a _____ or a _____ degree equation.	linear first
5. $4 + 2y = 8$ (*is/is not*) a linear equation.	is
6. To solve a linear equation, use the _____ of real numbers to simplify each expression.	properties
7. An important property which will help solve equations is the addition property of _____.	equality
8. This property says that for any expressions a, b, and c, the equations $a = b$ and $a + \_\_\_ = b + \_\_\_\_\_$ have the same _____.	c c; solution

9. Since the solution of x − 1 = 3 is 4, then the
    _____ of (x − 1) + 1 = 3 + ___ is also 4.   solution; 1

10. Using the addition property of equality, we can
    simplify x + 5 = 13 by subtracting ___ from   5
    both sides of the equation.

11. This gives x + 5 _____ = 13 _____.   − 5; − 5

12. Rewrite x + 5 − 5 as x + ____, or just ____.   0; x

13. Finally, x + 5 − 5 = 13 − 5 becomes x = ____.   8

14. The solution to the equation x = 8 is ____.   8
    Therefore, the solution to x + 5 = 13 is also
    _____.   8

15. The set of all solutions of an equation is called
    the _____ set for the equation.   solution

16. Therefore, the solution set for x + 5 = 13 is
    ____.   {8}

17. Another property which will help solve equations
    is the multiplication property of _____.   equality

18. This property says that for any expressions a,
    b, and c (with c ≠ 0), the equations a = b and
    _____ have the same _____.   ac = bc; solution

19. To solve 8x = 16, _____ both sides of the   divide
    equation by ____.   8

20. This gives

$$\frac{8x}{—} = \frac{16}{—}$$ 8; 8

    or   x = ____ .   2

    The solution set is _____ .   {2}

21. Using the _____ property of equality,   multiplication

    $\frac{3}{2}x = 9$ can be simplified by _____ both   multiplying

    sides of the equation by the _____ of $\frac{3}{2}$, reciprocal

    which is ___ . $\frac{2}{3}$

22. This gives $\left(\frac{2}{3}\right)\left(\frac{3}{2}x\right) = ($____$)9$. $\frac{2}{3}$

23. Rewrite $\left(\frac{2}{3}\right)\left(\frac{3}{2}x\right)$ as $\left(\frac{2}{3}\right)\left(\frac{3}{2}\right)x$, which equals 1x or

    ___ .   x

24. Therefore, $x = \left(\frac{2}{3}\right)9$, or $x = $ ___ .   6

25. The solution set of $\frac{3}{2}x = 9$ is _____ .   {6}

26. To simplify an equation involving parentheses,

    use the _____ property.   distributive

27. Simplify 3x + 5x as follows:

    3x + 5x = (___ + ___)x = ____ by the   3; 5; 8x

    _____ property. This process is called   distrubutive

    _____ like terms.   combining

28. Also, 5x − 2x + 8x − 14 = (__ − __ + __)x − 14   5; 2; 8

    = _____ − 14 by the _____ property.   11x; distributive

Use the distributive property and combine like terms
to simplify Frames 29-30.

29. $3(2x + 1) - 4x + 7 + x =$ _____ $3x + 10$

30. $4 - 2x + 4x - 8 + 4x - 2 =$ _____ $6x - 6$

**Complete the following guidelines in Frames 31-37 for
solving a linear equation in one variable.**

31. Eliminate _____ by multiplying both sides   fractions
     of the equation by the least common denominator.

32. Use the distributive property to clear _____.   parentheses

33. Simplify each side of the equation by _____   combining
     like terms.

34. Use the _____ property of equality to   addition
     simplify further.

35. Finally, use the _____ property of equal-   multiplication
     ity to write the equation in the form $x = k$.

36. The solution of the equation is then ___, and the   k
     the solution set is ____.   {k}

37. _____ the solution by substituting back into   Check
     the original equation.

38. To solve $6 - 3x + 5x = 9 + 4x + 3$, combine like
     terms on each side of the equation to get
     ___ + ___ = ___ + ___. The next step is to   6; 2x; 12; 4x
     use the _____ property of equality.   addition

39. Now we can subtract either 6 or 12. We choose
     to subtract 6 from both sides of the equation.
     This gives us ___ x = ___ + ___ x.   2; 6; 4

40. Again use the _____ property of equality and
    subtract 4x from both sides to get ___ x = ___.

    addition
    −2; 6

41. Finally, use the multiplication property of equal-
    ity and _____ both sides by ____ to get
    x = ____.

    divide; −2
    −3

42. The solution set of the original equation is
    _____.

    {−3}

43. To solve 3(2x − 4) = 20 − 2x, first use the
    _____ property to write _____ =
    _____.

    distributive;
    6x − 12
    20 − 2x

44. Now add 12 to both sides to get _____.

    6x = 32 − 2x

45. Next, add ____ to both sides to get _____.

    2x; 8x = 32

46. By dividing both sides by _____, we get _____.

    8; x = 4

47. Therefore, the solution set of the original equa-
    tion is _____.

    {4}

48. When solving a linear equation with fractions,
    eliminate the fractions by _____ both
    sides of the equation by the _____.

    multiplying
    least common
    denominator

49. In the equation $\frac{x}{3} + \frac{2x}{5} = \frac{11}{3}$ the least common
    denominator is ___.

    15

50. To solve $\frac{3}{4}(x - 2) = \frac{x}{3} - \frac{1}{4}$ multiply both sides by
    ____ to get ____ (x − 2) = _____.

    12; 9; 4x − 3

51. Then use the _____ property to clear the parentheses. This gives the equivalent equation _____.

    distributive

    $9x - 18 = 4x - 3$

52. Now add 18 to both sides to get _____.

    $9x = 4x + 15$

53. Next, subtract ___ from both sides to get _____.

    $4x$; $5x = 15$

54. Finally, divide both sides by ___ to get _____.

    5; $x = 3$

55. Thus, the solution set of the original equation is _____.

    $\{3\}$

56. When solving an equation with decimal coefficients, _____ both sides by the largest power of ____ necessary to obtain _____ coefficients.

    multiply

    10; integer

57. To solve $.006(x + 2) = .007x + .009$, first multiply both sides by _____ to clear the decimals.

    1000

58. The resulting equation is _____.

    $6(x + 2) = 7x + 9$

59. Now use the distributive property to clear the _____. The equation then becomes _____.

    parentheses

    $6x + 12 = 7x + 9$

60. Subtract _____ from both sides to get $12 = x + 9$.

    $6x$

61. Next, subtract ___ from both sides to get $3 = x$.

    9

62. Thus, the solution to the original equation is ___.

    $\{3\}$

Find the solution set for the equations in Frames 63–70.

63. $2x + 1 = 7$

    _____

    $\{3\}$

64. $-3x + 2 = 4x$ _____ $\{-2\}$

65. $-2x + 1 = 11$ _____ $\{-5\}$

66. $3x - 2x + 2 = 8$ _____ $\{6\}$

67. $3(2x - 1) = 2(4x - 3) + 1$ _____ $\{1\}$

68. $2x - 4(2x - 1) = -2(2x + 1)$ _____ $\{3\}$

69. $\frac{4x}{3} - \frac{1}{5}(12 - x) = \frac{x}{3}$ _____ $\{2\}$

70. $.05(2 - p) = .03p + .18$ _____ $\{-1\}$

71. A linear equation with a finite number of elements
    in its solution set is called a _____   conditional
    equation.

72. A linear equation that is true for every value of
    the variable is an _____, while an equation   identity
    with no solution is a _____.   contradiction

Write <u>conditional</u>, <u>identity</u>, or <u>contradiction</u> for each
equation in Frames 73–76.

73. $4(x - 2) + 3x = 7x - 10 + 2$ _____   identity

74. $2(m - 4) + 3m = 4(m - 1) + m$ _____   contradiction

75. $9p - 8 + 11p - 7p = 2p + 4 - 12$ _____   conditional

76. $12(a + 3) - 7a - 3 = -1 + 2[4 - (3 + a)] + 7a + 32$

    _____   identity

2.2 Formulas and Topics from Geometry

[1] Solve a formula for a specified variable. (Frames 1-19)

[2] Solve applied problems using formulas. (Frames 20-26)

[3] Solve problems about angle measure. (Frames 27-32)

All the formulas you need are given at the back of this Study Guide.

---

1. The fact that "distance equals the product of the rate and the time" can be expressed by the formula _____ .	$d = rt$
2. This formula is solved directly for ___ .	d
3. To solve directly for t, we would isolate ___ on one side of the equation.	t
4. This could be done by dividing each side of $d = rt$ by ___ .	r
5. This gives $$\frac{d}{\underline{}} = \frac{rt}{\underline{}}$$ or ____ = t.	$r; r$ $$\frac{d}{r}$$
6. If d = 800 miles, and r = 40 miles per hour, we can find t by using the formula t = ___ from Frame 5. Substitute the given values in the formula. $$t = \frac{d}{r}$$ $$t = \frac{\underline{}}{\underline{}}$$ $$t = \underline{}$$	$$\frac{d}{r}$$ 800 40 20 hours

7. The formula P = a + b + c says that the "perim-
    eter of a triangle equals the sum of the lengths
    of the three _____.   sides

8. To solve directly for b, subtract ___ and ___   a; c
    from both sides of the equation.

9. This gives _____ = b.   P − a − c

10. If P = 18, a = 5, and c = 6, use the formula
    _____ to find b.   b = P − a − c

    b = ____ − ____ − ____   18; 5; 6
    b = ____   7

11. The formula $A = \frac{1}{2}(b + B)h$ gives the area of a
    _____.   trapezoid

12. To solve directly for b, first multiply both
    sides by ___ to get ____ = _____.   2; 2A; (b + B)h

13. Now divide both sides by ____ to get   h

    ____ = _____. $\frac{2A}{h}$; b + B

14. Subtracting B from both sides gives _____ = b. $\frac{2A}{h}$ − B

15. If A = 50, h = 5, and B = 6, we can use the
    formula to find b.

    $b = \dfrac{2 \cdot \rule{1cm}{0.4pt}}{\rule{1cm}{0.4pt}} - \rule{1cm}{0.4pt}$ 50; 6
                                                       5

    b = ____   14

Solve the given formula directly for the given variable in Frames 16–19.

16. I = prt, for p

p = _____

$$\frac{I}{rt}$$

17. S = 2πrh + 2πr², for h

h = _____

$$\frac{S - 2\pi r^2}{2\pi r}$$

18. V = LWH, for W

W = _____

$$\frac{V}{LH}$$

19. A = ½bh, for b

b = _____

$$\frac{2A}{h}$$

Solve the problems in Frames 20–26.

20. Suppose $100 is deposited at simple interest at 5% for 3 years. Find the amount of interest earned by the money.

Here we use the formula _____. In this problem, p = _____, r = ____, and t = ____. Hence

I = (   )(   )(   )
I = _____

I = prt
$100; .05; 3

100; .05; 3
$15

21. How many years will it take for $1000 to earn $240 simple interest at 6% per year?

Again we use the formula for simple interest, _____. Here we know ___, ___, and ___, and we are looking for ___. Solve I = prt for t, to get

t = _____.

Substitute the numbers given in this problem.

$$t = \frac{\rule{2em}{0.4pt}}{\rule{1em}{0.4pt} \cdot \rule{1em}{0.4pt}}$$

t = _____

I = prt; I; p; r
t

$$\frac{I}{pr}$$

240
1000; .06

4 years

22. If the circumference of a circle is 36π inches, find the radius of the circle.

Use the formula _____. We know ___, and we are looking for ___. If we solve C = 2πr for r, we get

r = _____.

In our case, C = _____. Hence

r = _____

or   r = _____.

$C = 2\pi r$; C

r

$\dfrac{C}{2\pi}$

36π

$\dfrac{36\pi}{2\pi}$

18 inches

23. Suppose a right circular cylinder has a base with a radius of 5 inches. If the surface area is 90π, find the height.

Use the formula _____. We know ____ and ____, and we are looking for ____. Solving the formula for h, we get

h = _____.

Here S = _____ and r = ___. Substitute.

$$h = \dfrac{-2\pi(\quad)^2}{2\pi(\underline{\quad})}$$

$$= \dfrac{-}{\underline{\quad}}$$

$$= \dfrac{40\pi}{\underline{\quad}}$$

h = _____.

$S = 2\pi rh + 2\pi r^2$

S; r; h

$\dfrac{S - 2\pi r^2}{2\pi r}$

90π; 5

90π; 5
5

90π; 50π
10π

10π

4 inches

24. A box has a volume of 400 cubic centimeters. If the height is 2 centimeters and the width of the base is 10 centimeters, find the length of the base.

Formula: _____

length of base = _____

$V = LWH$

20 centimeters

25. The area of a trapezoid is 32 square centimeters. If one base is 6 and the height is 4, find the length of the other base.

    Formula: _____

    length of other base = _____

    $A = \frac{1}{2}(b + B)h$

    10 centimeters

26. What principal must be invested at 6% simple interest to earn $240 in 5 years?

    Formula: _____
    principle = _____

    $I = prt$
    $800

27. Using the diagram below in which lines p and q are parallel, name the pairs of angles that satisfy each description.

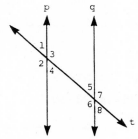

    Interior angles on the same side of the transversal: ____ and ____ (or _____ and _____).

    Alternate interior angles: ___ and ___
    (or ___ and ___)

    Alternate exterior angles: ___ and ___
    (or ___ and ___)

    Vertical angles: ___ and ___, (or ___ and ___, or
    ___ and ___, or ___ and ___)

    3; 5; 4; 6

    3; 6
    4; 5

    2; 7
    1; 8

    1; 4; 2; 3
    5; 8; 6; 7

28. In the diagram given below where lines a and b are parallel, the marked angles are _____ _____ angles.

    alternate
    interior

    $(4x - 3)°$

    $(7 + 3x)°$

29. Thus, their angle measures _____.	are equal
30. To solve for x, use the equation _____ = _____.	$4x - 3 = 7 + 3x$
31. Solving the equation gives x = ____.	10
32. Therefore, both angles measure _____.	37°

2.3 Applications of Linear Equations

1️⃣ Translate from word expressions to mathematical expressions. (Frames 1–17)

2️⃣ Write equations from given information. (Frames 18–20)

3️⃣ Solve problems about unknown numerical quantities. (Frames 21–24)

4️⃣ Solve problems about percents, simple interest and mixture. (Frames 25–27)

5️⃣ Solve problems about angles: supplementary, complementary, and sums of angles in triangles. (Frames 28–39)

Translate the verbal expressions in Frames 1–16 into mathematical expressions. Use x to represent the variable.

1 The sum of a number and 8: _____	$x + 8$
2. The sum of twice a number and −4: _____	$2x + (-4)$ or $2x - 4$
3. 9 more than a number: _____	$x + 9$
4. A number increased by 5: _____	$x + 5$
5. Three times a number, increased by 7: _____	$3x + 7$
6. 5 less than a number: _____	$x - 5$

7. A number decreased by 8: _____ $x - 8$

8. The difference between 5 times a number and 2:

    _____ $5x - 2$

9. 8 times a number: ____ $8x$

10. The product of a number and 3/4: ____ $\frac{3}{4}x$

11. The product of a number, and the number increased
    by 8: _____ $x(x + 8)$

12. 3/4 of a number: ____ $\frac{3}{4}x$

13. 5/8 of the difference between a number and 4:

    _____ $\frac{5}{8}(x - 4)$

14. The quotient of 7 and a number: ____ $\frac{7}{x}$

15. The quotient of the sum of a number and 4, and 5:

    _____ $\frac{x + 4}{5}$

16. The ratio of a number, and the sum of the number

    and 4: _____ $\frac{x}{x + 4}$

17. The word "is" often translates as ___.   =

**Write equations in Frames 18—20, but do not solve.
Use x as the variable.**

18. Five times a number, increased by 11, is 40.

    _____ $5x + 11 = 40$

19. If the product of -2 and a number is added to 4,
    the result is one more than the number.

    _____ $4 - 2x = x + 1$

20. If the quotient of 8 and a number is subtracted from the number, the result is 2.

_____

$$x - \frac{8}{x} = 2$$

Solve the applied problems in Frames 21–27.

21. **Find two consecutive integers such that the first increased by 2 is 17 less than their sum.**

    Let x represent the first integer. Then _____ represents the second. "The first increased by 2" is written _____. "Their sum" is written _____. "17 less than their sum" is written _____. "The first increased by 2 is 17 less than their sum" is written _____ = _____. Solve this equation, obtaining x = ___. x represents the _____ integer. Thus, the second integer is x + 1 = ____.

    x + 1

    x + 2

    x + (x + 1)

    x + (x + 1) − 17

    x + 2;
    x + (x + 1) − 17
    18

    first

22. **When a number is tripled, and this product increased by 5, the result is 6 less than four times the number. Find the number.**

    Let x be the _____. "A number is tripled" is written ____. "The product increased by 5" is written _____. "6 less than four times the number" is written _____. Write an equation representing the statement of the problem.

    _____

    Solve the equation. x = ____

    number

    3x

    3x + 5

    4x − 6

    3x + 5 = 4x − 6

    11

23. **Harry wishes to fence in a rectangular piece of land. He has 200 feet of fence, and he wants the length to be four times the width. Find the length and width of the land he can enclose.**

Let W = width. Then the length is given by ____.   4W
200 feet is the _____ of the rectangular   perimeter
area. The formula for the perimeter of a rectangle is P = _____. Use the formula to   2W + 2L
write the equation for this problem.

_____   200 = 2W + 2(4W)

Solve the equation. W = ___   20
Length = ___; width = ___   80; 20

24. **Joann is five years older than her sister Teresa.
Joann's age is one less than twice Teresa's age.
Find the ages of the two girls.**

Let x = Teresa's age. Then Joann's age is _____.   x + 5
Write "Joann's age is one less than twice Teresa's
age" in symbols, completing the equation for this
problem.

_____   x + 5 = 2x - 1

Solve the equation. x = ___   6
Teresa's age = ___; Joann's age = ____   6; 11

25. **Roy Gerard received an inheritance. Part of the
money was invested at 11%, and $6000 more than
this amount was invested at 9%. The total annual
interest from both investments is $3940. Find
the amount invested at each rate.**

_____ at 11% $17,000
_____ at 9% $23,000

26. **How many liters of 20% alcohol solution must be
mixed with 80 liters of a 50% solution to get a
mixture which is 40% alcohol?**

_____ liters   40

27. A grocer carries two grades of coffee; one sells
     for $4 per pound and the other sells for $7 per
     pound. How many pounds of the $7 coffee should
     be mixed with 125 pounds of the $4 coffee to get
     a mixture which will sell for $5.50 per pound?

                        _____ pounds   125

28. An angle that measures 90° is called a _____   right
     angle.

29. A straight angle measures _____.   180°

30. Two angles are _____ of each other if the   supplements
     sum of the measures of the two angles is 180°.

31. If the sum of the measures of two angles is 90°,
     the angles are called _____ angles.   complementary

32. The sum of the three angles in a triangle is ____.   180°

In the diagram below, if angle A is a right angle, find
the measures of angles B and C.

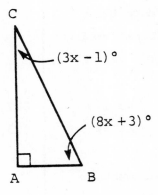

33. Since angle A is a right angle, its measure is
     ____.   90°

34. The sum of all three angles must be _____, so the   180°
     sum of angles B and C must be ____ - ____ - ____.   180°; 90°; 90°

35. Angles B and C are _____ angles.   complementary

36. Use the diagram to write an equation _____ + | $3x - 1$
     _____ = 90 | $8x + 3$

37. Solve the equation. x = ____ | 8

38. Angle B measures ____ and angle C measures ____. | 67°; 23°

39. Check: do the measures of all three angles add
     to 180°? | yes
     ____ + ____ + ____ = ____ | 90°; 67°; 23°; 180°

2.4 More Applications of Linear Equations

1 Solve problems about different denominations of money.
   (Frames 1-6)

2 Solve problems about uniform motion. (Frames 7-9)

3 Solve problems about geometric figures. (Frame 10)

---

1. When solving problems about denominations of
    money multiply the number of monetary units of
    the same kind by the _____ to get | denomination
    the total _____. | monetary value

Find the total monetary value of each of the following.

2. 10 nickels: _____ | $.50

3. 12 dimes: _____ | $1.20

4. x quarters: _____ | $.25x

5. (15 - y) pennies: _____ | $.01(15 - y)

6. **Le Thu has 18 coins, some dimes and some pennies. The total value of the coins is $.90. How many of each type of coin does she have?**

   Let x = the number of dimes she has.

   Then _____ = the number of pennies she has.   18 − x

   The total value of the dimes is _____ and the $.10x

   total value of the pennies is _____. $.01(18 − x)

   The sum of the values is ____. This leads to the $.90

   equation _____.   .10x + .01(18 − x)
                                                              = .90

   Solve the equation. x = ___   8

   Le Thu has ___ dimes and ____ pennies.   8; 10

7. When solving uniform motion problems use the
   formula _____.   d = rt

8. It is usually very helpful to draw a _____ in   sketch
   order to set up the equation properly.

9. **Two planes leave the same airport at the same time, flying in opposite directions. The speed of the first plane is 150 mph more than the speed of the second plane. After 3 hours they are 2850 miles apart. Find the speed of each plane. Let x represent the speed of the second plane.**

   Label the sketch.

                           airport
   •————————————————□————————————————•
   plane 1   plane 2
   _____ mph   _____ mph   x + 150; x

                total distance = _____   2850

Complete the chart.

	RATE	TIME	DISTANCE
plane 1 (faster)			
plane 2 (slower)			

x + 150; 3;
3(x + 150)

x; 3; 3x

After 3 hours they are _____ miles apart, so the sum of the distances is _____ miles.

Thus the equation is _____.

Solve the equation. x = ____

2850

2850

3(x+150)+3x=2850

400

The rate of the faster plane is _____; the rate of the slower plane is _____.

550 mph

400 mph

10. In an isosceles trapezoid, the pair of non-parallel sides are the same length. In this diagram, the length of the shorter base (the top) is 3 more than the length of each side. The length of the longer base is 1 less than twice the length of the shorter base. If the perimeter is 38 centimeters find the length of the sides and each base. Let x be the length of one side.

Label the diagram.

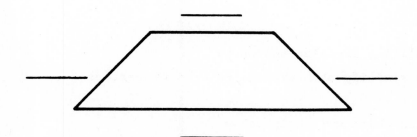

x + 3

x; x

2(x + 3) - 1

Write the equation. _____

Solve the equation. x = _____

	x+x+(x+3)+2(x+3)-1 = 38
	6

Each side is ____ centimeters, the shorter base is ___ centimeters and the longer base is ____ centimeters.

	6
	9; 17

2.5 Linear Inequalities on One Variable

[1] Solve linear inequalities using the addition property. (Frames 1–13)

[2] Solve linear inequalities using the multiplication property. (Frames 14–44)

[3] Solve linear inequalities having three parts. (Frames 45–49)

[4] Solve applied problems using linear inequalities. (Frames 50–51)

1 A linear inequality in one variable is of the form _____.	$ax + b < c$
2. x + 5 ≤ 13 is an example of an inequality. If x ≤ 8, then the inequality is (*true/false*).	true
3. Unlike an equation, an inequality usually has an _____ number of solutions.	infinite
4. (7 + 2) > 6 - 3x (*is/is not*) a linear inequality.	is
5. Like equations, inequalities are also solved by using _____ to simplify each expression.	properties
6. There are two properties which will help solve inequalities. The first is the _____ property of _____.	addition inequality

7. This property says that for any expressions a, b, and c, the inequalities a < b and _____ < _____ have the same _____.

a + c; b + c
solution

8. To solve x + 8 < 15, subtract ___ from both sides, getting

8

$$x + 8 - \underline{} < 15 - \underline{}$$

or $$x < \underline{}.$$

8; 8
7

9. Thus the solution set for the original inequality is {x| _____}.

x < 7

10. Graph this solution set.

is not

The parenthesis at 7 shows that 7 (*is/is not*) part of the graph.

is not

11. This solution set, {x|x < 7}, may be simplified in interval notation.

_____

(−∞, 7)

Solve the inequalities in Frames 12–13, and graph their solution sets.

12. x − 5 ≥ 11

$$x - 5 + \underline{} \geq 11 + \underline{}$$
$$x \geq \underline{}$$

5; 5
16

The solution set in interval notation is _____.

[16, ∞)

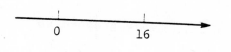

13.   −4m < −5m + 9

      Add ____ to each side, getting

                      −4m + ____ < −5m + 9 + ____

                              m < ____.

      Solution set: _____

      5m

      5m; 5m

      9

      (−∞, 9)

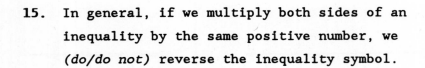

14.   If we multiply both sides of the inequality
      −3 < 4 by the positive number 7, we get
      (−3)7 ____ (4)7.

      <

15.   In general, if we multiply both sides of an
      inequality by the same positive number, we
      (do/do not) reverse the inequality symbol.

      do not

16.   If we divide both sides of 4x ≥ 20 by ____
      we get

                      $$\frac{4x}{\_\_} \geq \frac{20}{\_\_}$$

      or   _____.

      4

      4; 4

      x ≥ 5

17.   The _____ property of _____
      tells us that for any expressions a, b, and c,
      where c > 0, then a ≥ b and a • c ≥ b • c have
      the same _____ set.

      multiplication;
      inequality

      solution

18.   Therefore, in Frame 16, the solution set of
      4x ≥ 20 is _____.

      [5, ∞)

19.   This solution set, _____, is graphed as
      follows.

      [5, ∞)

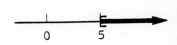

20. To solve $\frac{2}{3}x < -6$, multiply both sides by ____.

$\frac{3}{2}$

21. Since $\frac{3}{2}$ ____ 0, *(do/do not)* reverse the inequality symbol.

>; do not

22. This gives x < ____.

−9

23. This solution set, _____, is graphed as
follows.

(−∞, −9)

24. To solve $\frac{5}{4}x + 2 < -23$, first add ___ to both sides
sides to get $\frac{5}{4}x < $ ____.

−2

−25

25. Next, multiply both sides by ___ to get x < ____.

$\frac{4}{5}$; −20

26. This solution set, _____, is graphed as
follows.

(−∞, −20)

27. If we multiply both sides of the inequality
−3 < 4 by the negative number −7, we get
(−3)(−7) ___ (4)(−7).

>

28. In general, if we multiply both sides of an
inequality by the same negative number, we
(do/do not) reverse the inequality symbol.

do

29. The _____ property of _____
says that for any expressions a, b, and c, with
c < 0, a < b and a • c > b • c have the same
_____ set.

multiplication;
inequality

solution

30. To solve −4x < 20, divide both sides by ____, getting

$$\frac{-4x}{\underline{\quad}} \underline{\quad\quad} \frac{20}{-4}$$

or   x _____.

−4

−4; >

> −5

31. Since −4 < 0, the inequality symbol (*is/is not*) reversed.

is

32. Thus the solution set for the inequality −4x < 20 is _____.

(−5, ∞)

33. Graph this solution set.

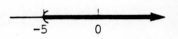

34. To solve 3 − 2(x + 4) ≤ 5(x − 4) + 1, first use the _____ property to write 3 − 2x − 8 ≤ _____ + 1.

distributive

5x − 20

35. By combining like terms we get _____ ≤ _____.

−2x − 5; 5x − 19

36. Subtracting 5x from both sides gives

_____ ≤ _____.

−7x − 5; −19

37. By adding ___ to both sides, we get ___ ≤ ___.

5; −7x; −14

38. Now divide both sides by ____ to get x ___ ___.

−7; ≥; 2

39. Note that since −7 ___ 0, we _____ the inequality symbol.

<; reversed

Solve each of the inequalities in Frames 40–44.

40. −6x ≥ −12   _____

(−∞, 2]

41. 2m − 3 < 3   _____

(−∞, 3)

42. 4r - 5 > 2r + 11   _____   (8, ∞)

43. -3x + 2 ≤ -5(x + 2)   _____   (-∞, -6]

44. -2(p + 5) ≥ 4(3 + p)   _____   (-∞, -11/3]

45. To solve 3 ≤ 2x + 5 ≤ 9, we want to get x by it-
     self in between the two inequality symbols. Start
     by subtracting ___ from each of the three parts of   5
     of the inequality to get ____ ≤ 2x ≤ ____.   -2; 4

46. Next divide each part by ___ to get   2
     ____ ≤ x ≤ ___.   -1; 2

47. This solution set, _____, is graphed as   [-1, 2]
     follows.

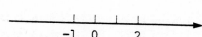   

Solve and graph the inequalities in Frames 48-49.

48. 7 ≤ 5x - 3 ≤ 12   _____   [2, 3]

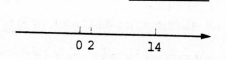   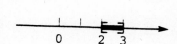

49. -2 < $\frac{1}{2}$m - 3 < 4   _____   (2, 14)

Solve the applied problems in Frames 50–51.

50. **If 1 is added to twice a number, the result is at least 4 more than the number. Find the number.**

Let x = the number.

"At least" translates as ___. Write an inequality using the information of the problem.

| | $\geq$ |

_____

| | $2x + 1 \geq x + 4$ |

Solve the inequality.

_____

| | $x \geq 3$ |

Our original inequality is satisfied by all
numbers _____ than or equal to ___.

| | greater; 3 |

51. **Tom cannot make more than $45 per day. Find
the least number of days he must work to earn
$495.**

_____ days

| | 11 |

2.6   Set Operations and Compound Inequalities

1 Find the intersection of two sets. (Frames 1–4)

2 Solve compound inequalities with the word <u>and</u>. (Frames 5–19)

3 Find the union of two sets. (Frames 20–24)

4 Solve compound inequalities with the word <u>or</u>. (Frames 25–36)

---

1 For any two sets A and B, the _____ of A and
B, symbolized _____, is defined as follows.

_____ = {x|x is an element of A **and** x is an
element of B}

	intersection
	$A \cap B$
	$A \cap B$

Find the intersections in Frames 2–4.

| | d |

2. $\{1, 2, 5, 7, 9\} \cap \{2, 5, 8, 11, 14\} = $ _____

| | $\{2, 5\}$ |

3. $\{a, b, c, d, f\} \cap \{a, c, f, h\} =$ _____ $\{a, c, f\}$

4. $\{2, 4, 6, 8\} \cap \{1, 3, 5, 7, 9\} =$ ___ $\emptyset$

5. A compound inequality is formed by joining two
    inequalities with a connective, such as "_____"   and
    or "___".   or

6. The solution set of a compound inequality having
    the connective "and" includes the _____   common
    elements from the solution sets of each part of
    the compound inequality.

7. To solve the compound inequality $2x \leq 8$ and
    $3x - 1 \geq 2$, solve each part, getting _____ $(-\infty, 4]$
    and _____. $[1, \infty)$

8. Graph these two solution sets.

    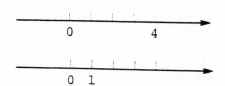   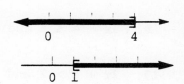

9. The common elements, or _____, of these   intersection
    two sets is the set _____. $[1, 4]$

10. This set is graphed as follows.

    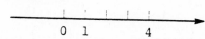   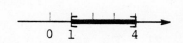

11. To indicate intersection, use the symbol ___. $\cap$

12. Using this symbol,
    $(-\infty, 4]$ ___ $[1, \infty) =$ _____. $\cap$; $[1, 4]$

13. Also, $[3, \infty)$ ___ $[-2, \infty) =$ _____. $\cap$; $[3, \infty)$

14. To solve $2x + 3 \leq 7$ and $-5x - 2 \leq 23$, solve each part, getting _____ and _____.

$(-\infty, 2]$; $(-5, \infty)$

15. Graph these two solution sets.

$2x + 3 \leq 7$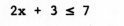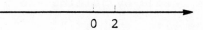

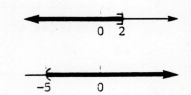

$-5x - 2 < 23$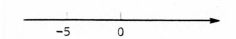

16. The solution set of the compound statement is the intersection of these two sets, or _____, which is graphed as follows.

$(-5, 2]$

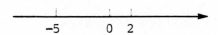

Solve the compound inequalities in Frames 17–19.

17. $x + 7 > 5$ and $x - 3 \leq 2$ _____

$(-2, 5]$

18. $3x - 4 \leq 2$ and $x + 4 > 8$ _____

$\varnothing$

19. $2x + 6 > 2$ and $-x - 3 \leq 2$ _____

$(-2, \infty)$

20. For any two sets A and B, the _____ of A and B, symbolized _____, is defined as follows:

_____ = $\{x \mid x$ is an element of A **or** x is an element of B$\}$

union

$A \cup B$

$A \cup B$

21. The solution set of a compound inequality having the connective "or" includes _____ elements from the solution sets of either (or both) parts of the compound inequality.

all

22. The set of all elements that belong to either (or both) of two sets is called the _____ of two sets.

union

23. To indicate union, use the symbol ___.

$\cup$

24. Using this symbol,

      $(-\infty, 4)$ ____ $[5, \infty)$ = _____ .

∪;
$(-\infty, 4) \cup (5, \infty)$

25. To solve $3x - 1 < 5$ or $2x \geq 10$, solve each part, getting _____ or _____ .

$(-\infty, 2)$; $[5, \infty)$

26. Graph these two solution sets.

    $x < 2$

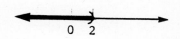

    $x \geq 5$

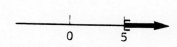

27. The solution set is then the _____ of these two sets, _____, which is graphed as follows.

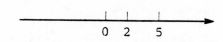

union
$(-\infty, 2) \cup [5, \infty)$

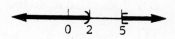

Solve the compound inequalities in Frames 28–30.

28. $x + 6 < 4$ or $x + 1 > 5$ _____

$(-\infty, -2) \cup (4, \infty)$

29. $2x - 3 < 3$ or $x + 4 < 5$ _____

$(-\infty, 3)$

30. $-x + 5 \leq 2$ or $5x - 2 < 13$ _____

$(-\infty, \infty)$

Let A = $\{1, 2, 3, 5, 7\}$, B = $\{2, 4, 6, 8\}$, and C = $\{3, 5\}$. List the elements in the sets of Frames 31–36.

31. $A \cap B$

    This set is made up of all the elements belonging to _____ set A and set B.

both

                $A \cap B$ = _____

$\{2\}$

32.   A ∪ B

This set is made up of all the elements belonging

to _____ set A or set B.   either

$$A \cup B = \{_____\}$$ {1,2,3,4,5,6,7,8}

33.   A ∩ C = {_____}   {3, 5}

34.   B ∪ ∅ = ___   B

35.   (A ∪ C) ∩ B = {___}   {2}

36.   Find the union of {1, 5, 7, 9} and {1, 7, 11, 12}.

_____   {1, 5, 7, 9, 11, 12]

2.7   Absolute Value Equations and Inequalities

[1]   Use the distance definition of absolute value. (Frames 1–2)

[2]   Solve equations of the form $|ax + b| = k$, $k > 0$. (Frames 3–20)

[3]   Solve inequalities of the form $|ax + b| < k$ and of the form $|ax + b| > k$, for $k > 0$. (Frames 21–60)

[4]   Solve absolute value equations that involve rewriting. (Frames 61–69)

[5]   Solve absolute value equations of the form $|ax + b| = |cx + d|$. (Frames 70–78)

[6]   Solve absolute value equations and inequalities that have only nonpositive constants on one side. (Frames 79–91)

---

1.   The absolute value of x represents the _____   distance

from x to ___. For example, $|2|$ represents the   0

distance from 2 to 0, or ___, and $|-7|$ repre-   2

sents the distance from −7 to 0 or ____.   7

2.   An equation which contains an absolute value

variable is called an _____ _____   absolute value

equation.

3. $|x| = 5$ and $|x - 2| = 5$ are both examples of
    _____ value _____.

    absolute;
    equations

4. The correspondence between distance and _____
    _____ suggests the equation $|x| = 5$ can be
    rewritten as _____ equations.

    absolute
    value
    two

5. These equations are _____ and _____.

    $x = 5$; $x = -5$

6. Therefore, the solution set of $|x| = 5$ is _____.

    $\{5, -5\}$

7. Recall from Section 1.1 that absolute value is
    never _____.

    negative

8. Therefore, there (is/is not) a solution to the
    equation $|x| = -5$.

    is not

9. The solution set is ___.

    $\emptyset$

10. Also, the solution set of $|x| = 0$ is _____.

    $\{0\}$

11. The equation $|x| = 8$ has _____ solutions.

    two

12. The solution set of $|x| = 8$ is _____.

    $\{8, -8\}$

13. To solve the equation $|x - 5| = 4$, use the
    correspondence between distance and
    _____ _____.

    absolute value

14. This gives
        $x - 5 =$ ___ or $x - 5 =$ ___.

    4; -4

15. Solving these two equations separately gives
        $x =$ ___ or $x =$ ___.

    9; 1

16. Therefore, the solution set of $|x - 5| = 4$ is

    _____ . $\{1, 9\}$

17. To solve $|2x - 1| = 9$, solve the two equations

    _____ or _____ . $2x - 1 = 9$;
                                                        $2x - 1 = -9$
    $2x =$ ____ or $2x =$ ____ 10; -8

    $x =$ ____ or $x =$ ____ 5; -4

18. Therefore, the solution set is _____ . $\{5, -4\}$

19. To solve $|3x + 1| = 4$, solve the two equations

    _____ or _____ . $3x + 1 = 4$;
                                                        $3x + 1 = -4$
    $3x =$ ____ or $3x =$ _____ 3; -5

    $x =$ ____ or $x =$ _____ 1; $-5/3$

20. The solution set is _____ . $\{1, -5/3\}$

21. To solve $|x| > 6$, we need to find all numbers
    whose distance from ___ is more than ___ . 0; 6

22. The solution would be made up of all numbers
    _____ than ____ , or greater than ____ .   less; -6; 6

23. This leads to the two inequalities

    _____ or _____ . $x < -6$; $x > 6$

24. Therefore, the _____ set of $|x| > 6$ is   solution

    _____ . $(-\infty, -6) \cup (6, \infty)$

25. This is graphed as follows.

    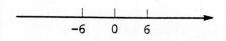

26. To solve $|x| \geq 12$, write x ___ ____ or $x \leq$ _____ . $\geq$; 12; -12

27. The solution set of $|x| \geq 12$ is

    _____.

    $(-\infty, -12] \cup [12, \infty)$

28. To solve $|x - 7| > 2$, write $x - 7$ ___ ___ or
    $x - 7$ ___ ___.

    $>$; 2
    $<$; -2

29. Solving each of these inequalities gives _____
    or _____.

    $x > 9$
    $x < 5$

30. The solution set is _____.

    $(-\infty, 5) \cup (9, \infty)$

31. To solve $|4 - 2z| \geq 8$, write $4 - 2z \geq 8$ or

    _____.

    $4 - 2z \leq -8$

32. Solving each of these inequalities gives

    $\quad\quad -2z \geq$ ___   or $-2z \leq$ ____

    $\quad\quad z$ ___ ___   or z ___ ___.

    4; -12
    $\leq$; -2; $\geq$; 6

33. The solution set is _____.

    $(-\infty, -2] \cup [6, \infty)$

Solve the absolute value inequalities in Frames 33-40.

34. $|x| \geq -5$ _____

    $(-\infty, \infty)$

35. $|x - 9| > 6$ _____

    $(-\infty, 3) \cup (15, \infty)$

36. $|4 - a| \geq 8$ _____

    $(-\infty, -4] \cup [12, \infty)$

37. $|x - 5| > 0$ _____

    $(-\infty, 5) \cup (5, \infty)$

38. $|-3p - 5| \geq 10$ _____

    $(-\infty, -5] \cup [5/3, \infty)$

39. $|2m - 5| \geq -2$ _____

    $(-\infty, \infty)$

40. $|m - 2| > 0$ _____

    $(-\infty, 2) \cup (2, \infty)$

41. $|a + 6| \geq 0$ _____ $(-\infty, \infty)$

42. Notice that the absolute value inequalities so
     far have all had ___ or ___ inequality symbols. $>;\ \geq$
     We will now consider absolute value inequalities
     having $<$ or $\leq$ inequality symbols.

43. To solve the absolute value inequality $|x| < 4$,
     we need all numbers whose _____ from 0 is   distance
     less than ___.   4

44. This includes all numbers between ___ and ___. $-4;\ 4$

45. Graph the numbers between -4 and 4.

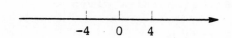   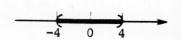

46. This is the graph of the set _____. $(-4, 4)$

47. This set is called the _____ set of   solution
     _____. $|x| < 4$

48. The solution set of $|x| < 7$ is _____. $(-7, 7)$

49. This is graphed as follows.

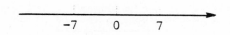   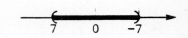

50. To solve $|x - 3| < 5$, rewrite it as
     ___ $< x - 3 <$ ___. $-5;\ 5$

51. By adding ___ to each part of the inequality,   3
     we get ___ $< x <$ ___. $-2;\ 8$

52. This gives the solution set _____. $(-2, 8)$

53. To solve $|2x - 7| < 9$, first write

    ___ < 2x - 7 < ___ . | -9; 9

54. Then add ___ to each part to get ___ < 2x < ___ . | 7; -2; 16

55. By dividing each part by __, we get ___ < x < ___ . | 2; -1; 8

56. Thus the solution set is _____ . | (-1, 8)

Solve the absolute value inequalities in Frames 57—60.

57. $|x| < 8$ _____ | (-8, 8)

58. $|x + 5| \le 13$ _____ | [-18, 8]

59. $|5 - 2x| \le 9$ _____ | [-2, 7]

60. $|x + 2| < -3$ _____ | ∅

61. To solve $|5x + 3| + 4 = 16$, first subtract ___ | 4
    from both sides to get $|5x + 3| = $ ___ . | 12

62. Then solve the two equations

    _____ or _____ . | 5x + 3 = 12;
                                | 5x + 3 = -12
    5x = _____ or 5x = _____ | 9; -15

    x = _____ or x = _____ | 9/5; -3

63. The solution set is _____ . | {9/5, -3]

Find the solution set of the absolute value equations in Frames 64—69.

64. $|x - 5| = 11$

    _____ or _____ | x - 5 = 11;
                              | x - 5 = -11
    x = _____ or x = _____ | 16; -6

    Solution set: _____ | {16, -6}

65. $|x - 6| = 14$

       _____ or _____ $x - 6 = 14;$
                                               $x - 6 = -14$

       $x =$ _____ or $x =$ _____ $20; -8$

       Solution set: _____ $\{20, -8\}$

66. $|2r + 3| = 9$

       _____ or _____ $2r + 3 = 9;$
                                               $2r + 3 = -9$

       $2r =$ _____ or $2r =$ _____ $6; -12$

       $r =$ _____ or $r =$ _____ $3; -6$

       Solution set: _____ $\{3, -6\}$

67. $\left|\dfrac{2}{3}m - 6\right| = 3$

       _____ or _____ $\dfrac{2}{3}m - 6 = 3;$

                                               $\dfrac{2}{3}m - 6 = -3$

       $\dfrac{2}{3}m =$ _____ or $\dfrac{2}{3}m =$ _____ $9; 3$

       $m =$ _____ or $m =$ _____ $\dfrac{27}{2}; \dfrac{9}{2}$

       Solution set: _____ $\left\{\dfrac{27}{2}, \dfrac{9}{2}\right\}$

68. $|3x + 2| - 4 = 7$

       First: $|3x + 2| =$ ____ 11

       _____ or _____ $3x + 2 = 11;$
                                               $3x + 2 = -11$

       $3x =$ _____ or $3x =$ _____ $9; -13$

       $x =$ _____ or $x =$ _____ $3; -\dfrac{13}{3}$

       Solution set: _____ $\left\{3, -\dfrac{13}{3}\right\}$

69. $|2q - 7| + 3 = 2$

       Solution set: _____ $\emptyset$

70. The equations we have solved so far have only one absolute value in them. However, $|x + 4| = |3x - 6|$ is also an _____ value _____.

absolute; equation

71. This equation can be solved if $x + 4$ and $3x - 6$ are _____, or if either one is the _____ of the other.

equal; negative

72. This is expressed as the following two equations.
    $$x + 4 = 3x - 6 \quad \text{or} \quad \underline{\hspace{1cm}} = -(\underline{\hspace{1.5cm}})$$

$x + 4$; $3x - 6$

73. This gives
    $$-2x = \underline{\hspace{1cm}} \quad \text{or} \quad 4x = \underline{\hspace{1cm}}.$$

-10; 2

$$x = \underline{\hspace{1cm}} \quad \text{or} \quad x = \underline{\hspace{1cm}}$$

5; $\frac{1}{2}$

Solution set: _____

$\{5, \frac{1}{2}\}$

Find the solution sets of the absolute value equations in Frames 74–78.

74. $|2x - 1| = |3x + 2|$

$2x - 1 = 3x + 2 \quad \text{or} \quad \underline{\hspace{3cm}}$

$2x - 1 = -(3x + 2)$

$-x = \underline{\hspace{1cm}} \quad \text{or} \quad 5x = \underline{\hspace{1cm}}$

3; -1

$x = \underline{\hspace{1cm}} \quad \text{or} \quad x = \underline{\hspace{1cm}}$

-3; $-\frac{1}{5}$

Solution set: _____

$\{-3, -\frac{1}{5}\}$

75. $|2b - 1| = |b + 4|$

$2b - 1 = b + 4 \quad \text{or} \quad \underline{\hspace{3cm}}$

$2b - 1 = -(b + 4)$

$b = \underline{\hspace{1cm}} \quad \text{or} \quad 3b = \underline{\hspace{1cm}}$

5; -3

$b = \underline{\hspace{1cm}} \quad \text{or} \quad b = \underline{\hspace{1cm}}$

5; -1

Solution set: _____

$\{5, -1\}$

76. $|3r + 2| = |r + 4|$

          $3r + 2 = r + 4$ or _____ | $3r + 2 = -(r + 4)$

               $2r =$ ____ or $4r =$ ____ | $2; -6$

                $r =$ ____ or $r =$ ____ | $1; -\dfrac{3}{2}$

    Solution set: _____ | $\{1, -\dfrac{3}{2}\}$

77. $|x + 6| = |x - 10|$

    This can be expressed as _____ or | $x + 6 = x - 10$
    _____. The equation $x + 6 = x - 10$ | $x + 6 = -(x - 10)$
    has _____ solution. The solution of $x + 6 =$ | no
    $-(x - 10)$ is ___. The solution set of $|x + 6| =$ | 2
    $|x - 10|$ is ____. | $\{2\}$

78. $|2z - 5| = |2z + 11|$

    Solution set: _____ | $\{-\dfrac{3}{2}\}$

79.   The absolute value of an expression can never
    be _____. | negative

80.   Therefore, the solution to the absolute value
    equation $|3x - 2| = -5$ is ___. | ∅

81.   The absolute value of an expression is _____ | always
    greater than or equal to zero.

82.   Thus, the solution to $|x| \geq -7$ is _____. | $(-\infty, \infty)$

83.   The absolute value of an expression will _____ be | only
    zero when the expression equals _____. | zero

84.   Therefore, the only equation that is derived from
    $|4t - 1| = 0$ is _____. | $4t - 1 = 0$

85.   The solution to this equation is _____. | $\{1/4\}$

Solve the absolute value equations and inequalities in Frames 86–91.

86. $|2 + 5x| = -1$ ∅

87. $|t + 2| < 0$ ∅

88. $|m + 3| \geq 0$ $(-\infty, \infty)$

89. $|2w - 3| > -6$ $(-\infty, \infty)$

90. $|x| \leq -5$ ∅

91. $|4x + 1| + 3 = 0$ ∅

Chapter 2 Test

The answers for these questions are at the back of this Study Guide.

Solve each of the following equations.

1. $9 + 3x = 2x$ 1. _____

2. $6z - 4 = 3 - 2z + 3z + 8$ 2. _____

3. $4x - (2 + 5x) = 2x + 6$ 3. _____

4. $|a| = 3$ 4. _____

5. $|1 + 2x| = 9$ 5. _____

6. $|3x + 8| = 16$ 6. _____

7. $|2m - 1| = |m + 1|$ 7. _____

8. $|3r - 2| = |4r - 5|$ 8. _____

9. $\frac{1}{2}(x + 2) = \frac{2}{3}x - 4$ 9. _____

10. $.5x + .07 = .01x - .42$ 10. _____

11. $|4x + 3| = 0$ 11. _____

12. $|7s + 2| + 5 = 0$ 12. _____

Solve each of the following, and graph the solution set on the number line.

13. $3x - 1 \geq 8$ 13. $\longrightarrow$

14. $-2 \leq 5q + 3 \leq 18$ 14. $\longrightarrow$

15. $2x - 4 \leq 6$ or $2x - 1 \geq 13$ 15. ⟶

16. $5x - 6 \geq 19$ and $-x + 4 \leq 13$ 16. ⟶

17. $|x - 4| \leq 6$ 17. ⟶

18. $|2z + 13| > 5$ 18. ⟶

19. $|3x + 4| > 0$ 19. ⟶

20. $|3 - 2b| \leq -4$ 20. ⟶

Solve each of the following for the indicated variable.

21. $7m + 2n = 8$, for m   21. _____

22. $3x - 5y + 2z = k$, for z   22. _____

23. Is the equation $3(r - 2) + 5r = 7 - [2 - (3 + 8r)]$
conditional, an identity, or a contradiction?   23. _____

Let $A = \{1, 2, 3\}$, $B = \{3, 6, 9, 12\}$ **and** $C = \{4, 5, 6, 7, 8, 9\}$. **List the elements in the following sets.**

24. $A \cap C$ 24. _____

25. $B \cup C$ 25. _____

26. $A \cap B$ 26. _____

27. $(B \cap C) \cup A$ 27. _____

Solve each of the following applied problems.

28.   The difference of two numbers is 11.   The
      larger number is 29 less than twice the
      smaller.   Find the two numbers.   28. _____

29.   The length of a rectangle is 2 less than
      twice the width.   The perimeter is 44.
      Find the length and width.   29. _____

30.   Margaret bought a suit worth $125, while the
      store was having a "25% off" sale. How much
      did she actually pay for the suit?

                                                          30. _____

31.   Chandra invests $18,000, some at 8% simple
      interest and some at 10% simple interest.
      If her total income from interest is $1580,
      how much did she invest at each rate?

                                                          31. _____

32.   How many liters of water must be added to 5
      liters of a 40% chemical solution to reduce
      the concentration to 10%?

                                                          32. _____

33.   Katy has 37 coins, some dimes and some quarters.
      The total value of the money is $5.05.   How many
      of each type of coin does she have?

                                                          33. _____

34.   Keith and Manoj have $12 to spend on a pizza.
      The pizzeria charges $7.50 for cheese only,
      and $1.25 for each additional topping.   Find
      the largest number of additional toppings
      they can choose.

                                                          34. _____

35. Bill leaves on a bicycle trip at 9:00 A.M.,
traveling at 4 miles per hour. Realizing
that his brother forgot his picnic lunch,
Paul leaves at 10:00 A.M.. How fast will
Paul have to ride to catch up to his brother
by 12:00 noon?

35. _____

36. Find the measure of each marked angle.   36. _____

$(8x - 10)°$

$3(x + 30)°$

CHAPTER 3 EXPONENTS AND POLYNOMIALS

3.1 Integer Exponents

1️⃣ Identify exponents and bases. (Frames 1–6)

2️⃣ Use the product rule for exponents. (Frames 7–13)

3️⃣ Define negative exponents. (Frames 14–21)

4️⃣ Use the quotient rule for exponents. (Frames 23–37)

5️⃣ Define a zero exponent. (Frames 38–43)

---

1. In the expression $(-4)^3$, 3 is called the

   _____, and -4 is called the _____.   **exponent; base**

2. To evaluate this expression, write $(-4)^3 =$

   ($\_\_\_$) $\cdot$ ($\_\_\_$) $\cdot$ ($\_\_\_$) $=$ _____.   **-4; -4; -4; -64**

Identify the exponent and base in Frames 3–6. Then evaluate each expression.

	Exponent	Base	Value
3. 5^3	_____	____	_____	**3; 5; 125**
4. 2^4	_____	____	_____	**4; 2; 16**
5. $\left(\frac{2}{3}\right)^3$	_____	____	_____	**3; $\frac{2}{3}$; $\frac{8}{27}$**
6. $\left(-\frac{1}{2}\right)^2$	_____	____	_____	**2; $-\frac{1}{2}$; $\frac{1}{4}$**

7. The _____ rule for exponents says that   **product**

   $a^m \cdot a^n =$ _____.   **a^{m+n}**

8. By this rule, $3^4 \cdot 3^5 = 3^{\overline{\quad}} =$ ____.   **$4 + 5$; 3^9**

Using the product rule, simplify the expressions in Frames 9–13. Write all answers with only positive exponents.

9. $(-2)^8 \cdot (-2)^3 =$ _____ = _____

$(-2)^{8+3}$; $(-2)^{11}$

10. $(-4)^3 \cdot (-4)^7 =$ _____

$(-4)^{10}$

11. $(-4m^3)(5m^4) = (-4)(5)m^{\overline{}} =$ _____$m^{\overline{}}$

$3 + 4$; -20; 7

12. $(-3x^4)(8x^2) = (\underline{})(\underline{})x^{\overline{}} =$ _____$x^{\overline{}}$

-3; 8; $4 + 2$; -24; 6

13. $(7p^3)(8p^5) = (\underline{})(\underline{})p^{\overline{}} =$ _____$p^{\overline{}}$

7; 8; $3 + 5$; 56; 8

14. To simplify work with quotients, we need to define a negative exponent. If $a \neq 0$, then

$$a^{-n} = \underline{} = \underline{}.$$

$\frac{1}{a^n}$; $\left(\frac{1}{a}\right)^n$

15. That is, if $a \neq 0$, a^{-n} equals the _____ of a raised to the nth power.

reciprocal

Write each expression in Frames 16–21 with only positive exponents and evaluate.

16. $2^{-5} = \dfrac{1}{\underline{}} =$ _____

2^5; $\frac{1}{32}$

17. $4^{-2} = \dfrac{1}{\underline{}} =$ _____

4^2; $\frac{1}{16}$

18. $8^{-1} = \dfrac{1}{\underline{}} =$ _____

8^1; $\frac{1}{8}$

19. $\left(\frac{3}{2}\right)^{-2} = ()^2 =$ _____

$\frac{2}{3}$; $\frac{4}{9}$

20. $2^{-1} + 3^{-1} = \dfrac{1}{\underline{}} + \dfrac{1}{\underline{}} =$ _____

2; 3; $\frac{5}{6}$

21. $(2k)^{-3} =$ _____ $1/(8k)^3$ or $(1/8k)^3$

22. Give the quotient rule for exponents: if $a \neq 0$,

$$\frac{a^m}{a^n} = \text{_____}.$$ a^{m-n}

**Use the quotient rule to write the expressions in
Frames 23–31 with only positive exponents.**

23. $\dfrac{9^5}{9^2} =$ _____ $=$ ___ 9^{5-2}; 9^3

24. $\dfrac{(-7)^6}{(-7)^5} =$ ___ -7

25. $\dfrac{4^5}{4^9} =$ _____ $=$ ___ 4^{5-9}; $\dfrac{1}{4^4}$

26. $\dfrac{(-11)^7}{(-11)^{10}} =$ _____ $\dfrac{1}{(-11)^3}$

27. $\dfrac{8m^{15}}{16m^9} = \left(\dfrac{8}{16}\right)m^{15-9} =$ _____ $\dfrac{m^6}{2}$

28. $\dfrac{3^5}{3^2} = 3^{\overline{\quad}} {}^{-} {}^{\overline{\quad}} = 3^{\overline{\quad}}$ 5; 2; 3

29. $\dfrac{2^{-4}}{2^{-6}} = 2^{\overline{\quad}} {}^{-} {}^{\overline{\quad}} = 2^{\overline{\quad}}$ -4; -6; 2

30. $\dfrac{3^5}{3^{-2}} = 3^{\overline{\quad}} {}^{-} {}^{\overline{\quad}} = 3^{\overline{\quad}}$ 5; -2; 7

31. $\dfrac{15^{-1}}{15} = 15^{-1} {}^{-} {}^{\overline{\quad}} = 15^{\overline{\quad}} =$ _____ 1; -2; $\dfrac{1}{15^2}$

**Solve the problems in Frames 32–37, which require both
the product rule and the quotient rule.**

32. $\dfrac{8^2 \cdot 8^{-5}}{8^{-4} \cdot 8^3} = \dfrac{8^{\overline{\quad}}}{8^{\overline{\quad}}} = 8^{\overline{\quad}} =$ ___ $2^{\begin{array}{l}-3;\\ \\ -1\end{array}}$ -2; $\dfrac{1}{8^2}$

33. $\dfrac{y^{-5} \cdot y^{-2}}{y^{-4}} = \dfrac{y^{\overline{\quad}}}{y^{-4}} =$ ___ -7; $\dfrac{1}{y^3}$

34. $\dfrac{z^{-6} \cdot z^4}{z^{-10} \cdot z^{-1}} =$ _____ | z^9

35. $\dfrac{8(-9m^2n)(2m^3n^{-4})}{m^6} =$ _____ | $-\dfrac{144}{mn^3}$

36. $\dfrac{r^s \cdot r^{2s+1}}{r^{3s}} =$ ___ | r

37. $\dfrac{z^{2y} \cdot z^{5y}}{6z} =$ _____ | $\dfrac{z^{7y-1}}{6}$

38. To be consistent with past work on exponents, a zero exponent must be defined as follows:

$$a^0 = \underline{\quad} \quad (\text{if } a \neq 0).$$

| 1

Simplify the expressions in Frames 2–6.

39. $9^0 =$ ___ | 1

40. $(-6)^0 =$ ___ | 1

41. $-(-2)^0 = -(\underline{\quad}) =$ ___ | 1; −1

42. $-(3^0 + 2^0 + 4^0) = -(\underline{\quad} + \underline{\quad} + \underline{\quad}) =$ ___ | 1; 1; 1; −3

43. $-(3 + 2 + 4)^0 =$ ___ | −1

3.2 Further Properties of Exponents

1. Use the power rule for exponents. (Frames 1–12)

2. Simplify exponential expressions. (Frames 13–20)

3. Use the rules of exponents with scientific notation. (Frames 21–44)

1. There are three power rules for exponents.

$$(a^m)^n = \underline{\hspace{1cm}}$$

$$(ab)^m = \underline{\hspace{1cm}}$$

$$\left(\frac{a}{b}\right)^m = \underline{\hspace{1cm}} \qquad (b \neq 0)$$

a^{mn}

$a^m b^m$

$\dfrac{a^m}{b^m}$

Use the power rules to simplify the expressions in Frames 8–18. Write all answers with only positive exponents.

2. $(7^3)^5 = \underline{\hspace{1.5cm}}$

7^{15}

3. $(2^4)^3 = \underline{\hspace{1.5cm}}$

2^{12}

4. $(5^2)^{-3} = 5^{\overline{\hspace{0.6cm}}} = \underline{\hspace{2cm}}$

$-6;\ 1/5^6$

5. $(13^{-1})^4 = 13^{\overline{\hspace{0.6cm}}} = \underline{\hspace{2cm}}$

$-4;\ 1/13^4$

6. $(7^{-2})^{-5} = 7^{\overline{\hspace{0.6cm}}}$

10

7. $(6x^2)^3 = (\underline{\hspace{0.6cm}})^3(\underline{\hspace{0.6cm}})^3 = \underline{\hspace{2cm}}$

$6;\ x^2;\ 6^3 x^6$ or $216x^6$

8. $(5m^3)^4 = (\underline{\hspace{0.6cm}})^4(\underline{\hspace{0.6cm}})^4 = \underline{\hspace{2cm}}$

$5;\ m^3;\ 5^4 m^{12}$ or $625m^{12}$

9. $\left(\dfrac{3}{8}\right)^4 = \underline{\hspace{2cm}}$

$\dfrac{3^4}{8^4}$

10. $\left(\dfrac{7^5}{2^2}\right)^3 = \underline{\hspace{1.5cm}}$

$\dfrac{7^{15}}{2^6}$

11. $\left(\dfrac{3}{x}\right)^4 = \underline{\hspace{1.5cm}}$

$\dfrac{3^4}{x^4}$ or $\dfrac{81}{x^4}$

12. $\left(\dfrac{9^{-2}}{m}\right)^{-1} = \underline{\hspace{1.5cm}}$

$9^2 m$ or $81m$

13. All the definitions and rules for exponents may be needed for some problems. For example, to simplify

$$\frac{(3^4)^{-2}}{3^2 \cdot 3^{-5}}$$

simplify the numerator with the _____ rule and the denominator with the _____ rule.

$$\frac{(3^4)^{-2}}{3^2 \cdot 3^{-5}} = \frac{3^{\overline{}}}{3^{\overline{}}}$$

Now use the _____ rule.

$$\frac{3^{-8}}{3^{-3}} = 3^{\overline{}} = \underline{}$$

| power |
| product |
| -8 |
| -3 |
| quotient |
| $-5; \dfrac{1}{3^5}$ |

Simplify each expression in Frames 20–27. Write all answers with only positive exponents.

14. $\dfrac{(m^{-3})^2}{m^4} = \dfrac{m^{\overline{}}}{m^4} = m^{\overline{}\,-\,\overline{}} = m^{\overline{}} = \dfrac{1}{\overline{}}$

$-6;\ -6;\ 4;\ -10;$ m^{10}

15. $\dfrac{(k^3r^{-3})^2 \cdot k^{-2}r^4}{r^{-1} \cdot k^2 \cdot k^{-4}} = \dfrac{k^{\overline{}}\,r^{\overline{}}}{k^{\overline{}}\,r^{\overline{}}} = k^{\overline{}}\,r^{\overline{}} = \underline{}$

$4;\ -2;\ 6;\ -1;\ \dfrac{k^6}{r}$

$-2;\ -1$

16. $\dfrac{(m^{-2}n^{-1})^{-2} \cdot m^{-3}n^5}{n^3 \cdot m^{-2} \cdot (mn)^{-2}} = \underline{}$

m^5n^6

17. $\dfrac{(3p^2q^4)(9p^3q)}{(6p^4q)(12p^5q^3)} = \dfrac{(\underline{})(p^{\overline{}})(q^{\overline{}})}{(\underline{})(p^{\overline{}})(q^{\overline{}})}$

$\qquad = \underline{}.$

$27;\ 5;\ 5$

$72;\ 9;\ 4$

$\dfrac{3q}{8p^4}$

18. $\dfrac{(-4m^3)^3(8m^2)^2}{(2m)^3(16m^2)^2} = \dfrac{(-4)^{\overline{}}(m^3)^{\overline{}} \cdot 8^{\overline{}}(\underline{})^{\overline{}}}{2^{\overline{}} \cdot m^{\overline{}} \cdot 16^{\overline{}} \cdot (\underline{})^{\overline{}}}$

$\qquad = \dfrac{(-64)(64)(\underline{})(\underline{})}{8 \cdot 256(\underline{})(\underline{})}$

$\qquad = \dfrac{\overline{} \cdot m^{\overline{}}}{m^{\overline{}}}$

$\qquad = \underline{}$

$3;\ 3;\ 2;\ m^2;\ 2$

$3;\ 3;\ 2;\ m^2;\ 2$

$m^9;\ m^4$

$m^3;\ m^4$

$-2;\ 13$
7

$-2m^6$

19. $\dfrac{(5z)^{-4}(y^{-2}z^5)}{25^{-1}z^3y^2}$ = _____ $\dfrac{1}{25z^2y^4}$

20. $\dfrac{(m^q)^2(m^4q^{-3})}{m^{-5}q}$ = _____ $m^{11}q^{-3}$

21. In scientific _____, a number is written   notation
     as the product of a number between ____ and ____   1; 10
     (or between −1 and −10, or equal to 1 or −1) and
     some _____ of 10.   power

22. Since 90,000 = ___ × 10,000, in scientific   9
     notation,

          90,000 = _____. 9×10^4

23. In general, count the number of digits from the
     _____ of the first nonzero digit to the   right
     _____ _____. This gives the absolute   decimal point
     value of the _____ on the ten.   exponent

24. The exponent is positive if the original number
     is _____ than 10. It is negative if the   greater
     original number is _____ than ____.   less; 1

**Write the numbers in Frames 25—30 in scientific
notation.**

25. 9,500,000 = 9.5 × 10‾‾   6

26. 253,000,000,000 = _____ 2.53×10^{11}

27. 79 = _____ 7.9×10^1

28. .0000052 = _____ × 10‾‾   5.2; −6

29. .00000000109 = _____ × 10‾‾   1.09; −9

30. −.0000000457 = _____ × 10‾‾   −4.57; −8

31. To convert from scientific to standard notation, move the decimal point as many places as the _____ on the 10. Move to the right for a _____ exponent, and to the left for a _____ exponent.

exponent

positive

negative

Write the numbers in Frames 32—36 without scientific notation.

32. $9.774 \times 10^3 = $ _____

9774

33. $7.63 \times 10^4 = $ _____

76,300

34. $6.42 \times 10^{-5} = $ _____

.0000642

35. $8.93 \times 10^{-7} = $ _____

.000000893

36. $-8.74 \times 10^5 = $ _____

−874,000

37. Scientific notation can also be used to simplify calculations. For example, to simplify

$$\frac{8 \times 10^6}{2 \times 10^5},$$

work as follows.

$$\frac{8 \times 10^6}{2 \times 10^5} = \frac{8}{2} \times 10^{\underline{\quad\quad}}$$

6 − 5

$$= \underline{\quad} \times 10^1 = \underline{\quad}$$

4; 40

Use scientific notation to compute in each of Frames 38—44.

38. $\dfrac{3 \times 10^{-2}}{1 \times 10^{-5}} = $ _____

3×10^3 or 3000

39. $\dfrac{125,000}{.005} = $ _____

2.5×10^7 or 25,000,000

40. $\dfrac{(7200)(.00016)}{(.008)(180,000)}$ = $\dfrac{(\underline{\quad} \times 10^{\overline{\quad}})(\underline{\quad} \times 10^{\overline{\quad}})}{(\underline{\quad} \times 10^{\overline{\quad}})(\underline{\quad} \times 10^{\overline{\quad}})}$

= $\underline{\quad} \times 10^{\overline{\quad}}$

= $\underline{\quad\quad\quad}$

7.2; 3; 1.6; −4

8; −3; 1.8; 5

.8; −3

.0008

41. $\dfrac{56,000}{.002 \times 1400}$ = $\underline{\quad\quad\quad\quad\quad\quad\quad}$

2×10^4 or 20,000

42. $\dfrac{-6}{1.2 \times 10^5}$ = $\underline{\quad\quad\quad\quad\quad\quad\quad}$

-5×10^{-5} or −.00005

(Round the numerical part of the answer to the nearest thousandth in Frames 43 and 44.)

43. $\dfrac{3.97 \times 28,000}{.000473 \times 25,421}$ = $\underline{\quad\quad}$ p' $\underline{\quad 8X \quad}$

9.245×10^3 or 9245

44. $\dfrac{.73 \times 96,000,000}{59,300 \times .00254}$ = $\underline{\quad\quad\quad\quad\quad}$

4.653×10^5 or 465,300

3.3   The Basic Ideas of Polynomials

[1] Know the basic definitions for polynomials. (Frames 1–12)

[2] Identify monomials, binomials, and trinomials. (Frames 13–20)

[3] Find the degree of a polynomial. (Frames 21–23)

[4] Add and subtract polynomials. (Frames 24–33)

[5] Use P(x) notation. (Frames 34–39)

1. An expression consisting of the product of a
   _____ and one or more _____ raised
   to nonnegative powers is called a _____.

   number; variables
   term

2. Thus, $-4x^2y^3$, $8x^2$, $9y^2$, and $6x^5$ are all examples
   of _____.

   terms

3. The number in the product is called the
    _____.

                                                          coefficient

Identify the coefficient in each of Frames 4–6.

4. $7x^5y^3$ ____

                                                          7

5. $-9m^2y^4$ ____

                                                          -9

6. x^5 ____

                                                          1

7. Any finite sum of terms is called a _____.

                                                          polynomial

8. $9x^3 - 4x^2 + 6$ is a _____ containing ___
    terms.

                                                          polynomial; 3

9. $-8x^5 - 6x^2 - 4x^3 + 3x$ is a polynomial of ___
    terms.

                                                          4

10. A polynomial containing only the variable x is
    called a polynomial _____. If the exponents

                                                          in x

    in x decrease from left to right, the polynomial
    is said to be in _____ _____ of the

                                                          descending powers

    variable.

11. $9x^4 - 5x^3 - 3x + 6$ is a polynomial in _____

                                                          descending

    powers of ___.

                                                          x

12. Write $7q - 9 + 4q^2 - 5q^4$ in descending powers
    of q.

                                       _____
                                                          $-5q^4 + 4q^2 + 7q - 9$

13. A polynomial containing exactly three terms is
    called a _____.

                                                          trinomial

14. A polynomial containing exactly two terms is
    called a _____.   binomial

15. A polynomial containing exactly one term is
    called a _____.   monomial

**Identify each of the polynomials in Frames 16—20 as a
trinomial, binomial, or monomial.**

16. $4x^9 - 8x^8 + 2x^3$ _____   trinomial

17. $3x^5 - 4x^2 + 11$ _____   trinomial

18. $8x^7 - 5x^4$ _____   binomial

19. $9x^{11}$ _____   monomial

20. -3 _____   monomial

21. The degree of a term with one variable is the
    _____ on the _____.   exponent; variable

22. The largest exponent on a polynomial with one
    variable is called the _____ of the poly-   degree
    nomial.

23. The degree of $4x^3 - 8x^4 - 3x^7 + 8$ is ___.   7

24. The _____ property can be used to sim-   distributive
    plify polynomials. Thus,

    $$9x + 14x = (\_\_\_ + \_\_\_)x = _____.$$ 9; 14; 23x

    Also,

    $$-7x^2y + 8x^2y - 4x^2y = (_____)x^2y$$ -7 + 8 - 4

    $$= _____.$$ $-3x^2y$

25. Terms such as $-7x^2y$ and $8x^2y$, which have exactly the same variables to the same powers, are called _____ terms. Only _____ terms can be combined.

like; like

26. To add $-4x^2 - 3x + 5$ and $7x^2 - 2x - 8$, use the _____ and associative properties to combine like terms. Then use the _____ property.

commutative
distributive

$(-4x^2 - 3x + 5) + (7x^2 - 2x - 8)$

= (_____) + (_____) + (_____)

= (_____)x^2 + (_____)x + (_____)

= _____

$-4x^2 + 7x^2$; $-3x - 2x$; $5 - 8$
$-4 + 7$; $-3 - 2$; $5 - 8$
$3x^2 - 5x - 3$

27. Add $-4x^4 + 3x^3 - 11x + 16$ and $3x^4 - 5x^3 + 8x - 6$.

_____

$-x^4 - 2x^3 - 3x + 10$

28. To subtract two polynomials, add the first polynomial and the _____ of the second polynomial. Use the fact that $-(a + b) =$ _____.
Therefore,

negative

$-a + (-b)$

$(9x^2 - 6x + 7) - (5x^2 - 3x + 5)$

= $9x^2 - 6x + 7 - ($___$) - ($___$) - ($___$)$

= $9x^2 - 6x + 7 + $_____$ + $_____$ + $_____

= _____

$5x^2$; $-3x$; 5
$(-5x^2)$; $3x$; (-5)
$4x^2 - 3x + 2$

Simplify each of the expressions in Frames 29–32.

29. $(4m^2 - 11m + 6) + (-9m^2 + 4m - 8)$
     answer: _____

$-5m^2 - 7m - 2$

30. $(11k^3 - 5k^2 + 6k - 2) - (8k^3 + 2k - 5)$
     answer: _____

$3k^3 - 5k^2 + 4k + 3$

31. $(9r^2 - 4 + 3r^2) - (5 - 2r^2)$
     answer: _____

$14r^2 - 9$

32. $(3m^4 - 5m^3) - 3m^3 + (2m^3 + 4m^4)$

answer: _____

$7m^4 - 6m^3$

33. Subtract: $-2y^3 + 4y^2 - 7y + 1$
$\underline{2y^3 - 2y^2 + 4y - 5}$

_____

$-4y^3 + 6y^2 - 11y + 6$

34. The symbol $P(x)$ is read _____.

P of x

35. If $P(x) = 2x^2 - 4x + 7$, then $P(3)$ is found by
replacing each ___ with ___.

x; 3

36. Evaluate $P(3)$.

$P(3) = 2(\underline{})^2 - 4(\underline{}) + 7$

3; 3

$= \underline{} - 12 + 7$

18

$= \underline{}$

13

Let $P(x) = 2 - 5x - x^2$, and evaluate the expressions
in Frames 37–39.

37. $P(-2) = 2 - 5(\underline{}) - (\underline{})^2$

-2; -2

$= 2 + 10 - \underline{}$

4

$= \underline{}$

8

38. $P(-4) = \underline{}$

6

39. $P(0) = \underline{}$

2

3.4 Multiplication of Polynomials

[1] Multiply terms. (Frames 1–6)

[2] Multiply any two polynomials. (Frames 7–10)

[3] Multiply binomials. (Frames 11–22)

[4] Find the product of the sum and difference of two terms. (Frames 23–30)

[5] Find the square of a binomial. (Frames 31–42)

1. The product of two terms can be found by using the
    rules for exponents from the earlier sections of
    this chapter. For example,

    $$9m^5(2m^3) = (\underline{} \cdot \underline{})m^{\overline{} + \overline{}}$$
    $$= \underline{}.$$

    9; 2; 5; 3
    $18m^8$

Find each product in Frames 2–6.

2. $(-3k^5)(4k^7) = (-3 \cdot \underline{})k^{\overline{} + \overline{}}$
    $$= \underline{}$$

    4; 5; 7
    $-12k^{12}$

3. $11y^3(7y^5) = \underline{}$

    $77y^8$

4. $-6p^5(-2p^9) = \underline{}$

    $12p^{14}$

5. $-5y^3(8y) = \underline{}$

    $-40y^4$

6. $7p^{11}(-4p^3) = \underline{}$

    $-28p^{14}$

7. To find the product of $3x^2$ and $5x^2 - 4x + 6$, use
    the \underline{} property to write

    distributive

    $$3x^2(5x^2 - 4x + 6)$$
    $$= (3x^2)(\underline{}) + (\underline{})(-4x) + (3x^2)(\underline{})$$
    $$= \underline{} + \underline{} + \underline{}$$
    $$= \underline{}.$$

    $5x^2$; $3x^2$; 6
    $15x^4$; $-12x^3$; $18x^2$
    $15x^4 - 12x^3 + 18x^2$

8. When working longer problems, it is often better
    to write the two factors vertically, as shown
    below.

    $$-3x^2 - 4x + 2$$
    $$\underline{ - 2x + 3}$$

    To find the product of these two polynomials,
    first multiply 3 times \underline{}.

    $-3x^2 - 4x + 2$

Step 1 $-3x^2 - 4x + 2$
                      $- 2x + 3$
_____

____ ____ ___

$-9x^2; - 12x; + 6$

Then multiply _____ times _____.

$-2x; -3x^2 - 4x + 2$

Step 2 $-3x^2 - 4x + 2$
                      $- 2x + 3$
_____
              $-9x^2 - 12x + 6$

____ ____ _____

$6x^3; 8x^2; -4x$

If we now add, we get the product.

Step 3 $-3x^2 - 4x + 2$
                      $- 2x + 3$
_____
              $-9x^2 - 12x + 6$
$6x^3 + 8x^2 - 4x$
_____

_____

$6x^3 - x^2 - 16x + 6$

Find each of the products in Frames 9–10.

9. $-3x^2 - x + 4$
                  $x - 3$
_____

____ ___ ___

____ ___ ____ _____

_____

$9x^2; 3x; -12$
$-3x^3; -x^2; 4x$

$-3x^3 + 8x^2 + 7x - 12$

10. $5x - 1$
           $3x + 4$
_____

____ ___

____ ____
_____

_____

$20x; -4$
$15x^2; -3x$

$15x^2 + 17x - 4$

11. The product of Frame 10 can be found in another
     way. Start by multiplying the first terms.

        $(5x - 1)(3x + 4)$ $(5x)(\_\_\_\_) = _____$
            ↑   ↑

$3x; 15x^2$

12. Multiply the outside terms.

$$(5x - 1)(3x + 4) \qquad (5x)(\underline{\quad}) = \underline{\quad\quad}$$

4; 20x

13. Multiply the inside terms.

$$(5x - 1)(3x + 4) \qquad (-1)(\underline{\quad}) = \underline{\quad\quad}$$

3x; −3x

14. Multiply the last terms.

$$(5x - 1)(3x + 4) \qquad (-1)(\underline{\quad}) = \underline{\quad\quad}$$

4; −4

15. The final product is found by adding the results from Frames 11–15.

$$(5x - 1)(3x + 4) = \underline{\quad} + 20x + (\underline{\quad}) + (-4)$$
$$= \underline{\quad\quad\quad}$$

15x²; −3x

15x² + 17x − 4

16. The shortcut procedure of Frames 11–15, where we multiply first terms, _____ terms, inside terms, and _____ terms, is called the _____ method.

outside

last

FOIL

Use the FOIL method to find the products in Frames 17–22.

17. $(7m - 2)(m + 4) = 7m^2 + \underline{\quad\quad} + \underline{\quad\quad} + (-8)$
$$= \underline{\quad\quad\quad}$$

28m; −2m

7m² + 26m − 8

18. $(3p + 1)(2p - 1) = \underline{\quad\quad} + (-3p) + 2p + \underline{\quad\quad}$
$$= \underline{\quad\quad\quad}$$

6p²; (−1)

6p² − p − 1

19. $(4q + 7)(8q - 3) = 32q^2 + \underline{\quad\quad} + \underline{\quad\quad} + \underline{\quad\quad}$
$$= \underline{\quad\quad\quad}$$

(−12q); 56q; (−21)

32q² + 44q − 21

20. $(11r + 5)(2r - 3) = \underline{\quad\quad\quad}$

22r² − 23r − 15

21. $(6y + 5z)(4y - z) = \underline{\quad\quad\quad}$

24y² + 14yz − 5z²

22. $(8k + 9m)(2k - 3m) =$ _____

$16k^2 - 6km - 27m^2$

23. One special product that comes up from time to time is called the product of the _____ and _____ of two terms:

sum

difference

$$(x + y)(x - y) = \text{_____}.$$

$x^2 - y^2$

24. The product of the sum and difference of two terms is the _____ of the _____ of the terms.

difference;
squares

Find each of the products in Frames 25—30.

25. $(8x - 3)(8x + 3) = (\underline{\quad})^2 - (\underline{\quad})^2$

$= \underline{\quad\quad\quad\quad}$

8x; 3
$64x^2 - 9$

26. $(7q + 4q)(7p - 4q) = (\quad)^2 - (\quad)^2$

$= \underline{\quad\quad\quad\quad}$

7p; 4q
$49p^2 - 16q^2$

27. $(11r + 2)(11r - 2) =$ _____

$121r^2 - 4$

28. $(5a + 9)(5a - 9) =$ _____

$25a^2 - 81$

29. $(4m + 3s)(4m - 3s) =$ _____

$16m^2 - 9s^2$

30. $(12k - 5r)(12k + 5r) =$ _____

$144k^2 - 25r^2$

31. One final type of binomial product is the square of a _____:

binomial

$$(x + y)^2 = \text{_____}$$

$x^2 + 2xy + y^2$

or $(x - y)^2 =$ _____.

$x^2 - 2xy + y^2$

32. In words: the square of a binomial is the square of the first _____, then _____ the product of both terms, plus the _____ of the second terms.

term; twice
square

Find the square of each binomial in Frames 33–38.

33. $(7y + 3)^2 = (\_\_\_)^2 + 2(\_\_\_)(3) + 3^2$

   $= 49y^2 + _____ + 9$

   7y; 7y

   42y

34. $(3r - 5)^2 = (3r)^2 - 2(\_\_\_\_)(\_\_\_) + \_\_\_$

   $= _____$

   3r; 5; 5^2

   $9r^2 - 30r + 25$

35. $(2a + 3b)^2 = (2a)^2 + 2(2a)(\_\_\_\_) + (\_\_\_\_)^2$

   $= _____$

   3b; 3b

   $4a^2 + 12ab + 9b^2$

36. $(11x + 5y)^2 = _____$

   $121x^2 + 110xy + 25y^2$

37. $(4m - 9n)^2 = _____$

   $16m^2 - 72mn + 81n^2$

38. $(2p - 13q)^2 = _____$

   $4p^2 - 52pq + 169q^2$

Use the special products to multiply the polynomials in Frames 39–42.

39. $[(4s + 3) - 2y]^2 = (\_\_\_\_\_)^2 - 2(_____)(\_\_\_) + (\_\_)^2$

   $= _____ - _____ + _____$

   $= _____$

   4s + 3; 4s + 3; 2y; 2y

   $(16s^2 + 24s + 9)$; $(16sy + 12y)$; $4y^2$

   $16s^2 - 16sy + 4y^2 + 24s - 12y + 9$

40. $[(5x - 2y) + 7][(5x - 2y) - 7] = _____$

   $25x^2 - 20xy + 4y^2 - 49$

41. $[8a + (5b - c)]^2 = _____$

   $64a^2 + 80ab - 16ac + 25b^2 - 10bc + c^2$

42. $[(3r + 5) - 2q][(3r + 5) + 2q] = _____$

   $9r^2 + 30r + 25 - 4q^2$

3.5 Greatest Common Factors; Factoring by Grouping

[1] Write integers in prime factored form. (Frames 1–4)

[2] Find the greatest common factor. (Frames 5–7)

[3] Factor out the greatest common factor. (Frames 8–21)

[4] Factor by grouping. (Frames 22–29)

---

1. A prime number is a positive _____ greater
    than ____ that can be _____ without re-
    mainder only by _____ and 1.

	integer
	1; divided
	itself

    Is 5 prime? (*yes/no*)

	yes

    Is 8 prime? (*yes/no*)

	no

2. Integers are often written in prime _____
    form, as a product of _____ numbers. For
    example, to write 72 in prime factored form,
    divide by the first prime, ___.

	factored
	prime
	2

    $$72 = 2 \cdot \underline{\quad}$$

	36

    Divide by 2 as many times as possible.

    $$72 = 2 \cdot 2 \cdot \underline{\quad}$$
    $$= 2 \cdot 2 \cdot 2 \cdot \underline{\quad}$$

	18
	9

    Now divide by the next prime, ___.

    $$72 = 2 \cdot 2 \cdot 2 \cdot \underline{\quad}$$

	$3 \cdot 3$

    Using exponents, the prime factored form of 72 is

    $$72 = \underline{\quad\quad}.$$

	$2^3 \cdot 3^2$

**Find the prime factored form of each integer in
Frames 3–4.**

3. $150 = 2 \cdot \underline{\quad}$
    $= 2 \cdot 3 \cdot \underline{\quad}$
    $= 2 \cdot 3 \cdot \underline{\quad}$
    $150 = \underline{\quad\quad}$ (Use exponents)

	75
	25
	$5 \cdot 5$
	$2 \cdot 3 \cdot 5^2$

4.   260 = _____ $2^2 \cdot 5 \cdot 13$

5.   To find the greatest _____ factor for a   common
     group of integers, find the largest integer
     that will _____ without _____   divide; remainder
     into every integer in the group.

6.   To find the greatest common factor of 24, 60,
     and 144, write the _____ factored form of   prime
     each number.

     24 = _____   60 = _____ $2^3 \cdot 3$; $2^2 \cdot 3 \cdot 5$
     144 = _____ $2^4 \cdot 3^2$

     The least number of times 2 appears is ____,   2
     the least number of times 3 appears is ____,   1
     and ____ does not appear in each factorization.   5
     Thus, the

     greatest common factor = _____ = ____. $2^2 \cdot 3$; 12

7.   To find the greatest common factor of terms with
     variables, write the prime factored form of the
     _____ coefficients of each term. The   numerical
     greatest common factor is the _____ of each   product
     different factor (_____ or _____) that   numerical;
     is common to all terms, with the exponent that is   variable
     the _____ exponent on the factor that appears   least
     in any term.

8.   The greatest common factor for the terms of
     $16m^2 + 4m + 12$ is ____ .   4

9.   The greatest common factor for $x^2 + x^3 + x^5$ is
     ____ . When working with variables, use the x^2
     _____ exponent on any of the variables.   least

10. We know that $2x(3x^2 - 2) =$ _____. The
process of starting with a polynomial like
$6x^3 - 4x$ and writing it as the product of ____
and _____ is called _____ the
polynomial.

$6x^3 - 4x$

$2x$

$3x^2 - 2$; factoring

11. The first step in factoring a polynomial is to
find the greatest common _____, the largest
term that is a factor of all the _____ of the
polynomial.

factor

terms

12. To factor $9m^3 - 18m^2 + 6m$, note that each of the
three terms of the polynomial contains ___ as a
factor. This means

$$9m^3 - 18m^2 + 6m$$

$$= (\underline{})(\underline{}) + (\underline{})(\underline{}) + (\underline{})(\underline{})$$

or, using the _____ property,

$$= (\underline{})(\underline{}).$$

The term 3m is the _____ _____
_____.

$3m$

$3m$; $3m^2$; $3m$;
$-6m$; $3m$; 2
distributive

$3m$; $3m^2 - 6m + 2$

greatest; common

factor

**Factor the greatest common factor from the polynomials
in Frames 13–21.**

13. $12p^4 - 18p^3 + 9p^2$

$$= (\underline{})(\underline{}) + (\underline{})(\underline{}) + (\underline{})(\underline{})$$

$$= (\underline{})(\underline{})$$

$3p^2$; $4p^2$; $3p^2$;
$-6p$; $3p^2$; 3
$3p^2$; $4p^2 - 6p + 3$

14. $-24r^5 + 36r^6 + 60r^7$

$$= (12r^5)(\underline{}) + (12r^5)(\underline{}) + (\underline{})(\underline{})$$

$$= (\underline{})(\underline{})$$

-2; $3r$; $12r^5$; $5r^2$

$12r^5$; $-2 + 3r + 5r^2$

We could also have worked this problem by factoring out _____, which would lead to an equivalent answer.

$-12r^5$

15. $64m^4n^3 - 48m^5n^2$

   = (_____)(____) + (_____)(____)

   = (_____)(_____)

   $16m^4n^2$; $4n$;
   $16m^4n^2$; $-3m$
   $16m^4n^2$; $4n - 3m$

16. $30p^4q^6 + 50p^3q^7$ = (_____)(_____)

   $10p^3q^6$; $3p + 5q$

17. $72r^5m^6 - 60r^6m^3 + 48r^7m^2$ = (____)(_____)

   $12r^5m^2$;
   $6m^4 - 5rm + 4r^2$

18. $50m^4a^3b^2 + 125m^7a^5b + 75m^3a^4b^3$ = _____

   $25m^3a^3b(2mb +$
   $5m^4a^2 + 3ab^2)$

19. $3(a + b)^2 - 6(a + b)$ has a common factor of

   _____.

   $3(a + b)$

   $3(a + b)^2 - 6(a + b)$ = [_____][_____]

   $3(a + b)$;
   $(a + b) - 2$

20. In the same way,

   $9(m + n)^3b^2 + 6(m + n)^4b$

   = [_____][_____].

   $3(m + n)^3b$;
   $3b + 2(m + n)$

   = $3(m + n)^3b(_____)$

   $3b + 2m + 2n$

21. $5a(2a - 3) + 7(2a - 3)$ = _____

   $(2a - 3)(5a + 7)$

22. Greatest common factors can be used to factor by

   _____. For example,

   grouping

   $$mx - my + 3x - 3y$$

   can be factored by _____ if we first write

   grouping

   $mx - my$ as _____, and then write $3x - 3y$ as

   $m(x - y)$

   _____.

   $3(x - y)$

   $mx - my + 3x - 3y = m(x - y) +$ _____

   $3(x - y)$

   = $(x - y)($_____$)$

   $m + 3$

   Here _____ is a common factor.

   $x - y$

Factor each of the polynomials in Frames 23—29. Use factoring by grouping.

23. $ma - mb + na - nb$ = (___)(_____) + (___)(_____)   m; a − b; n; a − b

                    = (_____)(_____)   a − b; m + n

24. $8ax - 6ay + 4xb - 3yb$

      = (___)(_____) + (___)(_____)   2a; 4x − 3y; b;
                                        4x − 3y

      = (_____)(_____)   4x − 3y; 2a + b

25. $5ka + 3ma - 5kb - 3mb$ = (_____)(_____)   5k + 3m; a − b

26. $21ab + 6ac + 14bd + 4cd$ = (_____)(_____)   7b + 2c; 3a + 2d

27. $35m^2 - 14m + 40m - 16$ = _____ $(5m - 2)(7m + 8)$

28. $3r^2s - 6r^2 + s - 2$ = _____ $(s - 2)(3r^2 + 1)$

29. $a^5b^2 + a^5 - 4b^2 - 4$ = _____ $(b^2 + 1)(a^5 - 4)$

3.6   Factoring Trinomials

1   Factor trinomials when the coefficient of the squared term is 1. (Frames 1—13)

2   Factor trinomials when the coefficient of the squared term is not 1. (Frames 14—23)

3   Use an alternative method for factoring trinomials.   (Frames 24—43)

4   Factor by substitution.   (Frames 44—47)

---

1.   We know that $(x + 2)(x - 3)$ = _____.   The $x^2 - x - 6$
     process of starting with $x^2 - x - 6$ and writing
     it as the product of _____ and _____   x + 2; x − 3
     is called _____.   In this section, we   factoring
     first discuss the factoring of polynomials like
     $x^2 + 3x + 2$, where the coefficient of x^2 is ___.   1

2. To factor $x^2 + 3x + 2$, look for two numbers whose product is ____ and whose sum is ____. These two numbers are ____ and ____. Thus

$$x^2 + 3x + 2 = (\underline{\hspace{1cm}})(\underline{\hspace{1cm}}).$$

This method of factoring trinomials works only when the _____ of x^2 is ____.

2; 3

1; 2

x + 1; x + 2

coefficient; 1

3. To factor $x^2 - 5x + 6$, look for two numbers whose product is ____ and whose sum is ____. These two numbers are ____ and ____. Thus,

$$x^2 - 5x + 6 = (\underline{\hspace{1cm}})(\underline{\hspace{1cm}}).$$

6; −5

−2; −3

x − 2; x − 3

Factor each of the trinomials in Frames 4–7.

4. $a^2 - 9a + 20 = (\underline{\hspace{1cm}})(\underline{\hspace{1cm}})$

a − 4; a − 5

5. $m^2 + 11m + 30 = (\underline{\hspace{1cm}})(\underline{\hspace{1cm}})$

m + 6; m + 5

6. $k^2 - 10k - 24 = (\underline{\hspace{1cm}})(\underline{\hspace{1cm}})$

k − 12; k + 2

7. $r^2 - 6r - 27 = (\underline{\hspace{1cm}})(\underline{\hspace{1cm}})$

r − 9; r + 3

8. To factor $3p^3 + 6p^2 - 24p$, first _____ out the _____ _____ factor. Here, this is ____. We have

$$3p^3 + 6p^2 - 24p = (\underline{\hspace{0.5cm}})(\underline{\hspace{2cm}})$$
$$= (\underline{\hspace{0.5cm}})(\underline{\hspace{1cm}})(\underline{\hspace{1cm}})$$

factor

greatest; common

3p

3p; $p^2 + 2p - 8$

3p; p + 4; p − 2

Factor each of the polynomials in Frames 9–13.

9. $9k^4 - 81k^3 - 90k^2 = (\underline{\hspace{0.5cm}})(\underline{\hspace{2cm}})$
$$= (\underline{\hspace{0.5cm}})(\underline{\hspace{1cm}})(\underline{\hspace{1cm}})$$

$9k^2$; $k^2 - 9k - 10$

$9k^2$; k − 10; k + 1

10. $r^3 + 7r^2 - 18r = (\underline{\hspace{0.5cm}})(\underline{\hspace{2cm}})$
$$= (\underline{\hspace{0.5cm}})(\underline{\hspace{1cm}})(\underline{\hspace{1cm}})$$

r; $r^2 + 7r - 18$

r; r + 9; r − 2

11. $a^2 - 10ab + 21b^2 = ($ _____ $)($ _____ $)$ a – 3b; a – 7b

12. $m^2 + mn - 72n^2 = ($ _____ $)($ _____ $)$ m + 9n; m – 8n

13. $y^2 + 5yz - 84z^2 = ($ _____ $)($ _____ $)$ y – 7z; y + 12z

14. To factor trinomials where the coefficient of the
     square term is not ____, we can use an extension   1
     of the method above.

15. For example, to factor $2m^2 - m - 10$, first find
     the product of the coefficient of the _____   squared
     term and the _____ term. The coefficient of   last
     the squared term is ___. The last term is ___.   2; –10
     The product of these numbers is (2)(-10) = ____.   –20
     The middle coefficient of $2m^2 - m - 10$ is ___.   –1

16. To factor $2m^2 - m - 10$, we need two numbers whose
     product is _____, and whose sum is ___. These   –20; –1
     numbers are ___ and ____.   –5; 4

17. Write the middle term, –m, as

              $-m =$ _____ $+$ ____.   –5m; 4m

18. Our trinomial, $2m^2 - m - 10$, now becomes

         $2m^2 - m - 10 = 2m^2 +$ _____ $- 10$.   –5m + 4n

19. Factor by grouping.

         $2m^2 - 5m + 4m - 10 = m($ _____ $) + 2($ _____ $)$ 2m – 5; 2m – 5
                             $=$ _____   (2m – 5)(m + 2)

20. To check, find the product $(2m - 5)(m + 2)$; you
     should get _____. $2m^2 - m - 10$

Factor in Frames 21–23.

21. $6a^2 + 7a - 5$

Multiply 6 and ____ to get ____. We need two numbers that multiply to give ____ and add to give ____. These numbers are ____ and ____. Write the trinomial as

$$6a^2 + 7a - 5 = 6a^2 + \underline{\hspace{2cm}} - 5.$$

Factor by grouping.

$$6a^2 + 10a - 3a - 5 = 2a(\underline{\hspace{1.5cm}}) - 1(\underline{\hspace{1.5cm}})$$
$$= \underline{\hspace{3cm}}$$

−5; −30
−30
7; 10; −3

10a − 3a

3a + 5; 3a + 5
(3a + 5)(2a − 1)

22. $12k^2 - 17k + 6$

We need two numbers that multiply to give $12(\underline{\hspace{0.5cm}}) = \underline{\hspace{1cm}}$ and add to give ____. The numbers are ____ and ____.

$$12k^2 - 17k + 6 = 12k^2 + \underline{\hspace{1.5cm}} + \underline{\hspace{1.5cm}} + 6$$
$$= 4k(\underline{\hspace{2cm}}) - 3(\underline{\hspace{2cm}})$$
$$= \underline{\hspace{4cm}}$$

6; 72; −17
−8; −9

(−8k); (−9k)
3k − 2; 3k − 2
(3k − 2)(4k − 3)

23. $15z^2 - 7z - 4 = \underline{\hspace{3cm}}$

(5z − 4)(3z + 1)

The next few frames show an alternate method of factoring.

24. To factor $2a^2 + 5a - 3$, first find the factors of $2a^2$, ____ and ____. The number −3 can be factored either as ____ and ____ or ____ and ____. Try various combinations of these factors to find the ones that work. First try

$$(2a + 1)(a - 3) = \underline{\hspace{2cm}},$$

which (*is/is not*) correct. Since only the middle sign of the product is off, try

$$(\underline{\hspace{2cm}})(\underline{\hspace{2cm}}) = 2a^2 + 5a - 3.$$

2a; a
3; −1; −3
1

$2a^2 - 5a - 3$

is not

2a − 1; a + 3

25. To factor $3b^2 - b - 2$, write the factors of $3b^2$, ____ and ____. The factors of -2 are ____ and ____ or ____ and ____. Try $(3b + 2)($ _____ $)$. This gives

$$(3b + 2)(b - 1) = \text{\underline{\hspace{3cm}}},$$

which (*is/is not*) correct.

3b; b; -2

1; 2; -1; b - 1

$3b^2 - b - 2$

is

26. Factor $6r^2 + 11r + 3$. Factors of $6r^2$ include ____ and ____ or ____ and ____. Factors of 3 include ____ and ____ or ____ and ____. Since the middle term of the polynomial that we are trying to factor has a ____ sign, we should use ____ and ____. Let us try $3r$ and ____ as factors of $6r^2$. If we try $2r + 3$ as one possible factor,

$$(2r + 3)(\text{\underline{\hspace{2cm}}}) = \text{\underline{\hspace{3cm}}},$$

which (*is/is not*) correct.

6r

r; 3r; 2r

3; 1; -3; -1

+; 3; 1

2r

$3r + 1$; $6r^2 + 11r + 3$

is

27. When factoring a trinomial of the form $ax^2 + bx + c$ with a negative, begin by factoring out ____.

$$-5x^2 - 13x + 6 = -1(\text{\underline{\hspace{3cm}}})$$
$$= -1(5x - 2)(\text{\underline{\hspace{1.5cm}}})$$

This answer may also be written correctly as $(\text{\underline{\hspace{2cm}}})(x + 3)$ or $(5x - 2)(\text{\underline{\hspace{1.5cm}}})$ when one of the binomial _____ is multiplied by the factor ____.

-1

$5x^2 + 13x - 6$

x + 3

-5x + 2; -x - 3

factors

-1

Factor the polynomials in Frames 28—38 by any method.

28. $12m^2 + 11m + 2 = (\text{\underline{\hspace{2cm}}})(\text{\underline{\hspace{2cm}}})$

3m + 2; 4m + 1

29. $15p^2 + 26p + 8 = (\text{\underline{\hspace{2cm}}})(\text{\underline{\hspace{2cm}}})$

5p + 2; 3p + 4

30. $-8p^2 - 10p + 3 = (\text{\underline{\hspace{2cm}}})(\text{\underline{\hspace{2cm}}})$

-4p + 1; 2p + 3
or 4p - 1; -2p - 3

31. $21k^2 + 22k - 8 = ($_____$)($_____$)$ $7k - 2;\ 3k + 4$

32. $36x^2 + x - 2 = ($_____$)($_____$)$ $9x - 2;\ 4x + 1$

33. $8y^2 + 2xy - 3x^2 = ($_____$)($_____$)$ $4y + 3x;\ 2y - x$

34. $6q^2 - 19pq - 20p^2 = ($_____$)($_____$)$ $6q + 5p;\ q - 4p$

35. $4r^2 - 25 = ($_____$)($_____$)$ $2r + 5;\ 2r - 5$

36. $9k^2 - 16 = ($_____$)($_____$)$ $3k + 4;\ 3k - 4$

37. $-9a^2 - 6a - 1 = ($_____$)($_____$)$ $-3a - 1;\ 3a + 1$

38. $30m^2 + 7mn - 15n^2 = ($_____$)($_____$)$ $5m - 3n;\ 6m + 5n$

First factor out a common factor in Frames 39–40. Then factor completely.

39. $4m^2n + 12mn - 72n = ($_____$)($_____$)$ $4n;\ m^2 + 3m - 18$

       $= ($_____$)($_____$)($_____$)$ $4n;\ m + 6;\ m - 3$

40. $6x^4z + 9x^3z^2 - 6x^2z^3 = ($_____$)($_____$)$ $3x^2z;$ $2x^2 + 3xz - 2z^2$

        $= ($_____$)($_____$)($_____$)$ $3x^2z;\ 2x - z;$ $x + 2z$

Factor.

41. $y^2(m + n)^2 - 5y(m + n)^2 - 14(m + n)^2$

   $= [$_____$]($_____$)$ $(m + n)^2;$ $y^2 - 5y - 14$

   $= [$_____$]($_____$)($_____$)$ $(m + n)^2;\ y - 7$ $y + 2$

42. $6(a + b)^2 + (a + b) - 2$

   $= [$_____$][$_____$]$ $3(a + b) + 2;$ $2(a + b) - 1$

43. $6(m - n)^2 - 5(m - n) - 4$

    = [_____][_____]

                                           $2(m - n) + 1;$
                                           $3(m - n) - 4$

44. To factor $2y^4 - 3y^2 - 9$, let x = _____. This y^2

converts the polynomial into _____, $2x^2 - 3x - 9$

which can be factored as (_____)(_____). $x - 3;$ $2x + 3$

Since x = _____, write y^2

    $2y^4 - 3y^2 - 9 =$ (_____)(_____). $y^2 - 3;$ $2y^2 + 3$

45. $6m^4 + 5m^2 - 25 =$ (_____)(_____) $3m^2 - 5;$ $2m^2 + 5$

46. $8k^4 + 10k^2 - 7 =$ (_____)(_____) $4k^2 + 7;$ $2k^2 - 1$

47. $8k^4 - 18k^2m - 5m^2 =$ (_____)(_____) $4k^2 + m;$ $2k^2 - 5m$

3.7 Special Factoring

[1] Factor the difference of two squares. (Frames 1–8)

[2] Factor a perfect square trinomial. (Frames 9–14)

[3] Factor the difference of two cubes. (Frames 15–22)

[4] Factor the sum of two cubes. (Frames 23–28)

1. $x^2 - y^2 =$ (_____)(_____) $x + y;$ $x - y$

    $x^2 - y^2$ is called the difference of two _____.   squares

Factor the polynomials in Frames 2–7.

2. $a^2 - 16b^2 =$ (_____)(_____) $a + 4b;$ $a - 4b$

3. $4m^2 - 9n^2 =$ (_____)(_____) $2m + 3n;$ $2m - 3n$

4. $25p^2 - 81q^2 =$ (_____)(_____) $5p + 9q;$ $5p - 9q$

5. $121a^2 - 16b^2 =$ (_____)(_____) $11a + 4b;$ $11a - 4b$

6. $a^4 - b^4 = ($ _____ $)($ _____ $)$ $a^2 + b^2$; $a^2 - b^2$

     $= ($ _____ $)($ _____ $)($ _____ $)$ $a^2 + b^2$; $a + b$;
     $a - b$

7. $m^2 - 16 = ($ _____ $)($ _____ $)$ $m^2 + 4$; $m^2 - 4$

     $= ($ _____ $)($ _____ $)($ _____ $)$ $m^2 + 4$; $m + 2$;
     $m - 2$

8. $m^2 + 4$ (*can/cannot*) be factored. In general,   cannot

     we cannot factor the _____ of two squares.   sum

9. $a^2 + 2ab + b^2 = $ _____ $(a + b)^2$

     The expression $a^2 + 2ab + b^2$ is called a _____   perfect
     square trinomial.

Factor the polynomials in Frames 10–14.

10. $k^2 + 6k + 9 = ($ _____ $)^2$ $k + 3$

11. $b^2 + 4b + 4 = ($ _____ $)^2$ $b + 2$

12. $9k^2 - 12kq + 4q^2 = ($ _____ $)^2$ $3k - 2q$

13. $4r^2 + 4rs + s^2 = ($ _____ $)^2$ $2r + s$

14. $4x^2 - 20xy + 25y^2 = ($ _____ $)^2$ $2x - 5y$

15. $x^3 - y^3 = ($ _____ $)($ _____ $)$ $x - y$; $x^2 + xy + y^2$

     $x^3 - y^3$ is called the _____ of two   difference
     _____ .   cubes

Factor each of the polynomials in Frames 16–22.

16. $8a^3 - 1 = ($ _____ $)($ _____ $)$ $2a - 1$;
     $4a^2 + 2a + 1$

17. $k^3 - 8 = ($ _____ $)($ _____ $)$ $k - 2$; $k^2 + 2k + 4$

18. $a^3 - 27 = ($ _____ $)($ _____ $)$

$a - 3;\ a^2 + 3a + 9$

19. $x^3 - 8y^3 = ($ _____ $)($ _____ $)$

$x - 2y;$
$x^2 + 2xy + 4y^2$

20. $8a^3 - 27b^3 = ($ _____ $)($ _____ $)$

$2a - 3b;$
$4a^2 + 6ab + 9b^2$

21. $64m^3 - n^3 = ($ _____ $)($ _____ $)$

$4m - n;$
$16m^2 + 4mn + n^2$

22. $x^9 - y^9 = ($ _____ $)^3 - ($ _____ $)^3$

$x^3;\ y^3$

  $= ($ _____ $)($ _____ $)$

$x^3 - y^3;$
$x^6 + x^3y^3 + y^6$

  $= ($ _____ $)($ _____ $)(x^6 + x^3y^3 + y^6)$

$x - y;\ x^2 + xy + y^2$

23. $x^3 + y^3 = ($ _____ $)($ _____ $)$

$x + y;\ x^2 - xy + y^2$

  $x^3 + y^3$ is called the _____ of two _____.

sum; cubes

Factor each of the polynomials in Frames 24–27.

24. $27b^3 + 1 = ($ _____ $)($ _____ $)$

$3b + 1;\ 9b^2 - 3b + 1$

25. $1000a^3 + 27 = (10a + 3)($ _____ $)$

$100a^2 - 30a + 9$

26. $a^3 + 27b^3 = ($ _____ $)($ _____ $)$

$a + 3b;$
$a^2 - 3ab + 9b^2$

27. $64x^3 + 27y^2 = (4x + 3y)($ _____ $)$

$16x^2 - 12xy + 9y^2$

28. $(m - 3)^3 + 8$

  $= (m - 3)^3 + ($ _____ $)^3$

2

  $= [(m - 3) + $ _____ $][(m - 3)^2 + $ _____ $+ 2^2]$

$2;\ 2(m - 3)$

  $= (m - $ _____ $)(m^2 - 6m + 9 + $ _____ $- 6 + $ _____ $)$

$1;\ 2m;\ 4$

  $= (m - 1)($ _____ $)$

$m^2 - 4m + 7$

3.8 General Methods of Factoring

[1] Know the first step in trying to factor a polynomial.
(Frames 1–8)

[2] Know the rules for factoring binomials. (Frames 9–16)

[3] Know the rules for factoring trinomials. (Frames 17–28)

[4] Know the rules for factoring polynomials of more than three terms.
(Frames 29–35)

---

1. A polynomial is in factored form if the polynomial
is written as a product of _____ polynomials
with _____ coefficients. None of the
polynomial factors can be further _____,
except that a _____ factor need not be
_____ completely.

prime
integer
factored
monomial
factored

2. The order of the factors (*does/does not*) matter.

does not

3. As the first step in factoring a polynomial,
always look for a _____ factor.

common

Factor out the greatest common factor in Frames 4–8.

4. $12m + 36 = $ _____

$12(m + 3)$

5. $8x^2 + 16x^3 = $ _____
(Be careful: don't forget the ____.)

$8x^2(1 + 2x)$
1

6. $10m^2n^3 + 15m^4n^2 = $ _____

$5m^2n^2(2n + 3m^2)$

7. $6x(x + 4) - 5(x + 4) = $ _____

$(x + 4)(6x - 5)$

8. $(a - 1)^3b^4 - 2(a - 1)^5b^2 = $ _____

$(a - 1)^3b^2 \cdot$
$[b^2 - 2(a - 1)^2]$

9. For a polynomial of two terms, (a _____),
    look for one of the following forms.

    binomial

    Difference of two squares:

    $$x^2 - y^2 = \underline{\hspace{4cm}}$$

    $(x + y)(x - y)$

    Difference of two cubes:

    $$x^3 - y^3 = (x - y)(\underline{\hspace{2.5cm}})$$

    $x^2 + xy + y^2$

    Sum of two cubes:

    $$x^3 + y^3 = \underline{\hspace{5cm}}$$

    $(x + y)(x^2 - xy + y^2)$

Factor the binomials in Frames 10–16.

10. $121m^2 - 25 = (\underline{\hspace{1.2cm}})^2 - (\underline{\hspace{1.2cm}})^2$

     $= \underline{\hspace{3cm}}$

     11m; 5
     $(11m + 5)(11m - 5)$

11. $64y^2 - 9z^2 = (\underline{\hspace{2cm}})(\underline{\hspace{2cm}})$

     $8y + 3z; \ 8y - 3z$

12. $8p^3 + 125 = (\underline{\hspace{1.5cm}})^3 + (\underline{\hspace{1.2cm}})^3$

     $= (\underline{\hspace{2cm}})[(2p)^2 - \underline{\hspace{1.5cm}} + 5^2]$

     $= (2p + 5)(\underline{\hspace{3cm}})$

     2p; 5
     2p + 5; 10p
     $4p^2 - 10p + 25$

13. $r^3 - 27y^3 = r^3 - (\underline{\hspace{1.5cm}})^3$

     $= (\underline{\hspace{2cm}})[r^2 + \underline{\hspace{1.5cm}} + (3y)^2]$

     $= (r - 3y)(\underline{\hspace{3cm}})$

     3y
     r - 3y; 3ry
     $r^2 + 3ry + 9y^2$

14. $1000a^3 - b^3 = (\underline{\hspace{1.5cm}})(\underline{\hspace{3.5cm}})$

     10a - b;
     $100a^2 + 10ab + b^2$

15. $27w^3 + 1 = \underline{\hspace{7cm}}$

     $(3w + 1) \cdot$
     $(9w^2 - 3w + 1)$

16. $125y^3 + 216z^6 = \underline{\hspace{6cm}}$

     $(5y + 6z^2) \cdot$
     $(25y^2 - 30yz^2 + 36z^4)$

17. If a polynomial has three terms (a _____),   trinomial

first look to see if it is a perfect _____   square

trinomial.

$$x^2 + 2xy + y^2 = \underline{\hspace{3cm}}$$ $(x + y)^2$

Decide if the trinomials in Frames 18—20 are perfect squares. If they are, factor them.

18. $p^2 - 16p + 64$ _____   perfect square;
$(p - 8)^2$

19. $25y^2 - 60y + 144$ _____   not perfect square

20. $49k^2 + 84km + 36m^2$ _____   perfect square;
$(7k + 6m)^2$

21. If a trinomial is not a perfect square, factor

by the methods of Section 3.7. To factor

$z^2 - 5z - 14$, look for two numbers whose product

is _____ and whose sum is _____. Complete the   -14; -5

factorization.

$$z^2 - 5z - 14 = (\underline{\hspace{2cm}})(\underline{\hspace{2cm}})$$ $z - 7$; $z + 2$

22. To factor $6m^2 - 13m - 5$, we need two numbers

whose product is _____, and whose sum is   -30

_____. The numbers are _____ and _____. Write   -13; -15; 2

$6m^2 - 13m - 5$ as $6m^2 +$ _____ $- 5$. $-15m + 2m$

Factor by _____.   grouping

$$6m^2 - 15m + 2m - 5 = 3m(\underline{\hspace{2cm}}) + (\underline{\hspace{1cm}})(2m - 5)$$ $2m - 5$; 1

$$= (2m - 5)(\underline{\hspace{2cm}}).$$ $3m + 1$

Factor each trinomial in Frames 23—28.

23. $m^2 - 2m - 8 =$ _____ $(m - 4)(m + 2)$

24. $p^2 + 4p - 21 = $ _____ $(p + 7)(p - 3)$

25. $2r^2 - r - 15 = $ _____ $(2r + 5)(r - 3)$

26. $3a^2 + 13a + 14 = $ _____ $(3a + 7)(a + 2)$

27. $10m^2 + 17mn + 6n^2 = $ _____ $(5m + 6n)(2m + n)$

28. $12k^2 - 7ks - 10s^2 = $ _____ $(3k + 2s)(4k - 5s)$

29. If a polynomial has four or more terms, try
     factoring by _____ .   grouping

Factor by grouping in Frames 30–35.

30. $ab + ac - 2b - 2c = $ _____ $(a - 2)(b + c)$

31. $m^2 + 4m - nm - 4n = $ _____ $(m - n)(m + 4)$

32. $yx - 2yz + zx - 2z^2 = $ _____ $(y + z)(x - 2z)$

33. $rt - rv + 3st - 3sv = $ _____ $(r + 3s)(t - v)$

34. $ks + 3kt + 6s + 18t = $ _____ $(k + 6)(s + 3t)$

35. $27xy^2 - 3x + 18y^2 - 2 = $ _____ $(3x+2)(3y+1)(3y-1)$

3.9   Solving Equations by Factoring

1   Learn the zero–factor property.   (Frames 1–3)

2   Use the zero–factor property to solve equations. (Frames 4–16)

3   Solve applied problems that require the zero–factor property.
    (Frames 17–18)

1. $x^2 - 4x - 5 = 0$ is an example of a _____   quadratic
    equation.

2. In general, if a, b, and c are real numbers, then
    $ax^2 + bx + c = 0$ is called a _____ equation.   quadratic
    (We assume a ___ 0.) $\neq$

3. One method of solving quadratic equations uses the
    the zero-factor property: if a and b are numbers
    such that ab = 0, then _____ or _____ .   a = 0; b = 0

4. To find the solutions of (x - 4)(x + 2) = 0,
    use the zero-factor property, which gives

    _____ or _____ .   x - 4 = 0;
                                            x + 2 = 0

    If we solve each of these equations, we get

    x = ____ or x = ____ .   4; -2

    The solution set of (x - 4)(x + 2) = 0
    is _____ . $\{4, -2\}$

5. To solve (2x - 1)(4x + 3) = 0, start with

    _____ or _____   2x - 1 = 0;
                                          4x + 3 = 0

    x = ____ or x = ____ . $\dfrac{1}{2}$; $-\dfrac{3}{4}$

    The solution set of this equation is _____ . $\{1/2, -3/4\}$

6. To solve (3r + 5)(2r - 7) = 0 start with

    _____ or _____   3r + 5 = 0
                                          2r - 7 = 0

    r = ____ or r = ____ . $-\dfrac{5}{3}$; $\dfrac{7}{2}$

    The solution set is _____ . $\{-5/3, 7/2\}$

7. To solve the equation $x^2 - 4x - 5 = 0$, we must

   first try to _____ it.   Factor here as

             (_____)(_____) = 0.

   This leads to

             _____ or _____

           x = _____ or x = _____.

   The solution set is _____.

factor
x − 5; x + 1
x − 5 = 0; x + 1 = 0
5; −1
{5, −1}

Use factoring to solve the quadratic equation in Frames 8–14.

8. $m^2 - 5m + 6 = 0$

         (_____)(_____) = 0

         _____ or _____

         m = _____ or m = _____

   The solution set is _____.

m − 3; m − 2
m − 3 = 0; m − 2 = 0
3; 2
{3, 2}

9. $r^2 - 9r - 10 = 0$

         (_____)(_____) = 0

         _____ or _____

         r = _____ or r = _____

             Solution set: _____

r − 10; r + 1
r − 10 = 0; r + 1 = 0
10; −1
{10, −1}

10. $p^2 + 3p - 10 = 0$

         (_____)(_____) = 0

         _____ or _____

         p = _____ or p = _____

             Solution set: _____

p + 5; p − 2
p + 5 = 0; p − 2 = 0
−5; 2
{−5, 2}

11. $6k^2 + 5k - 6 = 0$

    (_____)(_____) = 0 $3k - 2$; $2k + 3$

    _____ or _____ $3k - 2 = 0$;
                                             $2k + 3 = 0$

    k = _____ or k = _____ $\frac{2}{3}$; $-\frac{3}{2}$

    Solution set: _____ $\{\frac{2}{3}, -\frac{3}{2}\}$

12. $10m^2 + 13m - 3 = 0$

    m = _____ or m = _____ $\frac{1}{5}$; $-\frac{3}{2}$

    Solution set: _____ $\{\frac{1}{5}, -\frac{3}{2}\}$

13. Solve $6r^2 - 5r = 0$

    Factor out the common factor.

    _____ = 0 $r(6r - 5)$

    _____ = 0 or _____ = 0 r; $6r - 5$

    r = _____ or r = _____ 0; $\frac{5}{6}$

    Solution set: _____ $\{0, \frac{5}{6}\}$

14. Solve $6m(m + 2) = m + 10$.

    First get the right side equal to ___. 0

    _____ = 0 $6m^2 + 11m - 10$

    m = _____ or m = _____ $\frac{2}{3}$; $-\frac{5}{2}$

    Solution set: _____ $\{\frac{2}{3}, -\frac{5}{2}\}$

15. Solve $2y^3 + y^2 = 6y$.

    First subtract ____ from both sides to get the $6y$
    right side equal to ___. 0

    _____ = 0 $2y^3 + y^2 - 6y = 0$

Factor completely, remembering to factor out the common factor of ___.

         y

_____ = 0 $y(y + 2)(2y - 3)$

____ = 0   or   _____ = 0   or   _____ = 0 $y; \ y + 2; \ 2y - 3$

$y =$ ____   or $y =$ ____   or $y =$ ____ $0; \ -2; \ \dfrac{3}{2}$

Solution set: _____ $\left\{0, \ -2, \ \dfrac{3}{2}\right\}$

16.   Solve $4p^3 + 4p^2 - 3p = 0$.

Solution set: _____ $\left\{0, \ \dfrac{1}{2}, \ -\dfrac{3}{2}\right\}$

Solve each applied problem in Frames 17–18.

17.   **Two integers have a sum of 12. The sum of their squares is 90. Find the integers.**

Let $x =$ one integer. Since the sum of the integers is ____, the other integer is _____. $12; \ 12 - x$
The sum of the squares of the integers is ____, 90
giving the equation

_____ = 90. $x^2 + (12 - x)^2$

Solve the equation.

$x^2 +$ _____ = 90 $144 - 24x + x^2$
$2x^2 - 24x +$ _____ = 0 54
$x^2 - 12x +$ _____ = 0 27

Factor.

_____ = 0 $(x - 9)(x - 3)$

Set each factor equal to 0.

_____ = 0   or   _____ = 0 $x - 9; \ x - 3$
$x =$ ____   or $x =$ ____ $9; \ 3$

If $x = 9$, then $12 - x =$ ___. If $x = 3$, then 3
$12 - x =$ ___. In either case, the integers 9
are ___ and ___. $9; \ 3$

18. **Find two consecutive odd integers whose product is 35.**

                                    _____

5 and 7 or
−7 and −5

Chapter 3   Test

The answers for these questions are at the back of this Study Guide.

Simplify each of the following.

1. $\left(\frac{4}{3}\right)^2$ 1. _____

2. $9x^2y^3 \cdot 4xy^4$ 2. _____

3. $\dfrac{(3p^2q)^2(4pq^2)^3}{(27p^{-5})(8q^5)^2}$ 3. _____

4. $\dfrac{(5a^2b^3)(2a^{-2}b^2)^3}{10(a^2b^3)^2(b^2)^4}$ 4. _____

5. $\dfrac{x^m \cdot x^{1-m}}{x^3x^0}$ 5. _____

6. Write .000904 in scientific notation.   6. _____

Perform the following operations.

7. $3y^2x(4x^2y - 3xy^2)$ 7. _____

8. $(5y + 2z)(3y - 4z)$ 8. _____

9. $(4x + 9)(4x - 9)$ 9. _____

10. $(p - 3)(p^2 - 4p + 6)$ 10. _____

11. $(2a - 3)(4a^2 - 4a + 2)$ 11. _____

12. $(5x^2 - 6x + 3) - (2x^2 - 4x + 5) - (3 - x)$ 12. _____

13. $(5 - 3y^2) + (2y^2 - 4) - [(3 - y^2) - (-4 + 2y^2)]$ 13. _____

14. $[(2x + y) - 3]^2$ 14. _____

Determine the degree of each polynomial and write the coefficient of the 3rd term, if one exists.

15. $-7ab^2 + 2a^3 + 4$

15. _____

16. $2x^2y + 3x - 2y^4$

16. _____

17. $12 - 2abc - 7a^2$

17. _____

18. $-5x + 2$

18. _____

Identify each polynomial as a monomial, binomial or trinomial.

19. $2x + 3$

19. _____

20. $-8ab^3$

20. _____

Factor each of the following.

21. $6a^2 + 12a + 9$

21. _____

22. $7p^2q + 14pq$

22. _____

23. $x^2 - 9x - 10$

23. _____

24. $-p^2 - 6p - 5$

24. _____

25. $6k^2 + 4k - 3$

25. _____

26. $8a^2 + 6ak - 9k$

26. _____

27. $64m^2 - 9n^2$

27. _____

28. $121b^2 - 36$

28. _____

29. $x^3 - 8$

29. _____

30. $27p^3 - 8$

30. _____

31. $p^3 + 8$

31. _____

32. $8a^6 + 1$

32. _____

33. $-a^2 - 14a - 49$

33. _____

34. $m^2 - 8mn + 16n^2$

34. _____

35. $64r^2 + 48rs + 9s^2$

35. _____

36. $km - rm + kn - rn$

36. _____

37. $3ab + 2a - 3bc - 2c$

37. _____

38. $1 + x + y + xy$

38. _____

39. $32xy^2 - 18x + 16y^2 - 9$

39. _____

40. If $P(x) = x^3 + 2x^2 - 3x - 5$, find $P(-2)$.

40. _____

Solve each of the following by factoring.

41. $2x^2 + 7x - 4$

41. _____

42. $6x^2 - x - 15 = 0$

42. _____

43. $15a^3 + 7a^2 - 2a = 0$

43. _____

44. Find two consecutive odd integers whose product is 99.

44. _____

CHAPTER 4 RATIONAL EXPRESSIONS

4.1 Basics of Rational Expressions

1̄ Define rational expressions. (Frames 1–2)

2̄ Find the numbers that make a rational expression undefined. (Frames 3–8)

3̄ Write rational expressions in lowest terms. (Frames 9–24)

4̄ Write a rational expression with a specified denominator. (Frames 25–30)

---

1. A rational number is the quotient of two _____, | integers
   with the denominator not ___. | 0

2. A rational expression is the quotient of two
   _____, with the denominator not ____. | polynomials; 0

3. Any number can be used as a replacement for the
   _____ in a rational expression, except for | variable
   values that make the _____ equal ____. | denominator; 0

4. In the rational expression

   $$\frac{6}{m + 9}$$

   the value m – ____ cannot be used, since –9 makes | –9
   the denominator equal ____. The number –9 makes | 0
   the rational expression _____. | undefined

Find all numbers that make the rational expressions undefined in Frames 5–8.

5. $\dfrac{9p + 4}{3p - 7}$
   The denominator is 0 when p – _____, so _____ | 7/3; 7/3
   makes the rational expression undefined.

6. $\dfrac{-2}{2p^2 + 3p - 2}$

Set the denominator equal to ___, and _____ to get

_____ = 0.

The numbers _____ and _____ make the rational expression undefined.

0; factor

$(2p - 1)(p + 2)$

1/2; -2

7. $\dfrac{4 + r}{6r^2 + 11r - 10}$

The numbers that make the rational expression undefined are _____ and _____.

2/3; -5/2

8. $\dfrac{8k + 16}{9}$

The denominator can never be 0, so _____ can replace k in the rational expression.

any real number

9. According to the fundamental principle of rational numbers, if a/b is a rational number, and if c is any nonzero rational number, then

$$\dfrac{ac}{bc} = \text{\underline{}}.$$

$\dfrac{a}{b}$

10. By the fundamental principle of rational numbers,

$$\dfrac{(m - 3)^2(m + 2)^3}{(m - 3)^3(m + 2)^2} = \text{\underline{}}$$

$\dfrac{m + 2}{m - 3}$

Write the rational expressions in Frames 11–13 in lowest terms.

11. $\dfrac{k^3 r^2}{k r^3} = $ _____

$\dfrac{k^2}{r}$

12. $\dfrac{[(r + 2)^2]^2(r - 5)^4}{(r + 2)^6(r - 5)} = $ _____

$\dfrac{(r - 5)^3}{(r + 2)^2}$

13. $\dfrac{(mn^2)^3(m^2n)^4}{(m^5n)(m^2n^3)} = $ _____

m^4n^6

14. To write

$$\dfrac{a^2 + 2a - 8}{a^2 - 5a + 6}$$

in lowest terms, first _____ both numerator and denominator.

factor

$$\dfrac{a^2 + 2a - 8}{a^2 - 5a + 6} = \dfrac{(\_\_\_)(\_\_\_)}{(\_\_\_)(\_\_\_)}$$

$$= \_\_\_\_$$

$\dfrac{(a + 4)(a - 2)}{(a - 3)(a - 2)}$

$\dfrac{a + 4}{a - 3}$

Write the rational expressions in Frames 15–24 in lowest terms.

15. $\dfrac{a^2 - a}{a^2 + 2a} = \dfrac{(__)(___)}{(__)(___)} = $ _____

$\dfrac{(a)(a - 1)}{(a)(a + 2)} = \dfrac{a - 1}{a + 2}$

16. $\dfrac{2x - x^2}{10 - 5x} = \dfrac{(__)(___)}{(__)(___)} = $ _____

$\dfrac{(x)(2 - x)}{(5)(2 - x)} = \dfrac{x}{5}$

17. $\dfrac{2m^2 - 2n^2}{2m + 2n} = \dfrac{(\_\_)(\_\_\_)(\_\_\_)}{(\_\_)(\_\_\_)}$

$= $ _____

$\dfrac{(2)(m + n)(m - n)}{(2)(m + n)}$

$m - n$

18. $\dfrac{krm - 2kr}{kbm - 2kb} = \dfrac{(__)(___)}{(__)(___)} = $ _____

$\dfrac{(kr)(m - 2)}{(kb)(m - 2)} = \dfrac{r}{b}$

19. $\dfrac{2a^2 - 10a}{a^2 - 4a - 5} = \dfrac{(\_\_)(\_\_\_)}{(\_\_\_)(\_\_\_)}$

$= $ _____

$\dfrac{(2a)(a - 5)}{(a + 1)(a - 5)}$

$\dfrac{2a}{a + 1}$

20. $\dfrac{3b^2 + 12b}{b^2 - b - 20} = \dfrac{(\_\_)(\_\_\_)}{(\_\_\_)(\_\_\_)}$

$= $ _____

$\dfrac{(3b)(b + 4)}{(b - 5)(b + 4)}$

$\dfrac{3b}{b - 5}$

21. $\dfrac{x^2 - 9x + 18}{x^2 - 3x - 18} = \dfrac{(\_\_\_)(\_\_\_)}{(\_\_\_)(\_\_\_)}$

$= $ _____

$\dfrac{(x - 6)(x - 3)}{(x - 6)(x + 3)}$

$\dfrac{x - 3}{x + 3}$

22. $\dfrac{m^2 - 4n^2}{m^2 + 4mn + 4n^2} = \dfrac{(\underline{\hspace{1cm}})(\underline{\hspace{1cm}})}{(\underline{\hspace{1cm}})(\underline{\hspace{1cm}})}$

    $= \underline{\hspace{1.5cm}}$

$\dfrac{(m - 2n)(m + 2n)}{(m + 2n)(m + 2n)}$

$\dfrac{m - 2n}{m + 2n}$

23. $\dfrac{xm + 3ym - xn - 3yn}{xm + 3ym + xn + 3yn} = \dfrac{(\underline{\hspace{1cm}})(\underline{\hspace{1cm}})}{(\underline{\hspace{1cm}})(\underline{\hspace{1cm}})}$

    $= \underline{\hspace{1.5cm}}$

$\dfrac{(x + 3y)(m - n)}{(x + 3y)(m + n)}$

$\dfrac{m - n}{m + n}$

24. $\dfrac{y - 7}{7 - y} = \underline{\hspace{1.5cm}}$

-1

25. The fundamental principal can also be used to write rational expressions with an indicated _____. For example, to write

denominator

$$\dfrac{9}{5x}$$

    with a denominator of $20x^3$, divide $20x^3$ by ____ to get

$5x$

$$\dfrac{20x^3}{\underline{\hspace{0.8cm}}} = \underline{\hspace{1cm}}.$$

$4x^2$
$5x$

    Then multiply numerator and denominator of

    _____ by _____.

$\dfrac{9}{5x}$; $4x^2$

$$\dfrac{9}{5x} = \dfrac{9 \cdot (\underline{\hspace{0.8cm}})}{5x \cdot (\underline{\hspace{0.8cm}})}$$

$4x^2$
$4x^2$

$$= \underline{\hspace{1.5cm}}$$

$\dfrac{36x^2}{20x^3}$

Write the rational expressions in Frames 26–30 with the indicated denominator.

26. $\dfrac{a + 6}{5}$, $15a^2$ _____

$\dfrac{3a^2(a + 6)}{15a^2}$

27. $\dfrac{q - 7}{9}$, $27q^5$ _____

$\dfrac{3q^5(q - 7)}{27q^5}$

28. $\dfrac{7r}{r - 4}$, $r^2 - 4r$ _____

$\dfrac{7r^2}{r^2 - 4r}$

29. $\dfrac{12m}{m-9}$, $m^2 - 81$ _____

$\dfrac{12m(m+9)}{m^2 - 81}$

30. $\dfrac{2y}{3y-1}$, $6y^2 + 7y - 3$ _____

$\dfrac{2y(2y+3)}{6y^2 + 7y - 3}$

4.2 Multiplication and Division of Rational Expressions

1 Multiply rational expressions. (Frames 1–8)

2 Find reciprocals for rational expressions. (Frames 9–12)

3 Divide rational expressions. (Frames 13–22)

1. If P/Q and R/S are any two rational expressions, then

$$\frac{P}{Q} \cdot \frac{R}{S} = \underline{\qquad}.$$

$\dfrac{PR}{QS}$

Perform the multiplications in Frames 2–7.

2. $\dfrac{k^2 m}{k^2 m^3} \cdot \dfrac{k^3 m^2}{k^2 m^4} = \dfrac{k^{\overline{\quad}} \cdot m^{\overline{\quad}}}{k^{\overline{\quad}} \cdot m^{\overline{\quad}}} = \dfrac{k}{\underline{\quad}}$

5; 3
4; 7; m^4

3. $\dfrac{(6pq)^2}{9p^2 q} \cdot \dfrac{p^2 q^4}{p^3 q^2} = \underline{\qquad}$

$\dfrac{4q^3}{p}$

4. $\dfrac{(x-4)(x+3)}{(x-4)(x+5)} \cdot \dfrac{(x-3)(x+5)}{(x-3)(x+4)} = \underline{\qquad}$

$\dfrac{x+3}{x+4}$

5. $\dfrac{(2k-1)(k+4)}{(3k+2)(2k-1)} \cdot \dfrac{(3k+2)(k+6)}{(k-6)(k+6)} = \underline{\qquad}$

$\dfrac{k+4}{k-6}$

6. $\dfrac{a^2 - 5a}{a^2 + a - 2} \cdot \dfrac{a^2 - a - 6}{a^2 - 8a + 15}$

$= \dfrac{(\underline{\quad})(\underline{\quad})}{(\underline{\quad})(a+2)} \cdot \dfrac{(\underline{\quad})(\underline{\quad})}{(\underline{\quad})(\underline{\quad})}$

$= \underline{\qquad}$

a; a − 5; a + 2;
a − 3
a − 1; a − 3; a − 5
$\dfrac{a}{a-1}$

7. $\dfrac{3r^2 + r - 10}{3r^2 + 7r - 20} \cdot \dfrac{r^2 + 10r + 24}{2r^2 + 3r - 2}$

$= \dfrac{(\underline{})(\underline{})}{(\underline{})(\underline{})} \cdot \dfrac{(\underline{})(r + 6)}{(r + 2)(\underline{})}$

$= \underline{}$

$r + 2; \; 3r - 5; \; r + 4$
$3r - 5; \; r + 4; \; 2r - 1$

$\dfrac{r + 6}{2r - 1}$

8. $\dfrac{3r^2 - 5r - 2}{r^2 + 6r} \cdot \dfrac{r^2 + 2r - 24}{6r^2 - r - 1} \cdot \dfrac{10r^2 - 5r}{8r - 16}$

$= \underline{}$

$\dfrac{5(r - 4)}{8}$

9. Two rational expressions whose product is 1 are called _____ of each other.

reciprocals

Give the reciprocal of each rational expression in Frames 10–12.

10. $\dfrac{6}{m - 4}$ _____

$\dfrac{m - 4}{6}$

11. $\dfrac{8a^2 - 14a}{3a - 4}$ _____

$\dfrac{3a - 4}{8a^2 - 14a}$

12. $\dfrac{0}{k}$ _____

undefined (no reciprocal for 0)

13. To divide rational numbers, _____ the first by the _____ of the second.

multiply

reciprocal

14. That is, if a/b and c/d are rational numbers, with c/d $\neq$ ____, then

$$\dfrac{a}{b} \div \dfrac{c}{d} = \dfrac{a}{b} \cdot \underline{}.$$

0

$\dfrac{d}{c}$

Perform the operations indicated in Frames 15–22.

15. $\dfrac{x^2y}{xy^3} \div \dfrac{x^2y^2}{xy^3} = \dfrac{x^2y}{\underline{}} \cdot \underline{} = \underline{}$

$xy^3; \; \dfrac{xy^3}{x^2y^2}; \; \dfrac{1}{y}$

16. $\dfrac{r(r-2)}{(r-3)r} \div \dfrac{r^2(r-2)}{8(r-3)} = $ _____

$\dfrac{8}{r^2}$

17. $\dfrac{a^3(a+3)}{a^2(a-4)} \div \dfrac{a(a+3)}{a^4(a-4)} = $ _____

a^4

18. $\dfrac{k^2-5k}{4k+8} \div \dfrac{k^3-5k^2}{8k+16}$

$= \dfrac{(\quad)(\quad\quad)}{(\quad)(\quad\quad)} \div \dfrac{(\quad)(\quad\quad)}{(\quad)(\quad\quad)}$

$= \dfrac{(\quad)(\quad\quad)}{(\quad)(\quad\quad)} \cdot \dfrac{(\quad)(\quad\quad)}{(\quad)(\quad\quad)}$

$= $ _____

$k;\ k-5;\ k^2;\ k-5$
$4;\ k+2;\ 8;\ k+2$

$k;\ k-5;\ 8;\ k+2$
$4;\ k+2;\ k^2;\ k-5$

$\dfrac{2}{k}$

19. $\dfrac{x^2-4x-5}{x^2-2x-3} \div \dfrac{x^2-3x-10}{x^2+x-12}$

$= \dfrac{(\quad\quad)(\quad\quad)}{(\quad\quad)(\quad\quad)} \div \dfrac{(\quad\quad)(\quad\quad)}{(\quad\quad)(\quad\quad)}$

$= \dfrac{(x-5)(x+1)}{(x-3)(x+1)} \cdot \dfrac{(\quad\quad)(\quad\quad)}{(\quad\quad)(\quad\quad)}$

$= $ _____

$x-5;\ x+1;\ x-5;\ x+2$
$x-3;\ x+1;\ x+4;\ x-3$

$x+4;\ x-3$
$x-5;\ x+2$

$\dfrac{x+4}{x+2}$

20. $\dfrac{m^2-9m+18}{m^2+2m-15} \cdot \dfrac{m^2-m-6}{m^2-4m-12} \div \dfrac{m^2+m-12}{m^2+2m-8}$

$= $ _____

$\dfrac{m-2}{m+5}$

21. $\dfrac{x^2-x-20}{x^2+7x+12} \cdot \dfrac{x^2-2x-15}{x^2-2x-3} \div \dfrac{x^2-x-20}{x^2+5x+4}$

$= $ _____

$\dfrac{x-5}{x-3}$

22. $\dfrac{4x^2-y^2}{6x^2+5xy+y^2} \cdot \dfrac{3x^2+2xy-y^2}{2x^2+xy-y^2} \div \dfrac{3x^2+2xy-y^2}{2x^2-xy-3y^2}$

$= $ _____

$\dfrac{2x-3y}{3x+y}$

4.3   Addition and Subtraction of Rational Expressions

[1] Add and subtract rational expressions with the same denominator. (Frames 1–7)

[2] Find a least common denominator. (Frames 8–12)

[3] Add and subtract rational expressions with different denominators. (Frames 13–22)

---

1. To add or subtract two rational numbers with the same _____, just add or subtract the _____. Place the result over the common _____.

   That is, if a/b and c/b are rational numbers, then

   $$\frac{a}{b} + \frac{c}{b} = \underline{\qquad}$$

   and

   $$\frac{a}{b} - \frac{c}{b} = \underline{\qquad}.$$

   denominators
   numerators
   denominator

   $\dfrac{a + c}{b}$

   $\dfrac{a - c}{b}$

2. By the definition of addition,

   $$\frac{3}{x} + \frac{5}{x} = \underline{\qquad} = \underline{\quad}.$$

   $\dfrac{3 + 5}{x} = \dfrac{8}{x}$

3. Add: $\dfrac{a - 4}{2a} + \dfrac{6}{2a} = \underline{\qquad}$

   $\dfrac{a + 2}{2a}$

4. Subtract: $\dfrac{6}{5k} - \dfrac{11}{5k} = \underline{\qquad}$

   $-\dfrac{1}{k}$ (reduced)

5. Subtract: $\dfrac{9 - m}{m} - \dfrac{4 + m}{m} = \dfrac{9 - m - (\qquad)}{m}$

   $= \dfrac{\qquad}{m}$

   $4 + m$

   $5 - 2m$

6. Add: $\dfrac{4 + 3k}{5k} + \dfrac{-k - 3}{5k} = \underline{\qquad}$

   $\dfrac{2k + 1}{5k}$

7. Subtract: $\dfrac{m - n}{2m + n} - \dfrac{m + n}{2m + n} = \dfrac{(\underline{\hspace{1.5cm}}) - (\underline{\hspace{1.5cm}})}{(\underline{\hspace{1.5cm}})}$

$= \underline{\hspace{3cm}}$

$m - n; \ m + n$

$2m + n$

$\dfrac{-2n}{2m + n}$

8. If rational expressions to be added have different denominators, we must find a _____ _____ denominator. The least common denominator is a denominator that all denominators of the problem _____ into without a remainder.

least common

divide

9. To find a least common denominator, first _____ each denominator. Then take the _____ of all different factors, with each factor raised to the _____ power from any _____.

factor

product

greatest

denominator

10. To find the least common denominator for $10q^2$ and $75q^3$, factor 10 and 75 in each expression as

$10q^2 = \underline{\hspace{1.5cm}}$ and $75q^3 = \underline{\hspace{1.5cm}}$.

Then take each different factor with the _____ exponent. The least common denominator for $10q^2$ and $75q^3$ is

$\underline{\hspace{3cm}}$.

$2 \cdot 5q^2; \ 3 \cdot 5^2 q^3$

greatest

$2 \cdot 3 \cdot 5^2 \cdot q^3$
or $150q^3$

Find the least common denominator in Frames 11–12.

11. $6m - m^2, \ 9m^3$ \qquad $\underline{\hspace{4cm}}$

$9m^3(6 - m)$

12. $y^2 - 4, \ y^2 + 3y - 10$ \qquad $\underline{\hspace{4cm}}$

$(y + 2)(y - 2)(y + 5)$

13. To add $\dfrac{3}{2x}$ and $\dfrac{4}{5x}$, first find the least common

denominator, which is _____. Then write both   10x

fractions with a denominator of 10x.

$$\frac{3}{2x} + \frac{4}{5x} = \frac{}{10x} + \frac{}{10x} = \frac{}{10x}$$ 15; 8; 23

Add and subtract the rational expression in Frames 14–22.

14. $\dfrac{1}{a + b} + \dfrac{1}{a - b} = \dfrac{}{()()}$ $\dfrac{a - b}{(a + b)(a - b)}$

$$+ \frac{}{()()}$$ $\dfrac{a + b}{(a + b)(a - b)}$

$$= \frac{() + ()}{()()}$$ $\dfrac{(a - b) + (a + b)}{(a + b)(a - b)}$

$$= \frac{}{()()}$$ $\dfrac{2a}{(a + b)(a - b)}$

15. $\dfrac{3}{2m + 1} - \dfrac{4}{2(2m + 1)} = \dfrac{}{} - \dfrac{}{}$ $\dfrac{6}{2(2m + 1)} - \dfrac{4}{2(2m + 1)}$

$$= \frac{}{}$$ $\dfrac{2}{2(2m + 1)}$

$$= \frac{}{}$$ $\dfrac{1}{2m + 1}$

16. $\dfrac{2}{(x - 3)(x + 2)} + \dfrac{3}{(x + 2)(x - 1)}$

$$= \frac{()() + ()()}{()()()}$$ $\dfrac{2(x - 1) + 3(x - 3)}{(x - 3)(x + 2)(x - 1)}$

$$= \frac{}{}$$ $\dfrac{5x - 11}{(x - 3)(x + 2)(x - 1)}$

17. $\dfrac{1}{x - 2} + \dfrac{2}{x - 3} = \underline{}$ $\dfrac{3x - 7}{(x - 2)(x - 3)}$

18. $\dfrac{t + 2}{t} - \dfrac{t - 1}{t + 1} = \underline{}$ $\dfrac{4t + 2}{t(t + 1)}$

19. $\dfrac{2}{k^2 - k - 2} - \dfrac{3}{k^2 + 3k + 2}$

$$= \frac{2}{()()} - \frac{3}{()()}$$ k+1; k−2; k+1; k+2

$$= \underline{}$$ $\dfrac{-k + 10}{(k + 1)(k - 2)(k + 2)}$

20. $\dfrac{2}{r^2 + r - 12} - \dfrac{1}{r^2 - 2r - 3} =$ _____

$\dfrac{r - 2}{(r + 4)(r - 3)(r + 1)}$

21. $\dfrac{m}{m^2 - mn - 2n^2} + \dfrac{-3n}{m^2 + 4mn + 3n^2}$

= _____

$\dfrac{m^2 + 6n^2}{(m+n)(m-2n)(m+3n)}$

22. $\dfrac{3a}{a^2 + ab - 6b^2} - \dfrac{-9b}{a^2 - 5ab + 6b^2}$

= _____

$\dfrac{3a^2 + 27b^2}{(a-2b)(a+3b)(a-3b)}$

4.4 Complex Fractions

[1] Simplify complex fractions by simplifying numerator and denominator. (Frames 1–4)

[2] Simplify complex fractions by multiplying by a common denominator. (Frames 5–14)

[3] Simplify rational expressions with negative exponents. (Frames 15–20)

1. A complex fraction has a _____ in the numerator, the _____, or in both.

fraction

denominator

2. One way to simplify a complex fraction is to simplify the _____ and denominator separately. Then multiply the numerator by the _____ of the denominator.

numerator

reciprocal

3. To simplify the complex fraction

$$\dfrac{5 - \dfrac{3}{p}}{1 + \dfrac{4}{3p}},$$

first simplify the numerator: $5 - \dfrac{3}{p} = \dfrac{\rule{1cm}{0.4pt}}{p}$.

$5p - 3$

Then simplify the denominator: $1 + \dfrac{4}{3p} =$ _____.

$\dfrac{3p + 4}{3p}$

4. Then multiply the numerator by the reciprocal of the denominator.

$$\frac{5 - \dfrac{3}{p}}{1 + \dfrac{4}{3p}} = \frac{5p - 3}{p} \cdot \underline{\hspace{2cm}} = \underline{\hspace{2cm}}$$

$\dfrac{3p}{3p + 4}$; $\dfrac{3(5p - 3)}{3p + 4}$

5. An alternate method of solution is to multiply both the numerator and the denominator by the least _____ denominator of all the _____ .

common

denominators

6. To simplify the complex fraction

$$\frac{\dfrac{1}{k} + \dfrac{1}{k - 1}}{\dfrac{2}{k}},$$

we could multiply numerator and denominator by the common denominator _____ . This would give

$k(k - 1)$

$$\frac{k(k - 1)(\underline{\hphantom{xx}} + \underline{\hphantom{xxx}})}{k(k - 1)(\underline{\hphantom{xx}})}$$

$\dfrac{1}{k}$; $\dfrac{1}{k - 1}$

$\dfrac{2}{k}$

which equals

_____

$\dfrac{(k - 1) + k}{2(k - 1)}$

or

_____ .

$\dfrac{2k - 1}{2(k - 1)}$

Simplify the complex fractions in Frames 7–14.

7. $\dfrac{1 + \dfrac{2}{r}}{1 - \dfrac{2}{r}} = \dfrac{(\underline{\hphantom{xx}})\left(1 + \dfrac{2}{r}\right)}{(\underline{\hphantom{xx}})\left(1 - \dfrac{2}{r}\right)} = \underline{\hspace{2cm}}$

r; $\dfrac{r + 2}{r - 2}$

r

8. $\dfrac{\dfrac{1}{p} - \dfrac{1}{q}}{\dfrac{4}{pq}} = \dfrac{(\underline{\hphantom{xx}})\left(\dfrac{1}{p} - \dfrac{1}{q}\right)}{(\underline{\hphantom{xx}})\left(\dfrac{4}{pq}\right)} = \underline{\hspace{2cm}}$

pq; $\dfrac{q - p}{4}$

pq

9. $\dfrac{\dfrac{2}{s} - \dfrac{2}{r}}{1 - \dfrac{1}{sr}} = \dfrac{(\underline{\quad})\left(\dfrac{2}{s} - \dfrac{2}{r}\right)}{(\underline{\quad})\left(1 - \dfrac{1}{sr}\right)} = \underline{\hspace{3cm}}$

sr; $\dfrac{2r - 2s}{sr - 1}$

sr

10. $\dfrac{\dfrac{1}{r} + \dfrac{1}{r + 2}}{\dfrac{2}{r + 2}} = \dfrac{(\underline{\quad})(\underline{\quad})\left(\dfrac{1}{r} + \dfrac{1}{r + 2}\right)}{(\underline{\quad})(\underline{\quad})\left(\dfrac{2}{r + 2}\right)}$

$= \underline{\hspace{3cm}}$,

when can be simplified to

$= \underline{\hspace{3cm}}$.

r; r + 2

r; r + 2

$\dfrac{2r + 2}{2r}$

$\dfrac{r + 1}{r}$

11. $\dfrac{a - \dfrac{1}{a + 2}}{1 + \dfrac{1}{a + 2}} = \underline{\hspace{3cm}}$

$\dfrac{a^2 + 2a - 1}{a + 3}$

12. $\dfrac{m - 4}{m - \dfrac{4}{m - 4}} = \underline{\hspace{3cm}}$

$\dfrac{(m - 4)^2}{m^2 - 4m - 4}$

13. $\dfrac{5z - \dfrac{1}{z + 2}}{\dfrac{3}{z - 4} + z} = \underline{\hspace{4cm}}$

$\dfrac{(z-4)(5z^2+10z-1)}{(z+2)(z^2-4z+3)}$

14. $1 + \dfrac{1}{1 - \dfrac{1}{1 + \dfrac{1}{1 - \dfrac{1}{1 + 1}}}} = \underline{\hspace{2cm}}$

$\dfrac{5}{2}$

15. Complex fractions often involve negative

    $\underline{\hspace{3cm}}$.

exponents

16. To simplify $m^{-1} - 2n^{-1}$, use the definition of
    a negative exponent.

    $m^{-1} - 2n^{-1} = \underline{\hspace{2cm}} - \underline{\hspace{2cm}}$

    $= \underline{\hspace{3cm}}$

$\dfrac{1}{m}$; $\dfrac{2}{n}$

$\dfrac{n - 2m}{mn}$

Simplify the complex fractions in Frames 17–20.

17. $\dfrac{z^{-2}}{z^{-2}-1} = \dfrac{\underline{\quad}}{\underline{\quad}-1}$

$\dfrac{\dfrac{1}{z^2}}{z^2}$

$\dfrac{1}{z^2}$

$\dfrac{1}{z^2}$

$= \dfrac{\dfrac{1}{z^2}}{\underline{\qquad}}$

$\dfrac{1-z^2}{z^2}$

$= \underline{\qquad}$

$\dfrac{1}{1-z^2}$

18. $\dfrac{k^{-1}+p^{-2}\cdot}{k^{-1}-3p^{-2}} = \dfrac{\underline{\quad}+\underline{\quad}}{\underline{\quad}-\underline{\quad}}$

$\dfrac{1}{k};\ \dfrac{1}{p^2}$

$\dfrac{1}{k};\ \dfrac{3}{p^2}$

$= \underline{\qquad}$

$\dfrac{p^2+k}{p^2-3k}$

19. $(m^{-2}+n^{-1})^{-1} = (\underline{\quad}+\underline{\quad})^{-1}$

$\dfrac{1}{m^2};\ \dfrac{1}{n}$

$= (\underline{\qquad})^{-1}$

$\dfrac{n+m^2}{m^2n}$

$= \underline{\qquad}$

$\dfrac{m^2n}{n+m^2}$

20. $(4z)^{-2}-3k^{-1} = \underline{\qquad}-\underline{\quad}$

$\dfrac{1}{16z^2};\ \dfrac{3}{k}$

$= \underline{\qquad}$

$\dfrac{k-48z^2}{16kz^2}$

4.5 Division of Polynomials

1️⃣ Divide a polynomial by a monomial. (Frames 1–5)

2️⃣ Divide a polynomial by a polynomial of two or more terms. (Frames 6–22)

1. To divide a polynomial by a monomial (which has only _____ term), divide each _____ of the polynomial by the _____.

one; term
monomial

2. For example, to divide $5m^3 - 10m^2 + 15m$ by the monomial $5m$, divide each _____ by ____.

term; 5m

$$\frac{5m^3 - 10m^2 + 15m}{5m} = \frac{}{5m} + \frac{}{5m} + \frac{}{5m}$$

$5m^3$; $-10m^2$; $15m$

$$= \underline{\hspace{3cm}}$$

$m^2 - 2m + 3$

Divide the polynomials in Frames 3–5.

3. $\dfrac{8x^5 - 16x^4 + 12x^3}{8x^2} = \underline{\hspace{3cm}}$

$x^3 - 2x^2 + \dfrac{3x}{2}$

4. $\dfrac{15m^4p^3 - 9m^5p + 12m^3p^6}{6m^5p} = \underline{\hspace{3cm}}$

$\dfrac{5p^2}{2m} - \dfrac{3}{2} + \dfrac{2p^5}{m^2}$

This answer shows that the quotient of two polynomials need not be a _____.

polynomial

5. $\dfrac{-4a^3b^4c^5 + 6a^2bc^4}{12a^5b^2c^6} = \underline{\hspace{3cm}}$

$\dfrac{-b^2}{3a^2c} + \dfrac{1}{2a^3bc^2}$

6. To divide a _____ by a polynomial of more than one _____, use long division.

polynomial

term

7. To divide $6x^3 - 16x^2 + 11x - 2$ by $3x - 2$, first write the problem as follows.

$$\underline{\hspace{3cm}})\overline{6x^3 - 16x^2 + 11x - 2}$$

3x − 2

8. Then divide ____ into _____, obtaining _____. Put this number in the appropriate blank below. Then multiply _____ by _____ and subtract. Bring down the next term.

3x; $6x^3$; $2x^2$

2x²; 3x − 2

$$3x - 2)\overline{6x^3 - 16x^2 + 11x - 2}$$

$2x^2$

$6x^3 - 4x^2$

$-12x^2 + 11x$

9. _____ divides into _____, with a quotient of
    _____. Multiply _____ by _____, then
    subtract and bring down the next term.

$$\begin{array}{r} 2x^2 \quad \underline{\quad} \\ 3x-2\overline{)6x^3-16x^2+11x-2} \\ \underline{6x^3-\ 4x^2} \\ -12x^2+11x \\ \underline{\qquad} \\ \underline{\quad} \end{array}$$

3x; −12x²

−4x; −4x; 3x − 2

− 4x

−12x² + 8x

3x − 2

10. Divide _____ into _____, obtaining ___. Complete
    the problem.

$$\begin{array}{r} 2x^2-\ 4x \quad \underline{\quad} \\ 3x-2\overline{)6x^3-16x^2+11x-2} \\ \underline{6x^3-\ 4x^2} \\ -12x^2+11x \\ \underline{-12x^2+\ 8x} \\ 3x-2 \\ \underline{\qquad} \\ \underline{\quad} \end{array}$$

3x; 3x; 1

+ 1

3x − 2
0

11. Finally,

    $(6x^3 - 16x^2 + 11x - 2) \div (3x - 2) =$ _____.

2x² − 4x + 1

12. When dividing a polynomial with a missing term,
    use a _____ coefficient on the term that is
    missing.

zero

13. Remember to write the polynomial in _____
    powers of the variable.

descending

14. Divide $9x^2 - x^3 + 3x^4 - 1$ by $3x - 1$.

    First write $9x^2 - x^3 + 3x^4 - 1$ in _____
    powers of x. _____

descending

$3x^4 - x^3 + 9x^2 - 1$

Now, set up the long division, remembering to use
_____ for the missing term.   0x

$3x - 1\overline{)\underline{\quad} - \underline{\quad} + \underline{\quad} + \underline{\quad} - \underline{\quad}}$ $3x^4$; x^3; $9x^2$;
                                                                 0x; 1

Perform the division. The quotient is _____. $x^3 + 3x + 1$

15. Divide $6x^3 - 19x^2 + 18x - 15$ by $2x - 5$.

$$2x - 5\overline{)6x^3 - 19x^2 + 18x - 15}$$ $3x^2$; $-2x$; $+4$

                                                                 $6x^3 - 15x^2$

                                                                 $-4x^2 + 18x$

                                                                 $-4x^2 + 10x$

                                                                 $8x - 15$

                                                                 $8x - 20$

                                                                 5

16. The remainder here is ____, which is written in   5
the result as _____. Hence $5/(2x - 5)$

$$\frac{6x^3 - 19x^2 + 18x - 15}{2x - 5} = \text{_____} + \frac{\quad}{2x - 5}.$$ $3x^2 - 2x + 4$; 5

17. Divide $3x^3 - 4x^2 - 7x + 6$ by $3x - 6$.

$$3x - 6\overline{)3x^3 - 4x^2 - 7x + 6}$$ x^2

                                                                 $3x^3 - 6x^2$

$$2x^2 - 7x$$

At this point, ____ must be divided into _____. $3x$; $2x^2$
This requires a _____ coefficient in the   fractional
quotient.

$3x \cdot ? = 2x^2$ Divide both sides by ___. $3x$

$? = $ _____ Thus, the next term in the quotient $\frac{2}{3}x$

will be ___. $\frac{2}{3}x$

Continue the division process below.

$$
\begin{array}{r}
x^2 + \frac{2}{3}x \underline{\hspace{1.5cm}} \\
3x - 6 \overline{\smash{\big)}\ 3x^3 - 4x^2 - 7x + 6} \\
\underline{3x^3 - 6x^2} \\
\underline{\hspace{2cm}} \\
\underline{\hspace{2cm}} \\
\underline{\hspace{2cm}} \\
\underline{\hspace{2cm}} \\
\underline{\hspace{1.2cm}}
\end{array}
$$

-1

$2x^2 - 7x$

$2x^2 - 4x$

$-3x + 6$

$-3x + 6$

0

Perform the divisions indicated in Frames 18–22.

18. $(6x^3 + 19x^2 + 7x - 18) \div (2x + 3) =$ _____

$3x^2 + 5x - 4 + \dfrac{-6}{2x+3}$

19. $\dfrac{3x^4 - 10x^3 + 6x^2 + 8x - 11}{x - 2} =$ _____

$3x^3 - 4x^2 - 2x + 4 + \dfrac{-3}{x - 2}$

20. $(8x^3 - 8x + 15) \div (2x + 3) =$ _____

$4x^2 - 6x + 5$

21. $\dfrac{2a^4 - 2a^3 - 3a^2 + 23a - 8}{a^2 - 3a + 5} =$ _____

$2a^2 + 4a - 1 + \dfrac{-3}{a^2 - 3a + 5}$

22. $(9x^4 - x^2 + 18x - 6) \div (9x - 3) =$ _____

$x^3 + \dfrac{1}{3}x^2 + 2$

4.6 Synthetic Division

[1] Use synthetic division to divide by a polynomial of the form x − k.
(Frames 1–9)

[2] Use the remainder theorem to evaluate a polynomial. (Frames 10–13)

[3] Decide whether a given number is a solution of an equation.
(Frames 14–18)

1. Use the method of the last section to divide
   $3x^4 - 10x^3 + 6x^2 + 8x - 11$ by $x - 2$.

$$x - 2 \overline{)3x^4 - 10x^3 + 6x^2 + 8x - 11}$$

$$-4x^3 + 6x^2$$

$$-2x^2 + 8x$$

$$4x - 11$$

$3x^3$; $-4x^2$; $-2x$; 4
$3x^4 - 6x^3$
$-4x^3 + 8x^2$
$-2x^2 + 4x$
$4x - 8$
-3

2. The problem in Frame 1 can be worked by a shortcut
   method called _____ division. This shortcut
   can only be used when the divisor is of the form
   _____. To begin, write only the _____
   of $3x^4 - 10x^3 + 6x^2 + 8x - 11$ and only the number
   ____ from $x - 2$, as follows.

$$\underline{\quad} \overline{)\underline{\quad} \quad \underline{\quad} \quad \underline{\quad} \quad \underline{\quad} \quad \underline{\quad}}$$

synthetic
$x - k$; coefficients
2
2; 3; -10; 6; 8; -11

3. Bring down ___, multiply ___ and ___, and place
   the result in the next column.

$$2 \overline{)3 \quad -10 \quad 6 \quad 8 \quad -11}$$
$$\underline{\qquad}$$
$$3$$

3; 3; 2
6

4. Now add _____ and ____, multiply the sum by ____,
   and place the result in the next column.

$$2 \overline{)3 \quad -10 \quad 6 \quad 8 \quad -11}$$
$$6 \quad \underline{\quad}$$
$$3 \quad \underline{\quad}$$

-10; 6; 2

5. Add ___ and ___, multiply the sum by ____, and place the result in the next column.

$$2\overline{)3 \quad -10 \quad\quad 6 \quad\quad 8 \quad\quad -11}$$

```
2)3    -10      6      8     -11
        6      -8    ___
   _____
   3     -4    ___
```

6; -8; 2

-4
-2

6. Add ____ and ____. Multiply the sum by ____, and place the result in the next column. Then add _____ and ____.

```
2)3    -10      6      8     -11
        6      -8     -4    ____
   _____
   3     -4     -2    ____ ____
```

8; -4; 2

-11; 8

8

4; -3

7. The quotient and remainder are read from the _____ _____. The degree of the quotient is one _____ than the degree of the polynomial divided into. The result is

$$\underline{\hspace{5cm}} + \frac{}{x - 2}.$$

bottom row

less

$3x^3-4x^2-2x+4$; -3

Perform the divisions indicated in Frames 8–9.

8. $\dfrac{2x^4 + 3x^3 + x^2 - 4}{x + 2} = \underline{\hspace{4cm}}$

(Hint: the coefficient of x in the numerator is ____. Write x + 2 in the form x - k = x - (-2).)

$2x^3 - x^2 + 3x - 6 +$

$\dfrac{8}{x + 2}$

0

9. $\dfrac{3x^5 + 6x^4 - 5x^3 - 40x - 8}{x + 3} = \underline{\hspace{3cm}}$

$3x^4-3x^3+4x^2-12x$

$- 4 + \dfrac{4}{x + 3}$

10. One use of synthetic division is in evaluating a polynomial for a specific value of the variable. By the _____ theorem, if the polynomial $P(x)$ is divided by $x - k$, the remainder is equal to _____ .

    remainder

    $P(k)$

11. For example, suppose $P(x) = 2x^3 - 5x^2 + 7x - 3$. To find $P(-2)$, use synthetic division.

    $$\begin{array}{r|rrrr} -2 & 2 & -5 & 7 & -3 \\ & & & 18 & -50 \\ \hline & 2 & -9 & \underline{} & \underline{} \end{array}$$

    -4

    $25;\ -53$

    The remainder is _____ , so that $P(-2) = $ _____ .

    $-53;\ -53$

Let $P(x) = -x^4 + 2x^3 - 3x^2 + 4x - 2$. **Find each of the following.**

12. $P(2) = $ _____

    -6

13. $P(-1) = $ _____

    -12

14. To see if $x = -2$ is a solution of the equation $3x^4 - 2x^3 - 12x^2 + 10x + 4 = 0$, we could substitute ____ for ____ in the equation. It is easier to use synthetic division to divide _____ by $x + 2$. If the remainder is ____ , then $x = -2$ is a solution. Complete the following synthetic division.

    $-2;\ x$

    $3x^4-2x^3-12x^2+10x+4$

    0

    $$\begin{array}{r|rrrrr} -2 & 3 & -2 & -12 & 10 & 4 \\ & & \underline{} & \underline{} & \underline{} & \underline{} \\ \hline & \underline{} & \underline{} & \underline{} & \underline{} & \underline{} \end{array}$$

    $-6;\ 16;\ -8;\ -4$

    $3;\ -8;\ 4;\ 2;\ 0$

    Thus, $x = -2$ (*is/is not*) a solution of the given equation.

    is

Decide whether or not x = −3 is a solution of the equations in Frames 15–18.

15. $2x^3 + 4x^2 - 5x + 3 = 0$ _____   solution

16. $4x^4 + 9x^3 - 8x^2 + 5x - 6 = 0$ _____   not a solution

17. $-3x^5 - 8x^4 + 6x^3 + 10x^2 + 7x + 12 = 0$ _____   solution

18. Is 4 a solution of the equation

$$-2a^5 + 4a^4 + 13a^3 + 9a^2 + 48 = 0?$$ _____   yes

4.7   Equations with Rational Expressions

[1] Solve equations with rational expressions. (Frames 1–2 and 5–10)

[2] Know when potential solutions must be checked. (Frames 3–4)

---

1. To solve the equation

$$\frac{6}{x} - \frac{4}{x} = 5,$$

first multiply both sides of the equation by ____.   x
This gives

$$(\underline{}) \cdot \left(\frac{6}{x} - \frac{4}{x}\right) = (\underline{}) \cdot 5.$$ x; x

Then, by the distributive property,

$$(\underline{}) \cdot \frac{6}{x} - (\underline{}) \cdot \frac{4}{x} = (\underline{}) \cdot 5$$ x; x; x

or

$$\underline{} - \underline{} = \underline{}.$$ 6; 4; 5x

This simplifies to

$$\underline{} = 5x$$ 2

from which

$$x = \underline{}.$$ 2/5

Check by substituting _____ for x in the original   2/5
equation. The solution set is _____.   {2/5}

2. To solve the equation

$$\frac{5}{2m} = \frac{5}{m + 4},$$

first multiply both sides of the equation by
_____, the least common _____. | 2m(m + 4); denominator

This gives

$$[\underline{\quad\quad}](\underline{\quad}) = 2m(m + 4)\left(\frac{5}{m + 4}\right)$$ | $2m(m + 4);\ \frac{5}{2m}$

or

$$(\underline{\quad\quad})(\underline{\quad}) = (\underline{\quad})(\underline{\quad}),$$ | m + 4; 5; 2m; 5

which becomes _____ = _____ | 5m + 20; 10m
or, finally,   m = _____. | 4

To check, substitute ____ for ____ in the original | 4; m
equation.

$$\frac{5}{2(\underline{\ })} = \frac{5}{\underline{\ } + 4}$$ | 4; 4

$$\frac{5}{\underline{\ }} = \frac{5}{\underline{\ }}$$ | 8; 8

The solution (*is/is not*) correct. The solution | is
set is ____. | {4}

3. To solve the equation

$$\frac{3}{x + 3} = \frac{2}{x - 3} = \frac{12}{(x + 3)(x - 3)},$$

multiply both sides of the equation by
(_____)(_____). | x + 3; x - 3

This _____ expression cannot have the | variable
value _____. Since (x + 3)(x - 3) = 0 when | zero
x = _____ or x = _____, the solution cannot | -3; 3
be _____ or _____. Multiplying gives | -3; 3
_____ = _____ = 12. | 3(x - 3); 2(x + 3)

Solve this equation for x, obtaining x = _____.   3

But _____ is not a solution of the equation so   3

the solution set is _____. Checking the potential   Ø

solution x = _____ shows why _____ cannot be a   3; 3

solution.

or
$$\frac{3}{\underline{} + 3} = \frac{2}{\underline{} - 3} - \frac{12}{(\underline{} + 3)(\underline{} - 3)}$$ 3; 3; 3; 3

$$\frac{3}{6} = \frac{2}{0} - \frac{12}{0}.$$

However, division by ____ is undefined, which   0

confirms that the solution set of this equation

is _____.   Ø

4. In general, any time both sides of an equation

are _____ by an expression containing a   multiplied

_____, all potential solutions must be   variable

checked in the _____ equation.   original

In each of the equations of Frames 5–9, first list the least common denominator (LCD), and then find the solution.

5. $\dfrac{2}{x + 1} + \dfrac{1}{2} = \dfrac{5}{6}$ LCD = _____   6(x + 1)

$6(x + 1) \cdot \dfrac{2}{x + 1} + [\underline{}] \cdot \dfrac{1}{2} = [\underline{}] \cdot \dfrac{5}{6}$ 6(x + 1); 6(x + 1)

$\underline{} + \underline{} = \underline{}$ 12; 3(x + 1);
5(x + 1)

$12 + 3x + 3 = \underline{}$ 5x + 5

$10 = \underline{}$ 2x

$x = \underline{}$ 5

The solution (*is/is not*) correct. The solution   is

set is _____.   {5}

6. $\dfrac{4}{x-5} + \dfrac{1}{5} = 1$ LCD = _____ $5(x-5)$

$[\underline{\hspace{1.2cm}}] \cdot \dfrac{4}{x-5} + [\underline{\hspace{1.2cm}}] \cdot \dfrac{1}{5} = 5(x-5) \cdot 1$ $5(x-5);\ 5(x-5)$

$\underline{\hspace{1cm}} + \underline{\hspace{1cm}} = \underline{\hspace{1cm}}$ $20;\ x-5;\ 5x-25$

$\underline{\hspace{1cm}} = 4x$ 40

$x = \underline{\hspace{0.8cm}}$ 10

The solution (*is/is not*) correct. The solution   is

set is _____. $\{10\}$

7. $\dfrac{8}{a-4} + \dfrac{4}{a+4} = \dfrac{16}{a^2-16}$ LCD = _____ $(a+4)(a-4)$

$(a+4)(a-4) \cdot \dfrac{8}{a-4} + (a+4)(a-4) \cdot \dfrac{4}{a+4}$

$= [\underline{\hspace{3cm}}] \cdot \dfrac{16}{a^2-16}$ $(a+4)(a-4)$

$\underline{\hspace{1cm}} + \underline{\hspace{1cm}} = \underline{\hspace{0.8cm}}$ $8(a+4);\ 4(a-4);\ 16$

$12a + \underline{\hspace{0.8cm}} = \underline{\hspace{0.8cm}}$ $16;\ 16$

$12a = \underline{\hspace{0.8cm}}$ 0

$a = \underline{\hspace{0.8cm}}$ 0

The solution (*is/is not*) correct. The solution   is

set is _____. $\{0\}$

8. $\dfrac{2k}{k^2-4} + \dfrac{1}{k-2} = \dfrac{2}{k+2}$ LCD = _____ $(k+2)(k-2)$

$(k+2)(k-2) \cdot \dfrac{2k}{k^2-4} + (k+2)(k-2) \cdot \dfrac{1}{k-2}$

$= [\underline{\hspace{3cm}}] \cdot \dfrac{2}{k+2}$ $(k+2)(k-2)$

$\underline{\hspace{1cm}} + \underline{\hspace{1cm}} = \underline{\hspace{1.5cm}}$ $2k;\ k+2;$
        $2(k-2)$

$3k + 2 = \underline{\hspace{1.5cm}}$ $2k-4$

$k = \underline{\hspace{0.8cm}}$ -6

The solution (*is/is not*) correct. The solution   is

set is _____. $\{-6\}$

9. $\dfrac{4}{r+1} - \dfrac{r^2-2r+2}{r^2-2r-3} = \dfrac{r}{3-r}$ LCD = _____ $(r+1)(r-3)$

Solution set = _____ $\{2\}$

10. $2 - \dfrac{9}{x} - \dfrac{5}{x^2}$ LCD = _____ x^2

Solution set = _____ $\{5, \frac{1}{2}\}$

4.8 Applications

1. Find the value of an unknown variable in a formula. (Frames 1-5)

2. Solve a formula for a specified variable. (Frames 6-12)

3. Solve applications about numbers. (Frames 13-15)

4. Solve applications about work. (Frames 16-21)

5. Solve applications about distance. (Frames 22-23)

1. Use the formula

$$m = \frac{5pq}{8z},$$

and find z if m = 10, p = 24, and q = 2. Substitute into the formula.

$$\underline{\quad} = \frac{5(\quad)(\quad)}{8z}$$ 10; 24; 2

2. Simplify the numerator.

$$10 = \frac{\underline{\quad}}{8z}$$ 240

or $10 = \underline{\quad\quad}$ $\dfrac{30}{z}$

3. Solve the equation for z: z = ___.   3

In Frames 4-5, solve the formula

$$9p = \frac{6k + 2}{3r}$$

for r, using the given values of p and k.

4. p = 3, k = 1   r = _____ $\dfrac{8}{81}$

5. $p = -5$, $k = 0$ $r =$ _____ $-\dfrac{2}{135}$

6. Sometimes a formula must be solved for a speci-
    fied _____.   variable

7. For example, solve $F = \dfrac{Gm_1 m_2}{d^2}$ for m_2, first
    multiply both sides by ____ to get d^2

    _____ = _____. Fd^2; $Gm_1 m_2$

8. Then divide both sides by _____. Gm_1

    _____ = m_2 $\dfrac{Fd^2}{Gm_1}$

**Solve the formulas in Frames 9–12 for the specified
variable.**

9. $\dfrac{9}{k} + \dfrac{1}{m} = \dfrac{1}{4}$, for m

    Multiply both sides by _____. $4km$

    $(4km)\left(\dfrac{9}{k}\right) + (4km)(\underline{\ \ }) = 4km(\underline{\ \ })$ $\dfrac{1}{m}$; $\dfrac{1}{4}$

    _____ + _____ = _____ $36m$; $4k$; km

    Get all terms containing ___ together on one m
    side.

    _____ = $km -$ _____ $4k$; $36m$

    Factor on the right.

    $4k =$ _____ $m(k - 36)$

    Solve for m.

    $m =$ _____ $\dfrac{4k}{k - 36}$

10. $\dfrac{V_1 P_1}{T_1} = \dfrac{V_2 P_2}{T_2}$, for V_1 $V_1 =$ _____ $\dfrac{T_1 V_2 P_2}{P_1 T_2}$

11. $A = P + Prt$, for t $t =$ _____ $\dfrac{A - P}{Pr}$

12. $I = \dfrac{nE}{R + nr}$, for R   R = _____ | $\dfrac{nE - nrI}{I}$

In the rest of this section, we solve various word problems that produce equations with fractions.

13. **What number must be added to the numerator of 2/5 to make the result equal 19/40?**

Let x = the number to be added to the numerator. Then

$$\text{_____} = \frac{19}{40}.$$ | $\dfrac{2 + x}{5}$

Multiply both sides by ____ to get | 40

$$\text{_____} = \text{\_\_\_\_}$$ | $16 + 8x; 19$

or

$$x = \text{_____}.$$ | $\dfrac{3}{8}$

14. **Twice the reciprocal of a number is added to −4, giving an answer which reduces to −2/3. Find the number.**

Let x be the unknown number.

   "Twice the reciprocal" is _____. | $2(1/x)$ or $2/x$

Write the equation.

_____ | $-4 + \dfrac{2}{x} = -\dfrac{2}{3}$

Solve the equation: x = _____. | $\dfrac{3}{5}$

15. **If the reciprocal of the smaller of two consecutive integers is subtracted from twice the reciprocal of the larger, the result is 7 divided by the product of the integers. Find the integers.**

If x = the first integer, then _____ = the second integer. The equation here is | $x + 1$

$$\text{_____} = \text{_____}.$$ | $\dfrac{2}{x + 1} - \dfrac{1}{x};$ $\dfrac{7}{x(x + 1)}$

Solving this equation, we find that the first
integer = ____ and the second integer = ____.

8; 9

16. **Tom can do a job in 12 hours and Pam can do it in
9. How long will it take them if they work together?**

Let x represent the time it will take them if
they work together.

Since Tom can do the job in ___ hours, his rate
is ___ of the job per hour. Pam's rate is ____
of the job per hour. Use the chart below to help
organize the information.

12
1/12; 1/9

Worker	Rate	Time working together	Fractional part of the job done
Tom			
Pam			

$\frac{1}{12}$; x; $\frac{1}{12}$x

$\frac{1}{9}$; x; $\frac{1}{9}$x

The equation will be _____ + _____ = 1
           part done part done one whole
         by Tom   by Pam job done

$\frac{1}{12}$x; $\frac{1}{9}$x

Solve the equation. THe LCD is _____.

36

____ $\left(\frac{1}{12}x + \frac{1}{9}x\right)$ = ___ • 1

36; 36

____ + ____ = ____

3x; 4x; 36

____ = ____

7x; 36

x = _____

$\frac{36}{7}$

Thus, working together they complete the job in
_____ hours.

36/7

17. **An inlet pipe fills a sink in 24 minutes, while the drain will empty it in 30 minutes. If both the inlet and the drain are open, how long will it take to fill the sink?**

Let x = time to fill sink.

The inlet pipe fills the sink in ___ minutes, so   24

its rate is _____ of the sink per minute. The $\frac{1}{24}$

drain empties the sink in ___ minutes, so its   30

rate is ___ of the sink per minute. Use the $\frac{1}{30}$

chart below to help organize the information.

Pipe	Rate	Time working together	Fractional part of the job done
inlet			
drain			

$\frac{1}{24}$; x; $\frac{1}{24}$x

$\frac{1}{30}$; x; $\frac{1}{30}$x

Since the two pipes are working _____ each   against
other, subtract the work of the _____ from the   drain
work of the _____.   inlet

The equation is _____ − _____ − _____ . $\frac{1}{24}$x; $\frac{1}{30}$x; 1
     inlet   drain   one
   filling emptying full sink

Solve the equation. The LCD is ___.   120
x = _____   120
Thus, it takes _____ minutes for the sink to fill.   120

18. **Suppose, in Frame 17, that the inlet was opened for 4 minutes before the drain was opened. How long would it then take to fill the sink?**

Here (___)/24 = _____ of the sink was already   4; 1/6
filled when the drain was opened. This means
that _____ of the sink remains to be filled.   5/6

It took _____ minutes to fill the entire sink (Frame 17). Hence it would take

      (_____)(120) = _____ minutes

to complete the filling of the sink.

19. Fred can do a job in 6 hours and Sam in 10. If Fred does 1/2 of the job, and then Sam joins him, how long will it take them to complete the job?

First, find the time it would take them to do the entire job together: _____ hours. Since _____ of the job remains to be done, an additional

      (_____)(_____) = _____

hours will be required.

20. An outlet can empty a pool in 10 hours, while an inlet can fill it in 5 hours. How long will it take to fill the pool if both the inlet and outlet are open?

      _____ hours

21. A pipe can fill a vat in 15 hours, while an outlet can empty the vat in 20 hours. Suppose the inlet pipe is on for 10 hours, and then the outlet is opened. How much time would then be needed to fill the vat?

      _____ hours

22. George can drive 160 miles in the same time that Samantha can drive 280 miles. Samantha's speed is 30 miles per hour more than George's. What is Samantha's speed?

Let s = George's speed. Then, according to the problem, Samantha's speed = _____. The formula we need to use here is _____.

120

$\frac{5}{6}$; 100

15/4; 1/2

$\frac{1}{2}$; $\frac{15}{4}$; $\frac{15}{8}$

10

20

s + 30

d = rt

Since the problem says that both times are equal, it would be helpful to solve d = rt for t. If we do that, we get

$$t = \underline{\hspace{1cm}}.$$

George went _____ miles at a speed of _____. Hence his time is given by

$$t = \underline{\hspace{1cm}}.$$

In the same way, Samantha's time is

$$t = \underline{\hspace{1.5cm}}.$$

This information could be summarized in the following chart.

	d	r	t
George	160	s	$\frac{160}{s}$
Samantha	280	s + 30	$\frac{280}{s + 30}$

Since the times are equal,

$$\underline{\hspace{2cm}} = \underline{\hspace{3cm}}$$

Multiply both sides by _____, getting

$$\underline{\hspace{2.5cm}} = \underline{\hspace{1.5cm}},$$

or

$$s = \underline{\hspace{1cm}} \text{ miles per hour.}$$

Thus George's speed = _____ and Samantha's speed = _____.

Side notes:
$\frac{d}{r}$

160; s

$\frac{160}{s}$

$\frac{280}{s + 30}$

$\frac{160}{s}$; $\frac{280}{s + 30}$

s(s + 30)

160(s + 30); 280s

40

40 mph

70 mph

23. **Joan can go 30 miles downstream in the same time that she can go 10 miles upstream. Her boat goes 10 miles per hour in still water. Find the speed of the current.**

Let x = speed of the current.

Then the speed upstream = _____ and the speed $10 - x$

downstream = _____ . $10 + x$

Organize a chart to record the information.

(Remember $t = \dfrac{d}{r}$.)

	d	r	t
Upstream	10	10 - x	_____
Downstream	30	10 + x	_____

$\dfrac{10}{10 - x}$

$\dfrac{30}{10 + x}$

Since the time downstream is the same as the time

upstream, the equation here is

_____ = _____ . $\dfrac{10}{(10 - x)}; \dfrac{30}{(10 + x)}$

The solution is x = ___ miles per hour.   5

Chapter 4 Test

The answers for these questions are at the back of this Study Guide.

1. For what values is $\dfrac{8k - 7}{4k + 9}$ not defined?   1. _____

2. Write in lowest terms: $\dfrac{8p^2 - 25p + 3}{p^2 - 3p}$ 2. _____

Multiply or divide.

3. $\dfrac{9r^2}{8r} \cdot \dfrac{6r^3}{18r^5}$ 3. _____

4. $\dfrac{r^2 - r - 6}{r^2 - 3r - 10} \cdot \dfrac{r^2 - 4r - 5}{r^2 - 7r + 12}$ 4. _____

5. $\dfrac{4a^2 - 9}{4a^2 + 12a + 9} \div \dfrac{a^2 - a - 20}{2a^2 - 7a - 15}$ 5. _____

Find the least common denominator for each group of denominators.

6. $8y$, $12y^3$, $4y^2$ 6. _____

7. $a^2 - 4a + 3$, $a^2 - 3a$ 7. _____

Add or subtract.

8. $\dfrac{a}{1 + a} - \dfrac{a}{1 - a}$ 8. _____

9. $\dfrac{1}{r} + \dfrac{1}{r - 1}$ 9. _____

10. $\dfrac{2}{m + 1} - \dfrac{1}{m + 2}$ 10. _____

Simplify the following complex fractions.

11. $\dfrac{\dfrac{4r}{3r - 1}}{\dfrac{7r}{6r - 2}}$ 11. _____

12. $\dfrac{\dfrac{2}{k} - \dfrac{2}{k+1}}{\dfrac{2}{k} + \dfrac{2}{k+1}}$

12. _____

Divide.

13. $\dfrac{9p^4 - 6p^2 + 3p}{9p^2}$

13. _____

14. $(8x^3 + 18x^2 + 3x - 20) \div (2x + 5)$

14. _____

15. $\dfrac{6a^4 - 26a^3 + 49a^2 - 64a + 35}{3a - 7}$

15. _____

16. $(3x^4 + 4x^3 - 2x^2 + x - 9) \div (x + 2)$

16. _____

17. $(x^4 + x^3 - 6x^2 + 16) \div (x + 2)$

17. _____

18. $(10x^4 + x^3 - 13x^2 + 25x - 10) \div (10x - 5)$

18. _____

19. Is -3 a solution of the equation
$4x^4 + 10x^3 - 8x^2 - 5x + 3 = 0$?

19. _____

Solve each equation.

20. $2 - \dfrac{4}{m} = \dfrac{1}{2}$

20. _____

21. $\dfrac{5}{m-5} - \dfrac{1}{m+5} = \dfrac{30}{m^2 - 25}$

21. _____

22. $\dfrac{1}{r} - \dfrac{2}{r-1} = \dfrac{3}{5r}$

22. _____

Simplify.

23. $\dfrac{x^{-1} + y^{-2}}{x^{-2} - y^{-1}}$

23. _____

24. A formula from statistics says that $z = \dfrac{x - \mu}{\sigma}$
    Find σ if $z = .4$, $x = 56$ and $\mu = 52$.   24. _____

Solve for the specified variable.

25. $\dfrac{2}{p} - \dfrac{3}{q} = 1$, for p 25. _____

26. $\dfrac{m}{n} = \dfrac{a + 1}{a - 1}$, for a 26. _____

Solve each applied problem.

27. If twice the reciprocal of the larger of two
    consecutive integers is added to the recipro-
    cal of the smaller, the result is -17 divided
    by the product of the integers. Find the two
    integers.   27. _____

28. Fred can do a job in 12 hours and Kathleen in 8.
    Fred works 6 hours and then Kathleen joins him.
    How long will it take them to finish the job?   28. _____

29. A boat goes 7 miles per hour in still water. It
    can go 20 miles upstream in the same time as 50
    miles downstream. Find the speed of the current. 29. _____

CHAPTER 5 ROOTS AND RADICALS

5.1 Rational Exponents and Radicals

$\boxed{1}$ Define $a^{1/n}$. (Frames 1-12)

$\boxed{2}$ Use radical notation for nth roots. (Frames 13-40)

$\boxed{3}$ Define $a^{m/n}$. (Frames 41-50)

$\boxed{4}$ Write rational exponential expressions as radicals. (Frames 51-60)

---

1. The symbol $a^{1/n}$ is an _____ of _____.	nth root; a
2. If n is an even positive integer, and if $a > 0$, then $a^{1/n}$ is the _____ number whose nth power is _____: $$\left(a^{1/n}\right)^n = \text{\_\_\_\_\_}.$$	positive a a
3. Under these conditions, $a^{1/n}$ is called the _____ nth root of _____.	principal; a
Find each root in Frames 4-6.	
4. $100^{1/2} = $ ____, since $100 > 0$ and $10^2 = $ _____.	10; 100
5. $-169^{1/2} = $ _____	-13
6. $(-169)^{1/2} = $ _____	not a real number
7. If a is a real number, and n is an odd _____ integer, then $a^{1/n}$ is the real number (positive or _____) whose nth power is ____: $$\left(a^{1/n}\right)^n = \text{\_\_\_\_}.$$	positive negative; a a

Find each root in Frames 8–12.

8. $125^{1/3}$ = ____ since ____ = 125.

5; 5^3

9. $(-125)^{1/3}$ = ____ since ____ = -125.

-5; $(-5)^3$

10. $(-128)^{1/7}$ = ____

-2

11. $\left(\frac{27}{64}\right)^{1/3}$ = ____

$\frac{3}{4}$

12. $\left(\frac{1}{216}\right)^{1/3}$ = ____

$\frac{1}{6}$

13. The nth root of a is written ____.

$a^{1/n}$

14. As an alternative to $a^{1/n}$, the nth root of a can be written with radical notation as

$$a^{1/n} = \underline{\quad}.$$

$\sqrt[n]{a}$

15. The number a is the ____.

radicand

16. n is the ____ or ____.

index; order

17. $\sqrt[n]{a}$ is a ____.

radical

18. As a shortcut, the square root of a is written ____.

$\sqrt{a}$

Find each root in Frames 19–26.

19. $\sqrt{100}$ = ____ since ____ = 100.

10; 10^2

20. $-\sqrt{900}$ = ____

-30

21. $\sqrt{\dfrac{121}{144}} = $ _____

$\dfrac{11}{12}$

22. $\sqrt[3]{216} = $ ___

6

23. $\sqrt[3]{-216} = $ ___

−6

24. $\sqrt[4]{16} = $ ___

2

25. $\sqrt[5]{-32} = $ ____

−2

26. $\sqrt[6]{-64} = $ _____

not a real number

27. By definition, a square root of a^2 is a number that can be _____ to give a^2. This number is either a or ____.

squared
−a

28. The symbol $\sqrt{a}$ is used only for the _____ square root of ___.

nonnegative
a

29. The negative square root of a is written ____.

$-\sqrt{a}$

30. To make sure that the square root of a^2 is never _____, use absolute value bars:

$$\sqrt{a^2} = \text{_____}.$$

negative
$|a|$

31. For an even value of n,

$$\sqrt[n]{a^n} = \text{_____}.$$

$|a|$

32. For an odd value of n,

$$\sqrt[n]{a^n} = \text{\_\_\_}.$$

a

Find each root in Frames 33–40.

33. $\sqrt{7^2}$ = |___| = ___   7; 7

34. $\sqrt{(-11)^2}$ = |___| = ____   -11; 11

35. $\sqrt[4]{(-6)^4}$ = |___| = ___   -6; 6

36. $-\sqrt[6]{(-8)^6}$ = ____   -8

37. $\sqrt{25p^2}$ = _____ |5p| or 5|p|

38. $\sqrt{100m^4}$ = _____   10m²
     No absolute value bars are needed on m² since m²
     is never _____.   negative

39. $\sqrt[3]{(-12)^3}$ = _____   -12

40. $\sqrt[5]{(-25)^5}$ = _____   -25

41. We can now define $a^{m/n}$, where m and n are posi-
     tive _____, with m/n written in lowest   integers
     _____: terms

     $$a^{m/n} = (\text{\_\_\_\_\_})^m.$$ $a^{1/n}$

42. Using this definition,
     $$32^{2/5} = (32^{1/5})^{\underline{\quad}} = (\text{\_\_\_\_})^{\underline{\quad}} = \text{\_\_\_}.$$ 2; 2; 2; 4

43. The expression $a^{-m/n}$ = _____. $\dfrac{1}{a^{m/n}}$

Simplify the expressions in Frames 44–48.

44. $64^{2/3} = (64^{1/3})^{\underline{\quad}} = (\underline{\quad})^{\underline{\quad}} = \underline{\quad}$ 2; 4; 2; 16

45. $\left(\frac{9}{4}\right)^{3/2} = \left[(\quad)^{1/2}\right]^{\underline{\quad}} = (\underline{\quad})^{\underline{\quad}} = \underline{\quad}.$ $\frac{9}{4}$; 3; $\frac{3}{2}$; 3; $\frac{27}{8}$

46. $\left(\frac{1}{4}\right)^{-3/2} = \dfrac{1}{(\underline{\quad})^{3/2}}$ $\frac{1}{4}$

  $= \dfrac{1}{\dfrac{\underline{\quad}}{}}$ $\frac{1}{8}$

  $= \underline{\quad}$ 8

47. $\left(\frac{16}{81}\right)^{3/4} = \underline{\quad}$ $\frac{8}{27}$

48. $\left(\frac{1}{64}\right)^{-1/3} = \underline{\quad}$ 4

49. There is an alternative definition for $a^{m/n}$:

$$a^{m/n} = (\underline{\quad})^{1/n}.$$ a^m

50. In general, is $(a^{1/n})^m$ or $(a^m)^{1/n}$ the easier form to use? $\underline{\quad\quad\quad}$ $(a^{1/n})^m$

51. We know that $a^{m/n} = (\underline{\quad})^m$ and that $a^{m/n} = (\underline{\quad})^{1/n}.$ $a^{1/n}$ a^m

52. Using radicals,

$$a^{m/n} = (\underline{\quad})^m \text{ and } a^{m/n} = \sqrt[n]{\underline{\quad}}.$$ $\sqrt[n]{a}$; a^m

Write the expressions in Frames 53–56 with radicals.

53. $5^{2/3} = (\underline{\quad})^2$ $\sqrt[3]{5}$

54. $2p^{1/4} = 2\underline{\quad}$ (if $p \geq \underline{\quad}$) $\sqrt[4]{p}$; 0

55. $(6y)^{3/5}$ = _____ $\sqrt[5]{(6y)^3}$

56. $-5y^{4/7}$ = _____ $-5\sqrt[7]{y^4}$

**Replace each radical in Frames 57–60 with an equiv-
alent expression using rational exponents. Then
simplify if possible. Assume x is positive.**

57. $\sqrt{x^5}$ = $x^{\underline{\quad}/\underline{\quad}}$ 5; 2

58. $\sqrt[5]{x^3}$ = $x^{\underline{\quad}}$ 3/5

59. $\sqrt[8]{x^6}$ = $x^{\underline{\quad}}$ = $x^{\underline{\quad}}$ 6/8; 3/4

60. $\dfrac{\sqrt[4]{x^3} \cdot \sqrt{x}}{\sqrt{x^2}}$ = $\dfrac{x^{\underline{\quad}} \cdot x^{\underline{\quad}}}{x^{\underline{\quad}}}$ 3/4; 1/2

          1

          = $x^{\underline{\quad}}$ $x^{1/4}$ or $\sqrt[4]{x}$

5.2 More About Rational Exponents and Radicals

1️⃣ Use the rules of exponents with rational exponents.
    (Frames 1–20)

2️⃣ Use a calculator to find roots. (Frames 21–30)

---

1. All the familiar properties of exponents are also
    valid for _____ exponents.   rational

**Complete the rules for rational exponents r and s in
Frames 2–7.**

2. $a^r \cdot a^s$ = _____ a^{r+s}

3. a^{-r} = _____ $1/a^r$

4. $(a^r)^s =$ _____

a^{rs}

5. $\dfrac{a^r}{a^s} =$ _____

a^{r-s}

6. $(ab)^r =$ _____

$a^r b^r$

7. $\left(\dfrac{a}{b}\right)^r =$ ____

$\dfrac{a^r}{b^r}$

Use the rules of exponents to simplify the expressions in Frames 8–15.

8. $5^{1/3} \cdot 5^{2/3} = 5^{\underline{}+\underline{}} = \underline{}$

1/3; 2/3; 5

9. $49^{3/4} \cdot 49^{3/4} = 49^{\underline{}+\underline{}}$

3/4; 3/4

$= 49^{\underline{}}$

3/2

$= (49^{\underline{}})^{\underline{}}$

1/2; 3

$= \underline{}$

7^3 or 343

10. $9^{1/2} \cdot 9^{-3/2} = 9^{\underline{}+\underline{}} = 9^{\underline{}} = \underline{}$

1/2; −3/2; −1; 1/9

11. $16^{3/4} \cdot 16^{1/2} \cdot 16^{-1/4} = 16^{\underline{}} = \underline{}$

1; 16

12. $\dfrac{k^{2/3} \cdot k^{1/6}}{k^{-1/3}} = \dfrac{k^{\underline{}}}{k^{\underline{}}} = k^{\underline{}}$

5/6; 7/6
−1/3

13. $\dfrac{r^{-1/2} \cdot r^{1/4} \cdot r}{r^{-3/4} \cdot r^2} = \underline{}$

$\dfrac{1}{r^{1/2}}$

14. $\dfrac{(a^{3/4} \cdot a^{-1/2})^{-2}}{(a^{1/4} \cdot a^{3/2})^2} = \underline{}$

$\dfrac{1}{a^4}$

15. $\dfrac{x^{1/4} \cdot x^{3/8}}{(x^{-2/3})^{-3/4}} = x^{\underline{}}$

1/8

In Frames 16—20 replace all radicals with rational exponents, then apply the rules for rational exponents. Assume that all variables represent positive real numbers.

16. $\sqrt[5]{y^3} \cdot \sqrt[3]{y}$

     $= y^{---} \cdot y^{---}$ (convert to _____ exponents)   3/5; 1/3; rational

     $= y^{---+---}$ (use _____ rule for exponents)   3/5 + 1/3; product

     $= y^{---+---}$ (find a _____ _____)   9/15 + 5/15 <br> common denominator

     $= y^{---}$ 14/15

17. $\sqrt[5]{x^2} \cdot \sqrt[4]{x^3} =$ _____ $x^{23/20}$

18. $(\sqrt[3]{x^2})^4 =$ _____ $x^{8/3}$

19. $\dfrac{\sqrt[3]{y^4}}{\sqrt{y^5}} =$ _____ $\dfrac{1}{y^{7/6}}$

20. $\sqrt[3]{\sqrt[5]{y^4}} =$ _____ $y^{4/15}$

21. Many numbers, such as $\sqrt{5}$, or $\sqrt[3]{9}$, are not equal to any rational number. These numbers are called _____.   irrational
Decimal approximations for many irrational numbers can be found by using a calculator.

Find decimal approximations for the irrational numbers in Frames 22—24.

22. $\sqrt{71}$
Use a calculator.

$$\sqrt{71} \approx \text{_____}$$ 8.426

The symbol $\approx$ means "is _____ equal to." | approximately

23. $\sqrt{95} \approx$ _____ | 9.747

24. $-\sqrt{39} \approx$ _____ | -6.245

25. _____ calculators have a key marked $\boxed{y^x}$, called the _____ key. When this key is used with the _____ key, $\boxed{INV}$, we can find roots other than _____ roots. | Scientific

exponential

inverse

square

26. To find $\sqrt[6]{326}$ the typical keystroke sequence is

___ ___ ___ $\boxed{INV}$ $\boxed{y^x}$ ___ = . | 3; 2; 6; 6

The result is approximately _____. | 2.623

Find the decimal approximation for the expressions in Frames 27—30.

27. $\sqrt[5]{672}$ | 3.677

28. $\sqrt[3]{9417}$ | 21.117

29. $(743)^{1/6}$ | 3.010

30. $(34.7)^{4/3}$ | 113.181

5.3   Simplifying Radicals

1 Learn the product rule for radicals. (Frames 1–4)

2 Use the quotient rule for radicals. (Frames 5–8)

3 Simplify radicals. (Frames 9–22 and 27–30)

4 Simplify radicals by using different indexes. (Frames 23–26)

5 Use the Pythagorean formula to find the length of the side of a right triangle. (Frames 31–34)

---

1. For any real numbers a and b, and any natural number n,

$$\sqrt[n]{a} \cdot \sqrt[n]{b} = \underline{\qquad}.$$

This result is called the _____ rule for radicals.

$\sqrt[n]{ab}$

product

Use the product rule to simplify the expression in Frames 2–4.

2. $\sqrt[3]{9} \cdot \sqrt[3]{3} = \sqrt[3]{\underline{\quad}} = \underline{\quad}$

27; 3

3. $\sqrt{5} \cdot \sqrt{r} = \underline{\quad}$ (if r ≥ ___)

$\sqrt{5r}$; 0

4. $\sqrt[4]{3y} \cdot \sqrt[4]{7y^2} = \underline{\qquad}$ (if y ≥ 0)

$\sqrt[4]{21y^3}$

5. There is also a _____ rule; for real numbers a and b, with b ≠ ___, and any natural number n,

quotient

0

$$\frac{\sqrt[n]{a}}{\sqrt[n]{b}} = \underline{\qquad}.$$

$\sqrt[n]{\dfrac{a}{b}}$

For example,

$$\frac{\sqrt[5]{7}}{\sqrt[5]{9}} = \underline{\qquad}.$$

$\sqrt[5]{\dfrac{7}{9}}$

Simplify each expression in Frames 6—8.

6. $\dfrac{\sqrt[4]{32}}{\sqrt[4]{2}} = \sqrt[4]{\underline{\quad}} = \underline{\quad}$ $\qquad$ $\dfrac{32}{2};\ 2$

7. $\dfrac{\sqrt[3]{32}}{\sqrt[3]{4}} = \sqrt[3]{\underline{\quad}} = \underline{\quad}$ $\qquad$ $\dfrac{32}{4};\ 2$

8. $\sqrt[4]{\dfrac{y^{12}}{16}} = \underline{\qquad}$ (if $y \geq 0$) $\qquad$ $\dfrac{y^3}{2}$

9. A radical is in _____ form when $\qquad$ simplified

   (a) factored to prime factors, the radicand
        contains no factor to a _____ greater $\qquad$ power
        than or _____ to the index; $\qquad$ equal

   (b) the radicand has no _____; $\qquad$ fractions

   (c) no _____ contains a radical; $\qquad$ denominator

   (d) the exponents in the _____ and the $\qquad$ radicand
        _____ of the _____ have no common $\qquad$ index; radical
        factor.

Simplify each of the expressions in Frames 10—22.
Assume all variables represent positive real numbers.

10. $\sqrt{8} = \sqrt{\underline{\quad}\cdot\underline{\quad}} = \sqrt{\underline{\quad}}\cdot\sqrt{\underline{\quad}} = \underline{\quad}\,\sqrt{\underline{\quad}}$ $\qquad$ 4; 2; 4; 2; 2; 2

11. $\sqrt{18} = \sqrt{\underline{\quad}\cdot\underline{\quad}} = \sqrt{\underline{\quad}}\cdot\sqrt{\underline{\quad}} = \underline{\quad}\,\sqrt{\underline{\quad}}$ $\qquad$ 9; 2; 9; 2; 3; 2

12. $\sqrt{72} = \sqrt{\underline{\quad}\cdot\underline{\quad}} = \sqrt{\underline{\quad}}\cdot\sqrt{\underline{\quad}} = \underline{\quad}\,\sqrt{\underline{\quad}}$ $\qquad$ 36; 2; 36; 2; 6; 2

13. $\sqrt[3]{81} = \sqrt[3]{\underline{\quad}\cdot\underline{\quad}} = \sqrt[3]{\underline{\quad}}\cdot\sqrt[3]{\underline{\quad}}$ $\qquad$ 27; 3; 27; 3
     $= \underline{\quad}\,\sqrt[3]{\underline{\quad}}$ $\qquad$ 3; 3

14. $\sqrt{8x^2y^4} = \sqrt{\underline{}} \cdot \sqrt{\underline{}} \cdot \sqrt{\underline{}}$ 8; x^2; y^4

 $= \sqrt{\underline{} \cdot \underline{}}(\underline{})(\underline{})$ 4; 2; x; y^2

 $= \underline{} \sqrt{\underline{}}$ $2xy^2$; 2

15. $\sqrt{128x^5y^3} = \underline{} \sqrt{\underline{}}$ $8x^2y$; 2xy

16. $\sqrt{54k^8s^{10}} = \underline{} \sqrt{\underline{}}$ $3k^4s^5$; 6

17. $\sqrt{18x^3y^2} \cdot \sqrt{x^3y^4} = \sqrt{\underline{} \cdot \underline{}}$ $18x^3y^2$; x^3y^4

 $= \sqrt{(\underline{})x^{\underline{}}y^{\underline{}}}$ 18; 8 6

 $= \underline{} \sqrt{\underline{}}$ $3x^3y^3$; 2

18. $\sqrt{12p^5q^4} \cdot \sqrt{3p^7q^3} = \underline{} \sqrt{\underline{}}$ $6p^6q^3$; q

19. $\sqrt[4]{m^5n^6} \cdot \sqrt[4]{m^3n^9} = \underline{} \sqrt[4]{\underline{}}$ m^2n^3; n^3

20. $\sqrt[5]{a^7 \cdot b^5} \cdot \sqrt[5]{a^4b^7} = \underline{} \sqrt[5]{\underline{}}$ a^2b^2; ab^2

21. $\sqrt{\dfrac{3}{16}} = \dfrac{\sqrt{\underline{}}}{\sqrt{\underline{}}} = \underline{}$ 3; $\sqrt{3}$

          16; 4

22. $\sqrt[4]{\dfrac{r^3}{81}} = \underline{}$ $\dfrac{\sqrt[4]{r^3}}{3}$

23. For a radical to be simplified, an exponent
     in the _____ and the _____ on the   radicand; index
     radical must have no common _____.   factor

24. By this rule, $\sqrt[4]{7^2}$ (*is/is not*) simplified.   is not
     To simplify this radical, write it with
     rational exponents.

$$\sqrt[4]{7^2} = 7^{\overline{}}$$ 2/4

Since $7 > 0$, we can simplify the exponent to ____. Rewriting $7^{1/2}$ with radicals gives

$$\sqrt[4]{7^2} = \underline{\quad}.$$

1/2

$\sqrt{7}$

Simplify in Frames 25–26.

25. $\sqrt[9]{y^6}$ _____

$\sqrt[3]{y^2}$

26. $\sqrt[8]{625} = \sqrt[8]{\underline{\quad}} = \underline{\quad}$

5^4; $\sqrt{5}$

27. When multiplying radicals with different _____ use rational _____ to write each radical with the least common _____.

indexes

exponents

index

Multiply the radicals in Frames 28–30.

28. $\sqrt[5]{2} \cdot \sqrt[3]{4}$ The least common index is ____.

15

$\sqrt[5]{2} = 2^{\overline{\quad}} = 2^{\overline{15}} = \sqrt[15]{2^{\overline{\quad}}} = \sqrt[15]{\underline{\quad}}$

1/5; 3; 3; 8

$\sqrt[3]{4} = 4^{\overline{\quad}} = 4^{\overline{15}} = \sqrt[15]{4^{\overline{\quad}}} = \sqrt[15]{\underline{\quad}}$

1/3; 5; 5; 1024

$\sqrt[5]{2} \cdot \sqrt[3]{4} = \sqrt[15]{\underline{\quad}} \cdot \sqrt[15]{\underline{\quad}} = \sqrt[15]{\underline{\quad}}$

8; 1024; 8192

29. $\sqrt{5} \cdot \sqrt[3]{3} = \underline{\quad}$

$\sqrt[6]{1125}$

30. $\sqrt[4]{3} \cdot \sqrt[3]{6} = \underline{\quad}$

$\sqrt[12]{34,992}$

31. If c is the length of the longest side of a right triangle, the _____, and a and b are the lengths of the shorter sides, the _____, then _____. This result is called the _____ _____.

hypotenuse

legs

$c^2 = a^2 + b^2$

Pythagorean formula

Solving this equation for c, c = _____. This gives the length of the _____ directly.

$\sqrt{a^2 + b^2}$

hypotenuse

Use the Pythagorean formula to find the length of each missing side in Frames 32–34. Leave your answer as a simplified radical.

32.

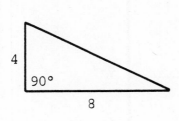

$c = \sqrt{\underline{\hphantom{x}}^2 + \underline{\hphantom{x}}^2}$ 4; 8

$= \sqrt{\underline{\hphantom{x}} + \underline{\hphantom{x}}}$ 16; 64

$= \sqrt{\underline{\hphantom{x}}}$ 80

$= \sqrt{\underline{\hphantom{x}}} \cdot \sqrt{\underline{\hphantom{x}}}$ 16; 5

$= \underline{\hphantom{xxx}}$ $4\sqrt{5}$

33.

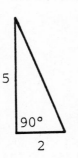

$c = \underline{\hphantom{xxxxx}}$ $\sqrt{29}$

34.

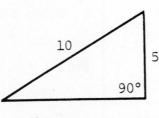

$a = \underline{\hphantom{xxxxx}}$ $5\sqrt{3}$

(Hint: use $a = \sqrt{c^2 - b^2}$.)

5.4 Adding and Subtracting Radical Expressions

[1] Define a radical expression. (Frames 1–2)

[2] Simplify radical expressions involving addition and subtraction.
(Frames 3–16)

1. A _____ expression is an _____
    expression that contains radicals.

    radical; algebraic

2. All of the following are _____ expressions.

    $\sqrt[3]{9} + 4$, $\dfrac{\sqrt{7} + \sqrt{5}}{11}$,   and $\sqrt{2} - \sqrt[4]{14}$

    radical

3. The distributive property can be used to simplify
    some expressions involving radical expressions
    with the same _____ and _____.   For
    example,

    indexes; radicands

    $$3\sqrt{5} - 7\sqrt{5} = (\underline{\hspace{1.5cm}})\sqrt{5} = \underline{\hspace{1cm}} \sqrt{5}.$$

    3 – 7; –4

4. In the same way,

    $2\sqrt{18} - 3\sqrt{32} = 2(\sqrt{\underline{\hspace{0.5cm}} \cdot \underline{\hspace{0.5cm}}} - 3(\sqrt{\underline{\hspace{0.5cm}} \cdot \underline{\hspace{0.5cm}}})$

    9; 2; 16; 2

    $= 2(\underline{\hspace{0.5cm}})(\sqrt{\underline{\hspace{0.5cm}}}) - 3(\underline{\hspace{0.5cm}})(\sqrt{\underline{\hspace{0.5cm}}})$

    3; 2; 4; 2

    $= \underline{\hspace{0.5cm}} \sqrt{\underline{\hspace{0.5cm}}} - \underline{\hspace{0.5cm}} \sqrt{\underline{\hspace{0.5cm}}}$

    6; 2; 12; 2

    $= \underline{\hspace{0.5cm}} \sqrt{\underline{\hspace{0.5cm}}}.$

    –6; 2

Simplify the expressions in Frames 5–16.

5. $7\sqrt{72} - 2\sqrt{50} = 7(\underline{\hspace{0.5cm}}\sqrt{\underline{\hspace{0.5cm}}}) - 2(\underline{\hspace{0.5cm}}\sqrt{\underline{\hspace{0.5cm}}})$

    6; 2; 5; 2

    $= \underline{\hspace{0.5cm}} \sqrt{\underline{\hspace{0.5cm}}} - \underline{\hspace{0.5cm}} \sqrt{\underline{\hspace{0.5cm}}}$

    42; 2; 10; 2

    $= \underline{\hspace{0.5cm}} \sqrt{\underline{\hspace{0.5cm}}}$

    32; 2

6. $3\sqrt{20} - 7\sqrt{45} = 3(\underline{\hspace{0.5cm}}\sqrt{\underline{\hspace{0.5cm}}}) - 7(\underline{\hspace{0.5cm}}\sqrt{\underline{\hspace{0.5cm}}})$

    2; 5; 3; 5

    $= \underline{\hspace{0.5cm}} \sqrt{5} - \underline{\hspace{0.5cm}} \sqrt{5}$

    6; 21

    $= \underline{\hspace{1.5cm}}$

    $-15\sqrt{5}$

7. $-3\sqrt{24} + 2\sqrt{54} - \sqrt{6} = -3(\underline{\quad}\sqrt{6}) + 2(\underline{\quad}\sqrt{6}) - \sqrt{6}$ 2; 3

   $= \underline{\quad}\sqrt{6} + \underline{\quad}\sqrt{6} - \sqrt{6}$ -6; 6

   $= \underline{\quad}$ $-\sqrt{6}$

8. $2\sqrt[3]{16} - 4\sqrt[3]{54} = 2(\underline{\quad}\sqrt[3]{2}) - 4(\underline{\quad}\sqrt[3]{2})$ 2; 3

   $= \underline{\quad}\sqrt[3]{2} - \underline{\quad}\sqrt[3]{2}$ 4; 12

   $= \underline{\quad\quad}$ $-8\sqrt[3]{2}$

9. $\sqrt{9x} - \sqrt{16x} = \underline{\quad}\sqrt{\underline{\quad}} - \underline{\quad}\sqrt{\underline{\quad}} = \underline{\quad}$ 3; x; 4; x; $-\sqrt{x}$

10. $\sqrt{18m^3} - \sqrt{72m^3} = \underline{\quad}\sqrt{2m} - \underline{\quad}\sqrt{2m}$ 3m; 6m

    $= \underline{\quad\quad}$ $-3m\sqrt{2m}$

11. $\sqrt{50p^3} - \sqrt{8p} = \underline{\quad\quad\quad\quad}$ $(5p - 2)\sqrt{2p}$

12. $3\sqrt{5x} - \sqrt{45x} - 4\sqrt{20x} = \underline{\quad\quad}$ $-8\sqrt{5x}$

13. $\dfrac{7\sqrt{2}}{\sqrt{9}} - \dfrac{\sqrt{2}}{6}$

    Simplify the denominator in $\underline{\quad}$. $\dfrac{7\sqrt{2}}{\sqrt{9}}$

    $$\frac{7\sqrt{2}}{\sqrt{9}} = \frac{7\sqrt{2}}{\underline{\quad}}$$

                                                                            3

    Therefore,

    $$\frac{7\sqrt{2}}{\sqrt{9}} - \frac{\sqrt{2}}{6} = \frac{7\sqrt{2}}{3} - \frac{\sqrt{2}}{6}$$

    Find the common denominator.

    $$= \frac{7 \cdot \sqrt{2}}{3 \cdot \underline{\quad}} - \frac{\sqrt{2}}{6}$$ 2
                                                                            2

    $$= \frac{14\sqrt{2}}{\underline{\quad}} - \frac{\sqrt{2}}{6}$$ 6

Simplify.

= _____ $\dfrac{13\sqrt{2}}{6}$

14. $\sqrt{\dfrac{3}{81}} - \sqrt{32} + \sqrt{\dfrac{3}{16}} =$ ___ − ___ + ___ $\dfrac{\sqrt{3}}{9}$; $4\sqrt{2}$; $\dfrac{\sqrt{3}}{4}$

= _____ − $4\sqrt{2}$ $\dfrac{13\sqrt{3}}{36}$

15. $\dfrac{2}{\sqrt{r^2}} - \dfrac{3\sqrt{r}}{3r} =$ ____ $\dfrac{2 - \sqrt{r}}{r}$

16. $5y\sqrt[3]{\dfrac{y}{8}} + 3\sqrt[3]{\dfrac{y^4}{27}} =$ _____ $\dfrac{7y\sqrt[3]{y}}{2}$

5.5 Multiplying and Dividing Radical Expressions

[1] Multiply radical expressions. (Frames 1–16)

[2] Rationalize denominators with one radical. (Frames 17–30)

[3] Rationalize denominators with binomials involving radicals. (Frames 31–36)

[4] Write radical quotients in lowest terms. (Frames 37–42)

---

1. Binomial expressions involving radicals can be multiplied using the _____ (First−Outer−Inner−Last) method of Chapter 4. | FOIL

2. To find $(\sqrt{7} - 5)(\sqrt{3} + \sqrt{2})$, work as follows.

$(\sqrt{7} - 5)(\sqrt{3} + \sqrt{2})$ First: $(\sqrt{7})(\_\_\_) =$ _____ $\sqrt{3}$; $\sqrt{21}$

$(\sqrt{7} - 5)(\sqrt{3} + \sqrt{2})$ Outer: $(\sqrt{7})(\_\_\_) =$ _____ $\sqrt{2}$; $\sqrt{14}$

$(\sqrt{7} - 5)(\sqrt{3} + \sqrt{2})$ Inner: $(-5)(\_\_\_) =$ _____ $\sqrt{3}$; $-5\sqrt{3}$

$(\sqrt{7} - 5)(\sqrt{3} + \sqrt{2})$ Last: $(-5)(\_\_\_) =$ _____ $\sqrt{2}$; $-5\sqrt{2}$

3. The product is _____.

$\sqrt{21} + \sqrt{14} - 5\sqrt{3} - 5\sqrt{2}$

4. Find the product $(\sqrt{2} + 3)\sqrt{3} - 4)$.

$(\sqrt{2} + 3)(\sqrt{3} - 4)$

   = $(\sqrt{2})(\underline{}) + (\sqrt{2})(\underline{}) + 3(\underline{}) + 3(\underline{})$

$\sqrt{3}$; −4; $\sqrt{3}$; −4

   = $\underline{} + \underline{} + \underline{} + \underline{}$

$\sqrt{6}$; $-4\sqrt{2}$; $3\sqrt{3}$; −12

5. In the same way,

   $(3 + \sqrt{5})(2 + \sqrt{3})$

   = $\underline{} + \underline{} + \underline{} + \underline{}$.

6; $3\sqrt{3}$; $2\sqrt{5}$; $\sqrt{15}$

6. To square a binomial such as x + y, write
$(x + y)^2 = \underline{}$. We can use this
formula to square a radical. For example,

$x^2 + 2xy + y^2$

   $(\sqrt{5} + 3)^2 = (\underline{})^2 + 2(\underline{})(\underline{}) + (\underline{})^2$

$\sqrt{5}$; $\sqrt{5}$; 3; 3

     = $\underline{} + \underline{} + \underline{}$

5; $6\sqrt{5}$; 9

     = $\underline{}$

$14 + 6\sqrt{5}$

7. Also,

   $(-4 + \sqrt{3})^2 = (\underline{})^2 + 2(\underline{})(\underline{}) + (\underline{})^2$

−4; −4; $\sqrt{3}$; $\sqrt{3}$

     = $\underline{} + \underline{} + \underline{}$

16; $-8\sqrt{3}$; 3

     = $\underline{}$.

$19 - 8\sqrt{3}$

Find each product in Frames 8–16.

8. $5(3 - \sqrt{2}) = (\underline{})(\underline{}) - (\underline{})(\underline{})$

5; 3; 5; $\sqrt{2}$

     = $\underline{}$

$15 - 5\sqrt{2}$

9. $\sqrt{3}(\sqrt{5} - \sqrt{2})$ = (___)(___) - (___)(___) $\sqrt{3}; \sqrt{5}; \sqrt{3}; \sqrt{2}$

    = _____ $\sqrt{15} - \sqrt{6}$

10. $\sqrt{2}(\sqrt{8} - \sqrt{18})$ = _____ - _____ = ___ $\sqrt{16}; \sqrt{36}; -2$

11. $\sqrt{5}(\sqrt{45} - \sqrt{3})$ = _____ $15 - \sqrt{15}$

12. $(2 + \sqrt{7})(2 - \sqrt{7})$ = ___ -3

13. $(4 + \sqrt{8})(4 - \sqrt{8})$ = ___ 8

14. $(-3 + \sqrt{11})(-3 - \sqrt{11})$ = ___ -2

15. $(\sqrt{2} + 3)^2$ = _____ $11 + 6\sqrt{2}$

16. $(-3 + \sqrt{5})^2$ = _____ $14 - 6\sqrt{5}$

17. To be simplified, an expression can have no
    radical in the _____.   denominator

18. The process of removing _____ from the   radicals
    denominator is called _____ the   rationalizing
    denominator.

19. To rationalize the denominator, we usually
    _____ both the numerator and   multiply
    denominator by some appropriate factor.

20. To rationalize the denominator of

    $$\frac{2}{\sqrt{5}},$$

    multiply numerator and denominator by _____. $\sqrt{5}$

21. This gives

$$\frac{2}{\sqrt{5}} = \frac{2(\underline{\hspace{1cm}})}{\sqrt{5}(\underline{\hspace{1cm}})}$$

$$= \underline{\hspace{2cm}} .$$

(Recall: $\sqrt{5} \cdot \sqrt{5} = \underline{\hspace{1cm}}$.)

$\sqrt{5}$
$\sqrt{5}$

$\frac{2\sqrt{5}}{5}$

5

22. To rationalize the denominator of

$$\frac{8}{\sqrt{32}},$$

we could multiply by $\sqrt{32}$. However, it is simpler
to simplify $\sqrt{32}$ first.

$$\sqrt{32} = \underline{\hspace{1.5cm}} .$$

$4\sqrt{2}$

23. Multiply the fraction by $\underline{\hspace{1.5cm}}$.

$$\frac{8}{\sqrt{32}} = \frac{8\sqrt{2}}{\sqrt{32} \cdot \sqrt{2}} = \underline{\hspace{2cm}}$$

$\frac{\sqrt{2}}{\sqrt{2}}$

$\frac{8\sqrt{2}}{8}$

24. Simplify the answer.

$$\frac{8}{\sqrt{32}} = \underline{\hspace{1.5cm}}$$

$\sqrt{2}$

**Rationalize the denominator in Frames 25–30. Assume
all variables represent positive real numbers.**

25. $\frac{6}{\sqrt{7}} = \underline{\hspace{2cm}}$

$\frac{6\sqrt{7}}{7}$

26. $\frac{12}{\sqrt{8}} = \frac{12}{2(\underline{\hspace{0.8cm}})} \cdot (\underline{\hspace{0.8cm}})$

$$= \frac{12\sqrt{2}}{\underline{\hspace{1cm}}}$$

$$= \underline{\hspace{1cm}}$$

$\sqrt{2}$

$\sqrt{2}$; $\sqrt{2}$

4

$3\sqrt{2}$

27. $\sqrt{\dfrac{12}{5}} = \dfrac{\sqrt{12}}{\rule{1cm}{0.4pt}}$

$\sqrt{5}$

$= \dfrac{2\sqrt{\rule{0.6cm}{0.4pt}}}{\sqrt{5}}$

3

$= \dfrac{2\sqrt{3}(\rule{1cm}{0.4pt})}{\sqrt{5}(\rule{1cm}{0.4pt})}$

$\sqrt{5}$

$\sqrt{5}$

$= \rule{2cm}{0.4pt}$

$\dfrac{2\sqrt{15}}{5}$

28. $\sqrt{\dfrac{8a^2}{b^3}} = \dfrac{\sqrt{8a^2}}{\rule{0.8cm}{0.4pt}} = \rule{2cm}{0.4pt}$

$\dfrac{\sqrt{8a^2}}{\sqrt{b^3}}$; $b\sqrt{b}$

Multiply both numerator and denominator by $\rule{1cm}{0.4pt}$.

$\sqrt{b}$

$= \dfrac{\sqrt{8a^2}(\rule{1cm}{0.4pt})}{b\sqrt{b}(\rule{1cm}{0.4pt})}$

$\sqrt{b}$

$\sqrt{b}$

$= \dfrac{\rule{2cm}{0.4pt}}{b \cdot b}$

$\sqrt{8a^2b}$

$= \dfrac{2a\sqrt{2b}}{\rule{1cm}{0.4pt}}$

b^2

29. $\dfrac{6}{\sqrt[3]{4}}$

We want a perfect $\rule{2cm}{0.4pt}$ in the denominator.

cube

Since 8 is a perfect cube, and $8 = 4 \cdot \rule{0.7cm}{0.4pt}$,

2

multiply both numerator and denominator by $\rule{1cm}{0.4pt}$.

$\sqrt[3]{2}$

$\dfrac{6}{\sqrt[3]{4}} = \dfrac{6\sqrt[3]{2}}{\sqrt[3]{4}(\rule{1cm}{0.4pt})}$

$\sqrt[3]{2}$

$= \dfrac{6\sqrt[3]{2}}{\rule{1cm}{0.4pt}}$

2

$= \rule{2cm}{0.4pt}$

$3\sqrt[3]{2}$

30. $\sqrt[3]{\dfrac{5}{36}} = \rule{2cm}{0.4pt}$

$\dfrac{\sqrt[3]{30}}{6}$

31. To rationalize a _____ that contains

a binomial involving radicals, such as

$$\frac{-2}{1 + \sqrt{7}},$$

multiply both the numerator and denominator by

_____, which is the _____ of

$$\frac{-2}{1 + \sqrt{7}} = \frac{-2(\quad\quad)}{(1 + \sqrt{7})(\quad\quad\quad)}$$

$$= \frac{\rule{3cm}{0.4pt}}{\rule{1.5cm}{0.4pt}}$$

$$= \rule{3cm}{0.4pt}.$$

denominator

$1 - \sqrt{7}$; conjugate

$1 - \sqrt{7}$
$1 - \sqrt{7}$

$\dfrac{-2(1 - \sqrt{7})}{-6}$

$\dfrac{1 - \sqrt{7}}{3}$

32. To rationalize the denominator or

$$\frac{3}{1 - \sqrt{5}},$$

multiply both the numerator and denominator by

_____.

This gives

$1 + \sqrt{5}$

$$\frac{3}{1 - \sqrt{5}} = \frac{3(\quad\quad\quad)}{(1 - \sqrt{5})(\quad\quad\quad)}$$

$$= \frac{\rule{3cm}{0.4pt}}{\rule{1.5cm}{0.4pt}}$$

$$= \rule{3.5cm}{0.4pt}.$$

$1 + \sqrt{5}$
$1 + \sqrt{5}$

$\dfrac{3(1 + \sqrt{5})}{-4}$

$\dfrac{-3(1 + \sqrt{5})}{4}$

Rationalize the denominators of each expression in Frames 33–36.

33. $\dfrac{\sqrt{5}}{\sqrt{5} - 1} = \dfrac{\sqrt{5}(\quad\quad\quad)}{(\sqrt{5} - 1)(\quad\quad\quad)} = \rule{3cm}{0.4pt}$

$\sqrt{5} + 1$; $\dfrac{5 + \sqrt{5}}{4}$

$\sqrt{5} + 1$

34. $\dfrac{3}{\sqrt{3} + \sqrt{2}} = \dfrac{3(\quad\quad\quad)}{(\sqrt{3} + \sqrt{2})(\quad\quad\quad)} = \rule{3cm}{0.4pt}$

$\sqrt{3} - \sqrt{2}$; $3\sqrt{3} - 3\sqrt{2}$

$\sqrt{3} - \sqrt{2}$

35. $\dfrac{\sqrt{8}}{\sqrt{2} + \sqrt{5}}$ = _____

$\dfrac{-4 + 2\sqrt{10}}{3}$

36. $\dfrac{2 - \sqrt{5}}{\sqrt{6} - \sqrt{3}}$ = _____

$\dfrac{2\sqrt{6} + 2\sqrt{3} - \sqrt{30} - \sqrt{15}}{3}$

37. To write

$$\dfrac{8 - 4\sqrt{7}}{6}$$

in lowest terms, factor out the common _____ in the _____.

factor

numerator

38. $\dfrac{8 - 4\sqrt{7}}{6}$ = $\dfrac{(\qquad)}{6}$

= _____

$4;\ 2 - \sqrt{7}$

$\dfrac{2(2 - \sqrt{7})}{3}$ or

$\dfrac{4 - 2\sqrt{7}}{3}$

Write the expressions in Frames 39–42 in lowest terms.

39. $\dfrac{12\sqrt{2} + 8\sqrt{5}}{2}$ = _____

$6\sqrt{2} + 4\sqrt{5}$

40. $\dfrac{6 - 10\sqrt{3}}{4}$ = _____

$\dfrac{3 - 5\sqrt{3}}{2}$

41. $\dfrac{8k - \sqrt{32k^2}}{4k}$ = $\dfrac{8k -}{4k}$

= _____

$4k\sqrt{2}$

$2 - \sqrt{2}$

42. $\dfrac{15p - \sqrt{50p^2}}{6p}$ = _____

$\dfrac{15 - 5\sqrt{2}}{6}$

5.6 Equations with Radicals

1. Learn the power rule. (Frame 1)

2. Solve radical equations, such as $\sqrt{3x + 4} = 8$. (Frames 2–6)

3. Solve radical equations, such as $\sqrt{m^2 - 4m + 9} = m - 1$, requiring the square of a binomial. (Frames 7–8)

4. Solve radical equations, such as $\sqrt{5m + 6} + \sqrt{3m + 4} = 2$, that require squaring twice. (Frames 9–10)

5. Solve radical equations with indexes greater than 2, that require squaring twice. (Frames 11–15)

---

1. To solve equations involving radicals, use the power rule: if both sides of an equation are raised to the same _____, all solutions of the _____ equation are also solutions of the new equation.

    Be careful: solutions to the new equation may or may not be solutions to the _____ equation.

    Every _____ to the new equation **must** be _____ in the original equation.

power
original
original
solution
checked

2. To solve $\sqrt{x - 4} = 3$, _____ both sides of the equation, getting

    $$\underline{\qquad} = \underline{\quad},$$
    or
    $$x = \underline{\quad}.$$

    Now verify that $x = \underline{\quad}$ is really the solution of the original equation. We have

    $$\sqrt{\underline{\quad} - 4} = 3$$
    $$\underline{\quad} = 3.$$

    Thus $x = 13$ (*is/is not*) correct. The solution set is _____.

square
x − 4; 9
13
13
13
3
is
{13}

Solve each of the equations in Frames 3—12.

3. $\sqrt{2x - 9} = \sqrt{x - 3}$

    _____ = _____ | 2x − 9; x − 3

            x = ____ | 6

    This solution (*does/does not*) check in the orig- | does
    inal equation. The solution set is _____. | {6}

4. $\sqrt{3p - 7} = \sqrt{p - 1}$

    _____ = _____ | 3p − 7; p − 1

            p = ____ | 3

    Now check:

        $\sqrt{3(\underline{\quad}) - 7} = \sqrt{(\underline{\quad}) - 1}$ | 3; 3

        _____ = _____. | $\sqrt{2}$; $\sqrt{2}$

    The solution (*does/does not*) check. The solu- | does
    tion set is _____. | {3}

5. $\sqrt{x + 4} - 3 = 0$. We first write

        _____ = ____, | $\sqrt{x + 4}$; 3

    which, when squared, gives

        _____ = ____ | x + 4; 9

            x = ____. | 5

    This solution (*does/does not*) check. The solu- | does
    tion set is _____. | {5}

6. $\sqrt{11x + 3} = -3\sqrt{x + 1}$

    _____ = _____ | 11x + 3; 9(x + 1)

            x = ____ | 3

This solution (*does/does not*) check. The solution set is ____ .

does not

∅

7. $\sqrt{x^2 - 4x + 5} = x + 1$

_____ = _____

_____ = ___

x = _____

$x^2 - 4x + 5$;
$x^2 + 2x + 1$
$-6x$; -4

$\dfrac{2}{3}$

This solution (*does/does not*) check. The solution set is _____ .

does

$\{2/3\}$

8. Solve $\sqrt{x^2 + 3x + 6} = 3x - 2$.

Solution set: _____

$\{2\}$

9. To solve the equation

$\sqrt{r + 6} - \sqrt{r + 1} = 1$,

it is necessary to _____ both sides _____ .

square; twice

To begin, add _____ on each side, getting

$\sqrt{r + 1}$

$\sqrt{r + 6} =$ _____ .

$1 + \sqrt{r + 1}$

Now square each side.

$r + 6 = 1 +$ _____ $+ r + 1$

$2\sqrt{r + 1}$

Simplify.

_____

$4 = 2\sqrt{r + 1}$

Divide both sides by 2.

_____

$2 = \sqrt{r + 1}$

Square both sides again.

$$4 = \underline{\hspace{2cm}}$$
$$\underline{\hspace{1cm}} - r$$

$r + 1$

3

Check this potential solution in the original equation. The solution set is _____.

$\{3\}$

10. Solve $\sqrt{2r + 5} + \sqrt{2 - r} = 3$.

    Solution set: _____

$\{2, -2\}$

11. $\sqrt[3]{x^2 + 7x + 2} = \sqrt[3]{x^2 + 6x + 1}$

    $x^2 + 7x + 2 = \underline{\hspace{2cm}}$

    $x = \underline{\hspace{1cm}}$

$x^2 + 6x + 1$

-1

This solution (*does/does not*) check. The solution set is _____.

does

$\{-1\}$

12. $\sqrt[3]{x^2 + 8x + 16} = \sqrt[3]{x^2}$

    $x^2 + 8x + 16 = \underline{\hspace{1cm}}$

    $x = \underline{\hspace{1cm}}$

x^2

-2

This solution (*does/does not*) check. The solution set is _____.

does

$\{-2\}$

13. $(3x - 2)^{1/4} = 2$

    $\underline{\hspace{3cm}} = \underline{\hspace{1cm}}$

    $x = \underline{\hspace{1cm}}$

$3x - 2;\ 2^4$ or 16

6

This solution (*does/does not*) check. The solution set is _____.

does

$\{6\}$

14. $(x - 5)^{1/3} + 1 = 0$

    $(x - 5)^{1/3} = \underline{\hspace{1cm}}$

    $x - 5 = \underline{\hspace{1cm}}$

    $x = \underline{\hspace{1cm}}$

-1

-1

4

This solution (*does/does not*) check. The solution set is _____.	does {4}

15. $(m - 2)^{1/4} + 6 = 0$
     The solution set is ____.

Ø

5.7   Complex Numbers

[1] Simplify numbers of the form $\sqrt{-b}$, b > 0.   (Frames 1–14)

[2] Recognize a complex number.   (Frames 15–19)

[3] Add and subtract complex numbers.   (Frames 20–29)

[4] Multiply complex numbers.   (Frames 30–38)

[5] Divide complex numbers.   (Frames 39–43)

[6] Find powers of i.   (Frames 44–52)

---

1. There (*is/is not*) a real number solution to the equation $x^2 + 1 = 0$. Any solution must be a number whose square is _____. In order to be able to find a solution for this equation, we need to discuss a new set of numbers.	is not −1
2. This new set of numbers is based on the number ____. By definition $$i^2 = \underline{\hspace{1cm}}$$	i −1
3. For any positive real number b, $\sqrt{-b} = \underline{\hspace{2cm}}$. Thus, $\sqrt{-16} = i \,\underline{\hspace{1cm}} = \underline{\hspace{1cm}}$.	$1\sqrt{b}$ $\sqrt{16}$; 4i
4. In the same way, $$\sqrt{-75} = i \,\underline{\hspace{1.5cm}} = i \,\underline{\hspace{1cm}}\,\underline{\hspace{1cm}}$$ $$= \underline{\hspace{1.5cm}}.$$	$\sqrt{75}$; $\sqrt{25}$; $\sqrt{3}$ $5i\sqrt{3}$

Simplify the expression in Frames 5–7.

5. $\sqrt{-121} =$ _____

11i

6. $\sqrt{-81} =$ _____

9i

7. $\sqrt{-27} =$ ___ $\sqrt{27} =$ ___ ___ $\sqrt{3} =$ _____

i; 3; i; $3i\sqrt{3}$

8. For $a \geq 0$, always change $\sqrt{-a}$ to the form _____, before performing any multiplication or division.

$i\sqrt{a}$

9. Using this, $\sqrt{-6} \cdot \sqrt{-6} =$ ___.

-6

10. Also, $\sqrt{-30} \cdot \sqrt{-30} =$ _____.

-30

11. Find the product $\sqrt{-8} \cdot \sqrt{-2}$.

$$\sqrt{-8} \cdot \sqrt{-2} = \underline{\quad} \sqrt{8} \cdot \underline{\quad} \sqrt{2}$$
$$= \underline{\quad} \sqrt{8 \cdot 2}$$
$$= \underline{\quad} \cdot \underline{\quad}$$
$$= \underline{\quad}$$

i; i

i^2

-1; 4

-4

Find the products in Frames 12–14.

12. $\sqrt{-8} \cdot \sqrt{-32} = \underline{\quad} \sqrt{8} \cdot \underline{\quad} \sqrt{32}$
$$= \underline{\quad} \cdot \sqrt{\underline{\quad\quad}}$$
$$= \underline{\quad} \cdot 16$$
$$= \underline{\quad}$$

i; i

i^2; $8 \cdot 32$

-1

-16

13. $\sqrt{-12} \cdot \sqrt{-6} = \underline{\quad\quad} \cdot \underline{\quad\quad}$
$$= \underline{\quad} \sqrt{\underline{\quad\quad}}$$
$$= \underline{\quad\quad}$$

$i\sqrt{12}$; $i\sqrt{6}$

i^2; $12 \cdot 6$

$-6\sqrt{2}$

14. $\sqrt{-32} \cdot \sqrt{-3}$ = _____ $-4\sqrt{6}$

15. A number of the form a + bi, where a and b are
     real numbers, is called a _____ number.   complex

16. Every real number is also a _____ number.   complex
     To see this, let ___ = 0 in the complex number   b
     a + bi.

17. Complex numbers with b ≠ 0 are called _____   imaginary
     numbers.

**Write real or imaginary for each of the numbers in
Frames 18–19.**

18. 2 – 5i   _____   imaginary

19. 25   _____   real

20. To add two _____ numbers, a + bi and   complex
     c + di, use the definition (a + bi) + (c + di)
     = (_____) + (_____)i.   a + c; b + d

Find each of the sums in Frames 21–23.

21. (7 + 4i) + (8 + 14i) = (_____) + (_____)i   7 + 8; 4 + 14
                         = ___ + ___i   15; 18

22. (6 – 3i) + (–2 + 5i) = ___ + ___   4; 2i

23. (–4 + 2i) + (3 – 2i) = ___   –1

24. Subtraction of two complex numbers is defined
     by writing

         (a + bi) – (c + di) = (a + bi) + (_____)   –c – di
                           = (_____) + (_____)i.   a – c; b – d

Find each of the following.

25. $(7 + 2i) - (4 + 4i) = $ _____

3 - 2i

26. $(-8 - 4i) - (2 - 3i) = $ _____

-10 - i

27. $(3 - 7i) - (4 + 2i) = $ _____

-1 - 9i

28. $(-5 + 3i) - (-2 - 7i) = $ _____

-3 + 10i

29. $(-8 + 3i) + (2 - 4i) - (6 - 2i) = $ _____

-12 + i

30. Multiply two complex numbers like 2 + 3i and
     4 - 5i just as if they were binomials. For
     example,

$$(2 + 3i)(4 - 5i)$$
$$= 2(\underline{\quad}) + 2(\underline{\quad}) + 3i(\underline{\quad}) + 3i(\underline{\quad})$$
$$= \underline{\quad} + \underline{\quad} + \underline{\quad} + \underline{\quad}$$
$$= 8 - 10i + 12i - 15(\underline{\quad})$$
$$= \underline{\qquad}.$$

4; -5i; 4; -5i

8; -10i; 12i;
-15i²

-1

23 + 2i

Perform the operations indicated in Frames 31-33.

31. $(3 - i)(-2 + 5i) = \underline{\quad} + \underline{\quad} + \underline{\quad} + \underline{\quad}$
$$= \underline{\qquad}$$

-6; 15i; 2i;
-5i²

-1 + 17i

32. $(-6 + 2i)(1 + i) = \underline{\quad} + \underline{\quad} + \underline{\quad} + \underline{\quad}$
$$= \underline{\qquad}$$

-6; -6i; 2i; 2i²

-8 - 4i

33. $(8 - i)(1 + 3i) = \underline{\qquad}$

11 + 23i

34. To square a complex number use the fact that

$$(x + y)^2 = \underline{\hspace{4cm}}.$$

$x^2 + 2xy + y^2$

Thus,

$$(2 + 3i)^2 = (\underline{})^2 + 2(\underline{})(\underline{}) + (\underline{})^2$$
$$= \underline{} + \underline{} + \underline{}$$
$$= \underline{\hspace{3cm}}.$$

2; 2; 3i; 3i

4; 12i; 9i²

−5 + 12i

35. In the same way,

$$(6 + i)^2 = (\underline{})^2 + 2(\underline{})(\underline{}) + (\underline{})^2$$
$$= \underline{} + \underline{} + \underline{}$$
$$= \underline{\hspace{3cm}}.$$

6; 6; i; i

36; 12i; i²

35 + 12i

36. Since $(x + y)(x - y) = \underline{\hspace{2cm}}$, we can find $(2 + 4i)(2 - 4i)$ by writing

$$(2 + 4i)(2 - 4i) = (\underline{})^2 - (\underline{})^2$$
$$= \underline{} - \underline{}$$
$$= \underline{} + \underline{}$$
$$= \underline{}.$$

$x^2 - y^2$

2; 4i

4; 16i²

4; 16

20

Find each of the products indicated in Frames 37–38.

37. $(2 - 3i)(2 + 3i) = (\underline{})^2 - (\underline{})^2$
$$= \underline{}$$

2; 3i

13

38. $(6 - 4i)(6 + 4i) = (\underline{})^2 - (\underline{})^2$
$$= \underline{}$$

6; 4i

52

39. The complex number 3 − 2i is called the conjugate of the complex number _____.

3 + 2i

40. To divide two complex numbers, multiply numerator and denominator by the _____ of the _____. For example, to find

$$\frac{1 - i}{2 + i},$$

conjugate

denominator

multiply numerator and denominator by _____ . $2 - i$

$$\frac{(1 - i)(\quad\quad)}{(2 + i)(\quad\quad)} = \frac{\rule{3cm}{0.4pt}}{\rule{1cm}{0.4pt}}$$

$2 - i; \; 1 - 3i$
$2 - i; \; 5$

Find each of the quotients indicated in Frames 41–43.

41. $\dfrac{3 + i}{2 + i} = \dfrac{(3 + i)(\quad\quad)}{(2 + i)(\quad\quad)} = \dfrac{\rule{3cm}{0.4pt}}{\rule{1cm}{0.4pt}}$ $2 - i; \; 7 - i$
$2 - i; \; 5$

42. $\dfrac{6 + 4i}{5 - 3i} = \rule{3cm}{0.4pt} = \rule{1cm}{0.4pt} + \rule{1cm}{0.4pt}$ $\dfrac{18 + 38i}{34}; \; \dfrac{9}{17}; \; \dfrac{19}{17}i$

43. $\dfrac{2 + 5i}{i} = \dfrac{(2 + 5i)(\quad\quad)}{i(\quad\quad)}$ $-i$
$-i$

$= \rule{3cm}{0.4pt}$ $5 - 2i$

44. We can use the definition of i to evaluate
powers of i. We know

$$i^2 = \rule{1cm}{0.4pt}.$$ -1

Thus

$$i^3 = \rule{1cm}{0.4pt} \cdot i = \rule{1cm}{0.4pt} \cdot i = \rule{1cm}{0.4pt}.$$ $i^2; \; -1; \; -i$

Also,

$$i^4 = \rule{1cm}{0.4pt} \cdot i^2 = \rule{1cm}{0.4pt} \cdot \rule{1cm}{0.4pt} = \rule{1cm}{0.4pt}.$$ $i^2; \; -1; \; -1; \; 1$

45. In summary,

$$i^1 = \rule{1cm}{0.4pt}, \; i^2 = \rule{1cm}{0.4pt}, \; i^3 = \rule{1cm}{0.4pt}, \; i^4 = \rule{1cm}{0.4pt}.$$ $i; \; -1; \; -i; \; 1$

46. Using this result,

$$i^9 = (i^4)^2 \cdot \rule{1cm}{0.4pt} = (\rule{1cm}{0.4pt})^2 \cdot \rule{1cm}{0.4pt} = \rule{1cm}{0.4pt}.$$ $i; \; 1; \; i; \; i$

Evaluate each of the powers of i in Frames 47–49.

47. $i^{12} = (i^4)^{\overline{\rule{1cm}{0.4pt}}} = \rule{1cm}{0.4pt}$ $3; \; 1$

48. $i^{11} = (i^4)^{\overline{\rule{1cm}{0.4pt}}} \cdot i^{\overline{\rule{1cm}{0.4pt}}} = \rule{1cm}{0.4pt}$ $2; \; 3; \; -i$

49. $i^{25} = $ ____

	i

50. To simplify i^{-1} use the definition of a negative exponent.

$$i^{-1} = \frac{1}{\underline{\quad}}$$

Then multiply numerator and denominator by ____.

$$i^{-1} = \frac{1 \cdot (\underline{\quad})}{i \cdot (\underline{\quad})}$$

$$= \underline{\quad\quad}$$

i

−i

−i
−i

−i

Simplify each of the expressions in Frames 51–52.

51. $i^{-5} = \dfrac{1}{i^5} = $ ____

−i

52. $i^{11} = $ ____

−i

Chapter 5 Test

The answers for these questions are at the back of this Study Guide.

Simplify each of the following.

1. $16^{1/2}$ 1. _____

2. $25^{-1/2}$ 2. _____

3. $32^{4/5}$ 3. _____

4. $\dfrac{m^{3/8} \cdot m^{5/8}}{m^2}$ 4. _____

5. $\dfrac{p^{-5/8}}{p^{-1/2}}$ 5. _____

6. $(p^{3/4} \cdot q^{1/4})^{-2}$ 6. _____

7. $\left(\dfrac{a^6 b^9}{p^{12}}\right)^{1/3}$ 7. _____

8. $\left(\dfrac{m^{-3} \cdot m^4}{m^{-1} \cdot m^2}\right)^{-2}$ 8. _____

9. Use the Pythagorean formula to find the length of the missing side.

                                        9. _____

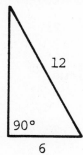

Simplify each of the following. Assume all variables represent positive real numbers.

10. $-\sqrt{75}$ 10. _____

11. $\sqrt{128}$ 11. _____

12. $\dfrac{m}{\sqrt{m}}$ 12. _____

13. $\sqrt[3]{27m^7n^9}$ 13. _____

14. $\sqrt{6} \cdot \sqrt{24}$ 14. _____

15. $2\sqrt{50} + 3\sqrt{72}$ 15. _____

16. $-3\sqrt{28} + 6\sqrt{63}$ 16. _____

17. $\dfrac{3}{2\sqrt{5}}$ 17. _____

18. $(\sqrt{2} - \sqrt{5})^2$ 18. _____

19. $(\sqrt{3} - \sqrt{2})(\sqrt{3} + \sqrt{2})$ 19. _____

20. $\dfrac{1}{\sqrt{2} - 3}$ 20. _____

21. $\sqrt{3} \cdot \sqrt[4]{5}$ 21. _____

Find the solution set of each of the following equations.

22. $\sqrt{x - 9} = \sqrt{2x - 20}$ 22. _____

23. $\sqrt{2x + 5} = x - 5$ 23. _____

24. $\sqrt[3]{x^2 + 5x + 10} = \sqrt[3]{x^2}$ 24. _____

25. $\sqrt{3r - 5} - \sqrt{r + 2} = 1$ 25. _____

Simplify each of the following. Write each answer in standard form.

26. $(-8 - 2i) - (-9 + 4i)$ 26. _____

27. $\sqrt{-8} \cdot \sqrt{-32}$ 27. _____

28. $(1 + 7i)(1 - 7i)$ 28. _____

29. $\dfrac{1}{i}$ 29. _____

30. $\dfrac{2 - i}{3 + i}$ 30. _____

CHAPTER 6 QUADRATIC EQUATIONS AND INEQUALITIES

6.1 Completing the Square

1⃞ Extend the zero-factor property to complex numbers.
(Frames 1-6)

2⃞ Learn the square root property. (Frames 7-8)

3⃞ Solve quadratic equations of the form $(ax - b)^2 = c$ using the square root property. (Frames 9-14)

4⃞ Solve quadratic equations by completing the square. (Frames 15-21)

---

1. An equation of the form $ax^2 + bx + c = 0$, where $a \neq 0$, is called _____. For example, $2y^2 - 7y + 4 = 0$ is a _____ equation.

> quadratic
> quadratic

2. In Chapter 3 to solve these equations, we used the _____ factor property: if a and b are real numbers, and if $ab = 0$, then _____ or _____.

> zero
> $a = 0$
> $b = 0$

3. To solve $2p^2 + 5p = 3$, first rewrite the equation so that one side is ___.

> 0

$$\text{\underline{\hspace{3cm}}} = 0$$

> $2p^2 + 5p - 3$

Now factor.

$$(\text{\underline{\hspace{2cm}}})(\text{\underline{\hspace{2cm}}}) = 0$$

> $2p - 1$; $p + 3$

Write two equations.

$$\text{\underline{\hspace{2cm}}} = 0 \ \text{ or } \ \text{\underline{\hspace{2cm}}} = 0$$

> $2p - 1$; $p + 3$

Solve each equation.

$$p = \text{\underline{\hspace{1.5cm}}} \ \text{ or } \ p = \text{\underline{\hspace{1.5cm}}}$$

> $\frac{1}{2}$; -3

By checking in the original equation, the solution set is _____

> $\{\frac{1}{2}, -3\}$

4. Solve $8r^2 + 14r = 15$.

    Solution set: _____

    $\{\frac{3}{4}, -\frac{5}{2}\}$

5. The _____ factor property can also be extended to
    the _____ numbers.

    zero

    complex

    If a and b are _____ numbers and if $ab = 0$

    complex

    then either _____ or _____ .

    $a = 0;\ b = 0$

6. To solve $x^2 + 9 = 0$, first rewrite the equation
    as a difference: _____ − (_____) = 0.

    $x^2;\ -9$

    Now factor. (x + ____)(x − ____) = 0

    $\sqrt{-9};\ \sqrt{-9}$

    Write two equations. _____ = 0 or _____ = 0

    $x + \sqrt{-9};\ x - \sqrt{-9}$

    Solve each equation. x = ____ or x = ____

    $-\sqrt{-9};\ \sqrt{-9}$

    Write the answer in a + bi form.

    x = ____ or x = ____

    $-3i;\ 3i$

    The solution set is _____ .

    $\{-3i,\ 3i\}$

7. Not all quadratic equations can be solved by
    factoring. To develop another method, we use
    a theorem which says that if a and b are com-
    plex numbers, and if $a^2 = b^2$, then a = ____ or

    b

    a = ____. Thus, if $b^2 = 25$, then b = ____ or

    $-b;\ 5$

    b = ____ .

    -5

8. Solve $y^2 = 7$.

    Solution set: _____

    $\{\sqrt{7},\ -\sqrt{7}\}$

9. If $(m - 1)^2 = 16$, then

    _____ = ____ or _____ = ____ ,

    $m - 1;\ 4;\ m - 1;\ -4$

    from which m = ____ or m = ____ .

    $5;\ -3$

Solve each of the quadratic equations in Frames 10–14.

10. $(a + 3)^2 = 25$

_____ = ____   or   _____ = ____   a + 3; 5; a + 3; −5

        a = ____   or   a = ____   2; −8

Solution set: _____ $\{2, -8\}$

11. $(p - 1)^2 = 7$

_____ = ____   or   _____ = ____ $p - 1$; $\sqrt{7}$; $p - 1$; $-\sqrt{7}$

p = _____   or   p = _____ $1 + \sqrt{7}$; $1 - \sqrt{7}$

Solution set: _____ $\{1 + \sqrt{7},\ 1 - \sqrt{7}\}$

12. $(3r + 2)^2 = 12$

3r + 2 = _____   or   3r + 2 = _____ $\sqrt{12}$; $-\sqrt{12}$

    3r = _____   or   3r = _____ $-2 + 2\sqrt{3}$; $-2 - 2\sqrt{3}$

     r = _____   or   r = _____ $\dfrac{-2 + 2\sqrt{3}}{3}$; $\dfrac{-2 - 2\sqrt{3}}{3}$

Solution set: _____ $\left\{\dfrac{-2 + 2\sqrt{3}}{3},\ \dfrac{-2 - 2\sqrt{3}}{3}\right\}$

13. $(k + 3)^2 = -4$

k + 3 = ____   or   k + 3 = ____   2i; −2i

    k = _____   or   k = _____   −3 + 2i; −3 − 2i

Solution set: _____ $\{-3 + 2i,\ -3 - 2i\}$

14. $(2a - 5)^2 = -16$

_____ = ____   or   _____ = ____   2a − 5; 4i; 2a − 5; −4i

   2a = _____   or   2a = _____   5 + 4i; 5 − 4i

    a = _____   or   a = _____ $\dfrac{5}{2} + 2i$; $\dfrac{5}{2} - 2i$

Solution set: _____ $\left\{\dfrac{5}{2} + 2i,\ \dfrac{5}{2} - 2i\right\}$

15. A quadratic equation such as $x^2 - 6x + 5 = 0$ can be converted to the form $(ax + b)^2 = c$ like those above by using the process of _____ the square. To do this in the given equation, first write $x^2 - 6x + 5 = 0$ as

completing

$$\underline{\hspace{3cm}} = \underline{\hspace{1cm}}.$$

$x^2 - 6x;\ -5$

Now take half of ____. This gives

−6

$$\frac{1}{2}(\underline{\hspace{1cm}}) = \underline{\hspace{1cm}}.$$

−6; −3

Then _____ this number: $(\underline{\hspace{1cm}})^2 = \underline{\hspace{1cm}}.$

square; −3; 9

Add ____ to both sides, giving

9

$$x^2 - 6x + \underline{\hspace{1cm}} = -5 + \underline{\hspace{1cm}}.$$

9; 9

The left—hand side should now be a _____ square. Factor on the left to get

perfect

$$(\underline{\hspace{2cm}})^2 = \underline{\hspace{1cm}},$$

$x - 3;\ 4$

which leads to

$$\underline{\hspace{2cm}} = \underline{\hspace{1cm}} \quad \text{or} \quad \underline{\hspace{2cm}} = \underline{\hspace{1cm}}.$$

$x - 3;\ 2;\ x - 3;\ -2$

Hence $\quad x = \underline{\hspace{1cm}} \quad$ or $\quad x = \underline{\hspace{1cm}}.$

5; 1

Solution set: $\underline{\hspace{3cm}}$

$\{5,\ 1\}$

16. Solve $k^2 - 5k + 4 = 0$.

$$\underline{\hspace{3cm}} = \underline{\hspace{1cm}}$$

$k^2 - 5k;\ -4$

Half of ____ is _____; squared, we have _____. Next,

$-5;\ -\dfrac{5}{2},\ \dfrac{25}{4}$

$$\underline{\hspace{2cm}} + \underline{\hspace{2cm}} = \underline{\hspace{1cm}} + \underline{\hspace{2cm}}.$$

$k^2 - 5k;\ \dfrac{25}{4};\ -4;$

$\dfrac{25}{4}$

The left—hand side is a perfect square.

$$(\underline{\hspace{2cm}})^2 = \underline{\hspace{2cm}}$$

$k - \dfrac{5}{2};\ \dfrac{9}{4}$

This leads to

$$\underline{\hspace{2cm}} = \underline{\hspace{1cm}} \quad \text{or} \quad k - \frac{5}{2} = \underline{\hspace{1cm}}.$$

$k - \dfrac{5}{2};\ \dfrac{3}{2};\ -\dfrac{3}{2}$

$$k = \underline{\hspace{1cm}} \quad \text{or} \quad k = \underline{\hspace{1cm}}$$

4; 1

Solution set: $\underline{\hspace{3cm}}$

$\{4;\ 1\}$

Solve the equations in Frames 17–21.

17. $x^2 - 2x + 5 = 0$

$$\underline{\hspace{3cm}} = \underline{\hspace{1.5cm}}$$

$$\underline{\hspace{3cm}} + \underline{\hspace{1cm}} = \underline{\hspace{1cm}} + \underline{\hspace{1.5cm}}$$

$$(\underline{\hspace{3cm}})^2 = \underline{\hspace{1.5cm}}$$

$$\underline{\hspace{2.5cm}} = \underline{\hspace{1cm}} \quad \text{or} \quad \underline{\hspace{2.5cm}} = \underline{\hspace{1.5cm}}$$

$$x = \underline{\hspace{2.5cm}} \quad \text{or} \quad x = \underline{\hspace{2.5cm}}$$

Solution set: $\underline{\hspace{3cm}}$

$x^2 - 2x$; -5

$x^2 - 2x$; 1; -5; 1

$x - 1$; -4

$x - 1$; $2i$; $x - 1$; $-2i$

$1 + 2i$; $1 - 2i$

$\{1 + 2i,\ 1 - 2i\}$

18. $3m^2 - 9m + 4 = 0$

First divide through by $\underline{\hspace{1.5cm}}$, obtaining

$$\underline{\hspace{4cm}} = 0.$$

Next, perform the steps mentioned above.

$$\underline{\hspace{2.5cm}} = \underline{\hspace{1.5cm}}$$

$$\underline{\hspace{3cm}} + \underline{\hspace{1cm}} = \underline{\hspace{1.5cm}} + \underline{\hspace{1cm}}$$

$$(\underline{\hspace{3cm}})^2 = \underline{\hspace{2cm}}$$

$$\underline{\hspace{2.5cm}} = \underline{\hspace{1.5cm}} \quad \text{or} \quad m - \frac{3}{2} = -\sqrt{\frac{11}{12}}$$

$$m = \underline{\hspace{2.5cm}} \quad \text{or} \quad m = \underline{\hspace{3cm}}$$

Since $\sqrt{\dfrac{11}{12}} = \underline{\hspace{2.5cm}}$, use a common denominator of 6 to write the solutions as

$$m = \underline{\hspace{2.5cm}} \quad \text{or} \quad m = \underline{\hspace{3cm}}.$$

Solution set: $\underline{\hspace{3cm}}$

3

$m^2 - 3m + \dfrac{4}{3}$

$m^2 - 3m$; $-\dfrac{4}{3}$

$m^2 - 3m$; $\dfrac{9}{4}$;
$-\dfrac{4}{3}$; $\dfrac{9}{4}$

$m - \dfrac{3}{2}$; $\dfrac{11}{12}$

$m - \dfrac{3}{2}$; $\sqrt{\dfrac{11}{12}}$

$\dfrac{3}{2} + \sqrt{\dfrac{11}{12}}$; $\dfrac{3}{2} - \sqrt{\dfrac{11}{12}}$

$\dfrac{\sqrt{33}}{6}$

$\dfrac{9 + \sqrt{33}}{6}$; $\dfrac{9 - \sqrt{33}}{6}$

$\left\{\dfrac{9 + \sqrt{33}}{6},\ \dfrac{9 - \sqrt{33}}{6}\right\}$

19. $2m^2 - 5m - 4 = 0$

Solution set: $\underline{\hspace{3cm}}$

$\left\{\dfrac{5 + \sqrt{57}}{4},\ \dfrac{5 - \sqrt{57}}{4}\right\}$

20. $2a^2 + 3a + 2 = 0$

Solution set: _____

$$\left\{\frac{-3 + i\sqrt{7}}{4}, \frac{-3 - i\sqrt{7}}{4}\right\}$$

21. $r^2 - 3r + 1 = 0$

Solution set: _____

$$\left\{\frac{3 + \sqrt{5}}{2}, \frac{3 - \sqrt{5}}{2}\right\}$$

6.2 The Quadratic Formula

1️⃣ Solve quadratic equations by using the quadratic formula. (Frames 1–7)

2️⃣ Solve quadratic equations with imaginary number solutions using the quadratic formula. (Frames 8–10)

3️⃣ Solve applications by using the quadratic formula. (Frames 11–12)

---

1. By the quadratic formula, the solutions of the equation $ax^2 + bx + c = 0$ $(a \neq 0)$ are given by

$$x = \underline{\hspace{3cm}}.$$

$$\frac{-b \pm \sqrt{b^2 - 4ac}}{2a}$$

2. To solve the equation $x^2 + 2x - 15 = 0$, first identify the letters a, b, and c. Here a = ____, b = ____, and c = _____. Thus

1; 2; −15

$$x = \frac{-(\quad) \pm \sqrt{(\quad)^2 - 4(\quad)(\quad)}}{2(\_\_)}$$

2; 2; 1; −15
1

$$= \frac{\pm \sqrt{\quad + \quad}}{\_\_\_}$$

−2; 4; 60
2

$$= \underline{\hspace{2.5cm}}.$$

$$\frac{-2 \pm 8}{2}$$

This leads to the two solutions

$$x = \underline{\hspace{2cm}} \quad \text{or} \quad x = \underline{\hspace{2cm}}$$

$$\frac{-2 + 8}{2}; \frac{-2 - 8}{2}$$

$$x = \underline{\hspace{1cm}} \quad \text{or} \quad x = \underline{\hspace{1cm}}.$$

3; −5

Solution set: _____

$$\{3, -5\}$$

3. To solve $6x^2 + 11x - 10 = 0$, first identify a,

    b, and c: a = ____ , b = ____ , and c = ____ .   6; 11; −10

    Thus

    $$x = \frac{-(\quad) \pm \sqrt{(\quad)^2 - 4(\quad)(\quad)}}{2(\quad)}$$ 11; 11; 6; −10
    6

    $$= \frac{\quad \pm \sqrt{\quad}}{\quad}.$$ −11; 361
    12

    Since $\sqrt{361}$ = ____ ,   19

    $$x = \underline{\quad} \quad \text{or} \quad x = \underline{\quad}.$$ $\frac{2}{3}$; $-\frac{5}{2}$

    Solution set: _____ $\{\frac{2}{3}, -\frac{5}{2}\}$

4. Solve $k^2 - 2k - 1 = 0$.

    We have a = ____ , b = ____ , and c = ____ .   Thus   1; −2; −1

    $$k = \frac{-(\quad) \pm \sqrt{(\quad)^2 - 4(\quad)(\quad)}}{2(\quad)}$$ −2; −2; 1; −1
    1

    $$= \frac{\quad \pm \sqrt{\quad}}{\quad}$$ 2; 8
    2

    $$= \underline{\qquad\qquad}$$ $\frac{2 \pm 2\sqrt{2}}{2}$

    $$k = \underline{\qquad\qquad} \quad \text{or} \quad k = \underline{\qquad\qquad}.$$ $1 + \sqrt{2}$; $1 - \sqrt{2}$

    Solution set: _____ $\{1 + \sqrt{2}, 1 - \sqrt{2}\}$

**For each of the quadratic equations in Frames 5–10,
first identify a, b, and c, and then use the quadratic formula to find the solutions of the equations.**

5. $9x^2 + 1 = 12x$

    a = ____ , b = ____ , c = ____   9; −12; 1

    $$x = \frac{-(\quad) \pm \sqrt{(\quad)^2 - 4(\quad)(\quad)}}{2(\quad)}$$ −12; −12; 9; 1
    9

    $$x = \frac{\quad \pm \sqrt{\quad}}{\quad}$$ 12; 108
    18

    Solution set: _____ $\{\frac{2 + \sqrt{3}}{3}, \frac{2 - \sqrt{3}}{3}\}$

6. $p^2 - p + 1$

$a =$ ___ , $b =$ ___ , $c =$ ___

$$p = \frac{-(\quad) \pm \sqrt{(\quad)^2 - 4(\quad)(\quad)}}{2(\quad)}$$

$$p = \frac{\pm \sqrt{\quad}}{\quad}$$

Solution set: _____ .

1; -1; -1
-1; -1; 1; -1
1
1; 5
2
$\{\frac{1+\sqrt{5}}{2}, \frac{1-\sqrt{5}}{2}\}$

7. $3k^2 + 2k = 2$

$a =$ ___ , $b =$ ___ , $c =$ ___

$$k = \frac{-(\quad) \pm \sqrt{(\quad)^2 - 4(\quad)(\quad)}}{2(\quad)}$$

$$= \frac{\pm \sqrt{\quad}}{\quad}$$

Since $\sqrt{28} =$ _____ , the solutions can be written as

$k =$ _____ or $k =$ _____ .

Solutions set: _____

3; 2; -2
2; 2; 3; -2
3
-2; 28
6
$2\sqrt{7}$
$\frac{-1+\sqrt{7}}{3}, \frac{-1-\sqrt{7}}{3}$
$\{\frac{-1+\sqrt{7}}{3}, \frac{-1-\sqrt{7}}{3}\}$

8. $4h^2 + 5 = 4h$

$a =$ ___ , $b =$ ___ , $c =$ ___

$$h = \frac{-(\quad) \pm \sqrt{(\quad)^2 - 4(\quad)(\quad)}}{2(\quad)}$$

$$h = \frac{\pm \sqrt{\quad}}{\quad}$$

Since $\sqrt{-64} =$ ___ , the solution set is

_____ .

4; -4; 5
-4; -4; 4; 5
4
4; -64
8
8i
$\{\frac{1}{2} + i, \frac{1}{2} - i\}$

9. $9m^2 - 12m + 13 = 0$

$a =$ ___ , $b =$ ___ , $c =$ ___

Solution set: _____

9; -12; 13
$\{\frac{2}{3} + i, \frac{2}{3} - i\}$

10. $2r^2 + 2ir = 5$

$$a = \underline{\quad}, \; b = \underline{\quad}, \; c = \underline{\quad}$$

$$r = \frac{-(\quad) \pm \sqrt{(\quad)^2 - 4(\quad)(\quad)}}{2(\underline{\quad})}$$

Since $(2i)^2 = \underline{\quad\quad}$,

$$r = \frac{\pm \sqrt{\overline{\quad\quad}}}{\underline{\quad}}$$

Solution set: $\underline{\quad\quad\quad\quad}$

2; 2i; −5

2i; 2i; 2; −5
2

−4

−2i; 36
4

$\left\{\frac{3}{2} - \frac{1}{2}i, \; -\frac{3}{2} - \frac{1}{2}i\right\}$

11. Jose and Miguel can paint a house together in 12/5 hours. Because Jose has a broken arm, it would take him 2 hours longer than Miguel if they each worked alone. How long would it take Miguel to paint the house working alone?

Let x = the time needed for Miguel to paint the house alone. Then ____ = the time needed for _____ to paint the house alone.

x + 2
Jose'

Miguel's rate is _____, while Jose's rate is _____.

1/x
1/(x + 2)

Recall that a worker's rate _____ by the time it takes to do the job together gives the _____ part of the job done by that worker.

multiplied

fractional

Use the chart below to organize the information.

Worker	Rate	Time working together	Fractional part of the job done
Miguel			
Jose			

$\frac{1}{x}$; $\frac{12}{5}$; $\frac{12}{5x}$

$\frac{1}{x+2}$; $\frac{12}{5}$
$\frac{12}{5(x+2)}$

Write the equation:

$$\frac{\rule{2cm}{0.4pt}}{\substack{\text{part done by}\\ \text{Miguel}}} + \frac{\rule{2cm}{0.4pt}}{\substack{\text{part done by}\\ \text{Jose}}} = \frac{\rule{2cm}{0.4pt}}{\substack{\text{one whole}\\ \text{job done}}}$$

$\dfrac{12}{5x}$; $\dfrac{12}{5(x + 2)}$; 1

Multiply both sides by the _____ _____

_____ , _____ .

least common denominator; $5x(x + 2)$

Simplify to get _____ = 0.

$5x^2 - 14x - 24$

Use the quadratic formula:

$$x = \frac{-(\quad) \pm \sqrt{(\quad)^2 - (\quad)(\quad)(\quad)}}{(\rule{0.8cm}{0.4pt})(\rule{0.8cm}{0.4pt})}$$

$$= \frac{\rule{1cm}{0.4pt} + \rule{1cm}{0.4pt}}{\rule{1.5cm}{0.4pt}}$$

-14; -14; 4; 5; -24
2; 5

14; $\sqrt{676}$
10

Simplify to get x = ____ or x = ____ .

4; -1.2

Do both answers make sense? ____

No

Reject the answer x = ____ because time cannot

-1.2

be _____ .

negative

Thus, it takes Miguel ____ hours to paint the house alone.

4

12. Russ and Bob can do a job in 3 3/5 hours when they work together. Bob alone would take 3 hours longer than Russ. How long would it take Russ working alone?

____ hours

6

6.3   The Discriminant; Sum and Product of the Solutions

$\boxed{1}$ Find the discriminant of a quadratic equation.   (Frames 1–5)

$\boxed{2}$ Use the discriminant to determine the number and type of solutions. (Frames 6–16)

$\boxed{3}$ Use the discriminant to decide whether a quadratic trinomial can be factored.   (Frames 17–21)

$\boxed{4}$ Find the sum and product of the solutions of a quadratic equation. (Frames 22–25)

$\boxed{5}$ Write a quadratic equation given its solutions.   (Frames 26)

---

1. The quantity $b^2 - 4ac$ in the quadratic formula is called the _____.

    discriminant

2. To find the discriminant for the equation $3x^2 - 4x + 2 = 0$, first identify a, b, and c.

    a = ____, b = ____, c = ____

    3; −4; 2

    Now calculate _____.

    $b^2 - 4ac$

    $b^2 - 4ac = (\underline{\quad})^2 - 4(\underline{\quad})(\underline{\quad}) = \underline{\quad}$

    −4; 3; 2; −8

Find the discriminant for the equations in Frames 3–5.

3. $4k^2 - 20k + 25 = 0$

    a = ____, b = ____, c = ____

    4; −20; 25

    $(\underline{\quad})^2 - 4(\underline{\quad})(\underline{\quad}) = \underline{\quad}$

    −20; 4; 25; 0

4. $6z^2 - 19z + 10 = 0$

    $(\underline{\quad})^2 - 4(\underline{\quad})(\underline{\quad}) = \underline{\quad}$

    −19; 6; 10; 121

5. $3r^2 - 6r - 1 = 0$

    $(\underline{\quad})^2 - 4(\underline{\quad})(\underline{\quad}) = \underline{\quad}$

    −6; 3; −1; 48

6. If the discriminant is the square of a nonzero integer, then the equation has exactly _____ different _____ solutions. (We assume throughout this section that all quadratic equations have _____ coefficients.)

two

rational

integer

7. If the discriminant is a positive number that is not the square of an integer, then the equation has exactly _____ different _____ solutions.

two; irrational

8. If the discriminant is zero, the equation has exactly _____ rational solution.

one

9. If the discriminant is negative, then the equation has _____ different _____ solutions.

two; complex

Find the discriminant for each of the equations in Frames 10–14. Use the discriminant to determine whether the given equations have solutions which are:

   (a) two different rational numbers
   (b) two different irrational numbers
   (c) exactly one rational number
   (d) two different imaginary numbers.

10. $25k^2 - 40k + 16 = 0$

   (_____)² − 4(_____)(_____) = _____

   Type of solutions: _____

-40; 25; 16; 0

(c)

11. $2m^2 - 3m - 1 = 0$

   (_____)² − 4(____)(____) = _____

   Type of solutions: _____

-3; 2; -1; 17

(b)

12. $6r^2 + 5r = 4$

(____)² – 4(____)(____) = _____   5; 6; –4; 121

Type of solutions: _____   (a)

13. $3k^2 - 4k + 2 = 0$

Type of solutions: _____   (d)

14. $7m^2 + 2m - 3 = 0$

Type of solutions: _____   (b)

15.   We can find a value of b so that the equation $4x^2 + bx + 25 = 0$ will have exactly one rational solution by using the discriminant. Here

(____)² – 4(____)(____) = _____.   b; 4; 25; $b^2 - 400$

The equation will have exactly one rational solution only if the discriminant equals _____.   0
Here

_____ = _____ $b^2 - 400$; 0

or

b = _____   or b = _____.   20; –20

16.   To find a value of a so that $ax^2 - 30x + 25 = 0$ will have only one rational solution, use the discriminant.

(____)² – 4(____)(____) = _____   –30; a; 25; 900 – 100a

_____ = _____   900 – 100a; 0

a = ____   9

17.   We can decide whether or not a trinomial $ax^2 + bx + c$ can be factored if its corresponding quadratic equation

$$ax^2 + bx + c = 0$$

has _____ _____.   rational; solutions

18. The quadratic equation

$$ax^2 + bx + c = 0$$

    has rational solutions, if its _____ is
    a perfect _____.

    discriminant
    square

19. To decide whether $10x^2 - 9x - 7$ can be factored,
    first write the corresponding quadratic equation.

    _____

    $10x^2 - 9x - 7 = 0$

    Then find the value of its _____.

    discriminant

    (_____)^2 − 4(_____)(_____) = _____

    −9; 10; −7; 361

    Since this value (*is/is not*) a perfect square,
    $10x^2 - 9x - 7$ (*can/cannot*) be factored. Its
    factors are (_____)(_____).

    is
    can
    5x − 7; 2x + 1

**Decide whether or not each trinomial in Frames 20–21
can be factored.**

20. $18y^2 + 8y - 5$

    Corresponding quadratic equation: _____

    $18y^2 + 8y - 5 = 0$

    Discrimiant: _____

    424

    $18y^2 + 8y - 5$ (*can/cannot*) be factored.

    cannot

21. $9q^2 - 15q + 4$

    Corresponding quadratic equation: _____

    $9q^2 - 15q + 4 = 0$

    Discriminant: _____

    81

    $9q^2 - 15q + 4$ (*can/cannot*) be factored.

    can

22. If x_1 and x_2 are the solutions of the quadratic
    equation

    $$ax^2 + bx + c = 0 \quad (a \neq 0)$$

    Then $x_1 + x_2 =$ _____ and $x_1 x_2 =$ _____.

    $-\dfrac{b}{a}; \dfrac{c}{a}$

In Frames 23–25, use the sum and products property to determine whether x_1 and x_2 of each equation are solutions of the equation.

23. $15x^2 + 2x - 8$; $x_1 = \frac{2}{3}$, $x_2 = -\frac{4}{5}$

If the solutions are 2/3 and −4/5, their sum should equal _____ and their product should equal _____. Here a = _____, b = _____, and c = _____.

$$-\frac{b}{a} = \underline{\hspace{1cm}}$$

$$x_1 + x_2 = \underline{\hspace{1cm}} + (\underline{\hspace{1cm}}) = \underline{\hspace{1cm}}$$

$$\frac{c}{a} = \underline{\hspace{1cm}}$$

$$x_1 x_2 = \underline{\hspace{1cm}} (\underline{\hspace{1cm}}) = \underline{\hspace{1cm}}$$

x_1 and x_2 (*are/are not*) solutions.

	−b/a
	c/a; 15; 2
	−8
	$-\frac{2}{15}$
	$\frac{2}{3}$; $-\frac{4}{5}$; $-\frac{2}{15}$
	$-\frac{8}{15}$
	$\frac{2}{3}$; $-\frac{4}{5}$; $-\frac{8}{15}$
	are

24. $28x^2 - 15x + 2$; $x_1 = \frac{2}{7}$, $x_2 = \frac{3}{4}$

a = _____, b = _____, c = _____

$$-\frac{b}{a} = \underline{\hspace{1cm}}$$

$$x_1 + x_2 = \underline{\hspace{1cm}}$$

$$\frac{c}{a} = \underline{\hspace{1cm}}$$

$$x_1 x_2 = \underline{\hspace{1cm}}$$

x_1 and x_2 (*are/are not*) solutions.

	28; −15; 2
	$\frac{15}{28}$
	$\frac{29}{28}$
	$\frac{1}{14}$
	$\frac{3}{14}$
	are not

25. $30x^2 + 53x + 21 = 0$; $x_1 = -\frac{3}{5}$, $-\frac{7}{6}$

x_1 and x_2 (*are/are not*) solutions.   are

26. To write a quadratic equation with the solution set $\{-5/4, -1/2\}$ write two equations.

x = _____   or   x = _____

x + _____ = _____   or   x + _____ = _____

	$-\frac{5}{4}$; $-\frac{1}{2}$
	$\frac{5}{4}$; 0; $\frac{1}{2}$; 0

By the zero—factor property, we have

$$(\underline{\hspace{2cm}})(\underline{\hspace{1.5cm}}) = 0,$$

and, using FOIL, we have

$$\underline{\hspace{4cm}} = 0.$$

Multiply by the least common denominator.

$$\underline{\hspace{4cm}} = 0$$

Use the sum and product properties.

a = _____, b = _____, c = _____

$$-\frac{b}{a} = \underline{\hspace{1cm}} \qquad x_1 + x_2 = \underline{\hspace{1.5cm}}$$

$$\frac{c}{a} = \underline{\hspace{1cm}} \qquad x_1 x_2 = \underline{\hspace{1.5cm}}$$

An equation with the solution set $\{-5/4, -1/2\}$ is _____.

$x + \frac{5}{4};\ x + \frac{1}{2}$
$x^2 + \frac{7}{4}x + \frac{5}{8}$
$8x^2 + 14x + 5$
$8;\ 14;\ 5$
$-\frac{7}{4};\ -\frac{7}{4}$
$\frac{5}{8};\ \frac{5}{8}$
$8x^2 + 14x + 5 = 0$

6.4 Equations Quadratic in Form

[1] Solve an equation with fractions by writing it in quadratic form. (Frames 1–6)

[2] Use quadratic equations to solve applied problems. (Frame 7)

[3] Solve an equation with radicals by writing it in quadratic form. (Frames 8–13)

[4] Solve an equation that is quadratic in form by substitution. (Frames 14–23)

---

1. To solve $1 - \frac{11}{k} + \frac{30}{k^2} = 0$, first multiply through by ____, obtaining

$$\underline{\hspace{3cm}} = 0,$$

which factors as

$$(\underline{\hspace{2cm}})(\underline{\hspace{2cm}}) = 0.$$

The solution set of the original equation is _____.

k^2
$k^2 - 11k + 30$
$k - 5;\ k - 6$
$\{5,\ 6\}$

Solve each of the equations in Frames 2–4.

2. $1 - \dfrac{8}{p} + \dfrac{15}{p^2} = 0$

    _____ = 0 $p^2 - 8p + 15$

    (_____)(_____) = 0 $p - 5;\ p - 3$

    Solution set: _____ $\{5,\ 3\}$

3. $2 - \dfrac{3}{p} - \dfrac{35}{p^2} = 0$

    _____ = 0 $2p^2 - 3p + 35$

    (_____)(_____) = 0 $p - 5;\ 2p + 7$

    Solution set: _____ $\{5,\ -\dfrac{7}{2}\}$

4. $2 + \dfrac{5}{a} - \dfrac{12}{a^2} = 0$

    _____ = 0 $2a^2 + 5a - 12$

    (_____)(_____) = 0 $a + 4;\ 2a - 3$

    Solution set: _____ $\{-4,\ \dfrac{3}{2}\}$

5. To solve $\dfrac{1}{m} + \dfrac{2}{m - 3} = \dfrac{9}{4}$, multiply both sides by

    _____. $4m(m - 3)$

    $4m(m - 3)(\underline{\ \ }) + 4m(m - 3)(\underline{\hspace{2cm}})$ $\dfrac{1}{m};\ \dfrac{2}{m - 3}$

    $= 4m(m - 3)\left(\dfrac{9}{4}\right)$

    _____ + _____ = 9m(m - 3) $4(m - 3);\ 8m$

    Simplify.

    _____ = $9m^2 - 27m$ $4m - 12 + 8m$

    _____ = $9m^2 - 27m$ $12m - 12$

    $0 =$ _____ $9m^2 - 39m + 12$

    Divide both sides by 3.

    $0 =$ _____ $3m^2 - 13m + 4$

    Factor.

    $0 =$ _____ $(3m - 1)(m - 4)$

Set each factor equal to 0.

$3m - 1 = 0$ or   _____ $= 0$ $m - 4$

$m =$ _____   or $m =$ _____ $\frac{1}{3}$; 4

Check each answer in the original equation; the

solution set is _____ . $\{\frac{1}{3}, 4\}$

6. Solve $\frac{3}{k + 1} - \frac{1}{k - 1} = \frac{1}{4}$.

Solution set: _____ $\{3, 5\}$

7. A boat goes 15 miles per hour in still water.
It can go 24 miles upstream, and return, in
3 hours, 20 minutes. Find the speed of the
current.

Let x = speed of the current. The speed going

upstream is _____, while the speed down- $15 - x$

stream is _____ . $15 + x$

From the distance formula, d = rt, we have

t = _____ . $\frac{d}{r}$

The time upstream is $\frac{d}{r} =$ _____ . $\frac{24}{15 - x}$

The time downstream is _____ . $\frac{24}{15 + x}$

Use this information to construct a chart.

	Distance	Rate	Time
Upstream	24	$15 - x$	_____
Downstream	24	$15 + x$	_____

$\frac{24}{15 - x}$

$\frac{24}{15 + x}$

The total time is 3 hours, 20 minutes, or

$3\frac{20}{60} =$ _____ hours. $3\frac{1}{3}$

Write $3\frac{1}{3}$ as an improper fraction: _____.

$\dfrac{10}{3}$

The time upstream plus the time downstream is

_____ hours, so

$\dfrac{10}{3}$

$$\underline{\hspace{4cm}} = \dfrac{10}{3}.$$

$\dfrac{24}{15 - x} + \dfrac{24}{15 + x}$

Solve this equation to find that the current
has a speed of _____ miles per hour.

3

8. To solve an equation with a square root, such as

$$p = \sqrt{8p - 7},$$

_____ both sides, getting

square

$$p^2 = \underline{\hspace{3cm}}$$

$8p - 7$

or _____ = 0.

$p^2 - 8p + 7$

Factor.

_____ = 0

$(p - 7)(p - 1)$

Place each factor equal to 0 to get

$$p = \underline{\hspace{1cm}} \quad \text{or} \quad p = \underline{\hspace{1cm}}.$$

7; 1

Check each answer in the original equation.

If $p = 7$, $\sqrt{8p - 7} = \sqrt{8(\underline{\hspace{1cm}}) - 7} = \underline{\hspace{1cm}} = \underline{\hspace{1cm}}$.

7; $\sqrt{49}$; 7

If $p = 1$, $\sqrt{8p - 7} = \sqrt{8 \cdot 1 - 7} = \underline{\hspace{1cm}}$.

1

Both answers check, so that the solution set is

_____.

$\{7, 1\}$

9. To solve $x = \sqrt{5x + 6}$, first _____ both sides.
This gives

square

$$\underline{\hspace{2cm}} = \underline{\hspace{2cm}}$$

x^2; $5x + 6$

or _____ = 0,

$x^2 - 5x - 6$

which factors as

$$(\underline{\hspace{2cm}})(\underline{\hspace{2cm}}) = 0.$$

$x - 6$; $x + 1$

The solutions of this last equation are

$$x = \underline{\hspace{1cm}} \quad \text{or} \quad x = \underline{\hspace{1cm}}.$$

6; −1

Check both these potential solutions in the original equation. Verify that x = 6 (*is/is not*) a solution, and x = −1 (*is/is not*) a solution. The solution set of the original equation is thus

_____.

is

is not

{6}

10. Solve $m = \sqrt{7m - 10}$.

$$\underline{\hspace{1.5cm}} = \underline{\hspace{2cm}}$$

$$\underline{\hspace{3cm}} = 0$$

$$(\underline{\hspace{1.5cm}})(\underline{\hspace{1.5cm}}) = 0$$

$$m = \underline{\hspace{1cm}} \quad \text{or} \quad m = \underline{\hspace{1cm}}$$

m^2; $7m - 10$

$m^2 - 7m + 10$

$m - 5$; $m - 2$

5; 2

After checking both these answers in the original equation, we find that the solution set of the original equation is _____.

{5, 2}

11. Solve $k = \sqrt{\dfrac{11k + 20}{3}}$.

$$\underline{\hspace{1.5cm}} = \underline{\hspace{2cm}}$$

$$3k^2 = \underline{\hspace{2cm}}$$

$$\underline{\hspace{3cm}} = 0$$

$$(\underline{\hspace{1.5cm}})(\underline{\hspace{1.5cm}}) = 0$$

$$k = \underline{\hspace{1cm}} \quad \text{or} \quad k = \underline{\hspace{1cm}}$$

k^2; $\dfrac{11k + 20}{3}$

$11k + 20$

$3k^2 - 11k - 20$

$3k + 4$; $k - 5$

$-\dfrac{4}{3}$; 5

After checking, we find that the solution set of the original equation is _____.

{5}

12. Solve $r = \sqrt{\dfrac{6 - r}{2}}$.

Solution set: _____

$\left\{\dfrac{3}{2}\right\}$

13. Solve $m = \sqrt{\dfrac{7m + 20}{3}}$.

Solution set: _____

{4}

14. To solve $3m^4 - 14m^2 + 8 = 0$, let $y = $ _____, so that the equation becomes

m^2

_____ $= 0$,

or   (_____)(_____) $= 0$.

$3y^2 - 14y + 8$
$3y - 2;\ y - 4$

From this

$$y = \underline{\quad} \quad \text{or} \quad y = \underline{\quad}.$$

$\frac{2}{3};\ 4$

Since $y = m^2$,

$$m^2 = \underline{\quad} \quad \text{or} \quad m^2 = \underline{\quad}.$$

$\frac{2}{3};\ -4$

If $m^2 = \frac{2}{3}$, then $m = $ _____. The solutions of $m^2 = 4$ are ____ and _____.

$\pm\sqrt{\frac{2}{3}} = \pm\frac{\sqrt{6}}{3}$
$2;\ -2$

The solution set is _____.

$\{2, -2, \frac{\sqrt{6}}{3}, -\frac{\sqrt{6}}{3}\}$

15. To solve $4k^4 - 12k^2 + 5 = 0$, let $x = $ ____. This gives

k^2

_____ $= 0$,

$4x^2 - 12x + 5$

from which $x = $ _____ or $x = $ _____.

$\frac{5}{2};\ \frac{1}{2}$

Since $x = k^2$,

$$k^2 = \underline{\quad} \quad \text{or} \quad k^2 = \underline{\quad}.$$

$\frac{5}{2};\ \frac{1}{2}$

If $k^2 = \frac{5}{2}$, then $k = $ _____ or $k = $ _____.

$\frac{\sqrt{10}}{2};\ -\frac{\sqrt{10}}{2}$

If $k^2 = \frac{1}{2}$, then $k = $ _____ or $k = $ _____.

$\frac{\sqrt{2}}{2};\ -\frac{\sqrt{2}}{2}$

The solution set is _____.

$\{\frac{\sqrt{10}}{2}, -\frac{\sqrt{10}}{2}, \frac{\sqrt{2}}{2}, -\frac{\sqrt{2}}{2}\}$

16. Solve $3p^4 - 16p^2 + 5 = 0$.

Solution set: _____

$\{\sqrt{5}, -\sqrt{5}, \frac{\sqrt{3}}{3}, -\frac{\sqrt{3}}{3}\}$

17. Solve $2(3r - 1)^2 - 7(3r - 1) + 5 = 0$.

Let $x = $ _____. This gives

$3r - 1$

_____ $= 0$

$2x^2 - 7x + 5$

Factor.

$$(\underline{\quad})(\underline{\quad}) = 0$$

$(x - 1)(2x - 5)$

Solve for x.

$$x = \underline{\quad} \quad \text{or} \quad x = \underline{\quad}$$

$1; \dfrac{5}{2}$

Replace x with \underline{\quad}.

$3r - 1$

$$3r - 1 = \underline{\quad} \quad \text{or} \quad 3r - 1 = \underline{\quad}$$

$1; \dfrac{5}{2}$

Solve for r to get the solution set

$$\{\underline{\quad}, \underline{\quad}\}.$$

$\dfrac{2}{3}; \dfrac{7}{6}$

18. The quadratic formula can also be used to solve equations quadratic in \underline{\quad}.

form

19. For example, to solve

$$m^4 - 4m^2 + 2 = 0,$$

let $x = \underline{\quad}$, getting

m^2

$$\underline{\quad} = 0.$$

$x^2 - 4x + 2$

Use the quadratic formula to get

$$x = \underline{\quad} \quad \text{or} \quad x = \underline{\quad}.$$

$2 + \sqrt{2}; \ 2 - \sqrt{2}$

Since $m^2 = x,$

$$m^2 = 2 + \sqrt{2} \quad \text{or} \quad m^2 = \underline{\quad}.$$

$2 - \sqrt{2}$

Take square roots to get the solution set

$$\{\sqrt{2 + \sqrt{2}}, \underline{\quad}, \sqrt{2 - \sqrt{2}}, \underline{\quad}\}.$$

$-\sqrt{2 + \sqrt{2}}; \ -\sqrt{2 - \sqrt{2}}$

20. To solve $p = \sqrt{\dfrac{4 - 7p}{2}}$, first \underline{\quad} both sides. This gives

square

$$p^2 = \underline{\quad}$$

$\dfrac{4 - 7p}{2}$

or

$$\underline{\quad} = \underline{\quad}.$$

$2p^2; \ 4 - 7p$

Solve this equation by the quadratic formula.

$$a = \underline{\quad}, \ b = \underline{\quad}, \ c = \underline{\quad}$$

2; 7; −4

Solving the equation gives

$$p = \underline{\qquad} \quad \text{or} \quad p = \underline{\qquad}.$$

$\frac{1}{2}$; −4

We must now _____ these potential solutions
in the original equation. Doing that, we find
that the solution set of the original equation
is

check

$$\{\underline{\quad}\}.$$

$\frac{1}{2}$

21. Solve the equation $\sqrt{13k} = \sqrt{6k^2 + 6}$.

$$\{\underline{\qquad\qquad}\}$$

$\frac{2}{3}; \frac{3}{2}$

22. Solve $\sqrt{x - 1} = \sqrt{3x + 1} - 2$.

Square both sides.

$$(\sqrt{x - 1})^2 = (\underline{\qquad\qquad})^2$$

$\sqrt{3x + 1} - 2$

$$x - 1 = 3x + 1 + \underline{\qquad\qquad}$$

$-4\sqrt{3x + 1} + 4$

Simplify.

$$\underline{\qquad\qquad} = -4\sqrt{3x + 1}$$

$-2x - 6$

Square both sides again.

$$(\underline{\qquad\qquad})^2 = (-4\sqrt{3x + 1})^2$$

$-2x - 6$

$$\underline{\qquad\qquad} = 16(3x + 1)$$

$4x^2 + 24x + 36$

$$\underline{\qquad\qquad} = 0$$

$4x^2 - 24x + 20$

Divide both sides by 4.

$$\underline{\qquad\qquad} = 0$$

$x^2 - 6x + 5$

Factor to get $x = \underline{\quad}$ or $x = \underline{\quad}$. Check each
of these answers in the original equation to
get the solution set _____.

5; 1

$\{5, 1\}$

23. Solve $\sqrt{2 - x} = \sqrt{7 - x} - 1$.

Solution set: _____

$\{-2\}$

6.5 Formulas and Applications

[1] Solve second-degree formulas for a specified variable. (Frames 1-6)

[2] Solve applied problems about motion along a straight line. (Frames 7-8)

[3] Solve applied problems using the Pythagorean formula. (Frames 9-12)

[4] Solve applied problems using formulas for area. (Frames 13-16)

---

1. To solve $m^2 - 4m = 6p$ for m, think of p as a

_____. Write the equation equal to 0.

_____ = 0

constant

$m^2 - 4m - 6p$

Solve this equation by using the quadratic formula. Here a = ____, b = ____, and c = ____.

1; -4; -6p

$$m = \frac{-(\quad) \pm \sqrt{(\quad)^2 - 4(\quad)(\quad)}}{2(\quad)}$$

-4; -4; 1; -6p
1

$$= \frac{4 \pm \sqrt{}}{2}$$

16 + 24p

2. Simplify $\sqrt{16 + 24p}$.

$$\sqrt{16 + 24p} = \sqrt{\quad(\quad\quad)}$$

4; 4 + 6p

$$= \underline{}$$

$2\sqrt{4 + 6p}$

3. Substitute this in the result for m.

$$m = \frac{4 \pm }{2}$$

$2\sqrt{4 + 6p}$

$$= \frac{2()}{2}$$

$2 \pm \sqrt{4 + 6p}$

$$m = \underline{}$$

$2 \pm \sqrt{4 + 6p}$

Solve the formulas in Frames 4–6 for the specified variable.

4. $R = \frac{1}{2}pq^2$, for q

q = _____ $\pm\dfrac{\sqrt{2Rp}}{p}$

5. $y = \dfrac{mv^2r}{6}$, for v   v = _____ $\pm\dfrac{\sqrt{6mry}}{mr}$

6. $z = \dfrac{\sqrt{ma}}{3}$, for m   m = _____ $\dfrac{9z^2}{a}$

7. The position of a particle moving in a straight line is given by

$$s = 8t^2 + 14t,$$

where s is in feet and t is time in seconds the particle has been in motion. How many seconds will it take for the particle to move 15 feet?

To find out, replace s with ____.   15

____ $= 8t^2 + 14t$ 15

Solve this equation (reject any _____   negative

answers) to get _____ seconds.   3/4

8. The position of a particle moving in a straight line is given by

$$s = 3t^2 - 8t,$$

where s is in feet and t is time in seconds the particle has been in motion. How many seconds will it take for the particle to move 35 feet?

_____ $= 3t^2 - 8t$ 35

Solve this equation (reject any negative answers) to get _____ seconds.   5

9. We can also use our knowledge of the quadratic formula to solve problems involving the Pythagorean formula from geometry: if a and b represent the two short sides of a right triangle, and if c represents the longest side, then

   _____ = _____ .

   $a^2 + b^2$; c^2

Solve the applied problems in Frames 10–16.

10. **On side of a right triangle is 2 more than twice the shortest side, and the longest side is 3 more than twice the shortest side. Find the three sides.**

    If x represents the shortest side, then the medium side is _____ and the longest side is _____ . Then the Pythagorean formula gives

    $2x + 2$

    $2x + 3$

    _____ + _____ = _____ .

    x^2; $(2x + 2)^2$; $(2x + 3)^2$

    Upon solving this equation, we find that the sides of the triangle are _____, _____, and _____. (When solving the equation above, we got two values for x: ____ or ____. The value x = ____ cannot be a side of a triangle, and is rejected.)

    5; 12

    13

    5; −1

    −1

11. **The medium side of a right triangle is 3 more than three times as long as the shortest side, and the longest side is 4 more than three times the shortest side. Find the three sides.**

    The sides are ____, ____, and ____.

    7; 24; 25

12. At a certain time, the shadow of a tower cast by the sun is 17 meters longer than the tower itself. The distance between the top of the tower and the end of the shadow is 1 meter longer than the length of the shadow. Find the height of the tower.

_____   7 meters

13. The length of a rectangle is 9 centimeters more than the width. The area is 22 square centimeters. Find the length and width of the rectangle.

    Let x = the width of the rectangle. The length is then _____. The area of a rectangle is $x + 9$
    given by the product of the length and _____,   width
    or

$$x(x + 9) = \underline{\hspace{1in}}.$$ 22

    Solve this equation.

$$\underline{\hspace{1.5in}} = 22$$ $x^2 + 9x$
$$\underline{\hspace{1.5in}} = 0$$ $x^2 + 9x - 22$

    Factor.

$$\underline{\hspace{1.5in}} = 0$$ $(x + 11)(x - 2)$
$$x = \underline{\hspace{0.5in}} \quad \text{or} \quad x = \underline{\hspace{0.5in}}$$ $-11; \, 2$

    Reject x = _____, giving the answer x = ____. $-11; \, 2$
    The width of the rectangle is ____ centimeters,   2
    and the length is $2 +$ ____ = ____ centimeters.   9; 11

14. The length of a rectangle is 3 more than the width. The area is 108. Find the length and width.

        length: ____, width: ____   12; 9

15. **Mary has a garden that is 15 feet by 18 feet. She wants to put a strip of bark of uniform width around it, using enough bark to cover 70 square feet. How wide a strip can she use?**

Let x represent the width of the bark strip. Complete the following figure.

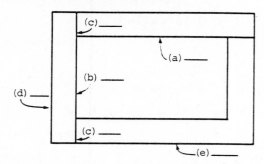

(a) 18
(b) 15
(c) x
(d) 15 + 2x
(e) 18 + 2x

From the figure, we get the equation

(_____)(_____) − (____)(____) = ____.

15 + 2x; 18 + 2x; 15; 18; 70

Solve this equation.

x = ____ (feet/foot)

1 foot

16. **A rectangle has a length that is 6 inches more than the width. If a square 2 inches on a side is cut from each corner, and the sides then turned up to form a box, the volume is 224 cubic inches. Find the dimensions of the original rectangle.**

Complete this figure.

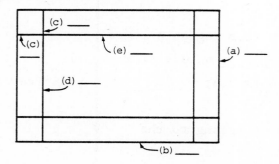

(a) x
(b) x + 6
(c) 2
(d) x − 4
(e) x + 2

length: ____ inches, width: ____ inches

18; 12

6.6 Nonlinear Inequalities

1 Solve quadratic inequalities. (Frames 1–19)

2 Solve polynomial inequalities of degree 3 or more. (Frames 20–21)

3 Solve rational inequalities. (Frames 22–27)

---

1. An inequality such as $x^2 + 2x - 15 < 0$ is called
    a _____ inequality.

    quadratic

2. To solve the inequality $x^2 + 2x - 15 < 0$, we first
    solve the quadratic equation $x^2 + 2x - 15 = $ ___ 0.

    =

3. We can factor $x^2 + 2x - 15$ as (_____)(_____),
    which leads to the solutions x = ____ or x = ____.

    x + 5; x – 3
    –5; 3

4. Using the factorization from Frame 3, we can
    write the original problem as

    (_____)(_____) ___ 0.

    x + 5; x – 3; <

5. The two solutions of the equation, x = ____ or
    x = ____, can be used to divide the number line
    into _____ regions, as shown below.

    –5
    3
    three

    Region   Region   Region
      A   B   C

    –5; 3

6. We know that if one value of x from region A makes
    $(x + 5)(x - 3)$ positive, then _____ values of x
    from Region A will make $(x + 5)(x - 3)$ _____.

    all
    positive

7. To see if the points of Region A belong to the
    solution set of $(x + 5)(x - 3) < 0$, we select
    any _____ from Region A and try it. Let us
    select x = –7. We then have

    number

$$(x + 5)(x - 3) = (\underline{\quad} + 5)(\underline{\quad} - 3)$$
$$= (\underline{\quad})(\underline{\quad})$$
$$= \underline{\quad}$$
$$20 \underline{\quad} 0.$$

-7; -7

-2; -10

20

≮

8. Thus the points of Region A (*do/do not*) belong to the solution set.

do not

9. If we select x = 0 from Region B, we have

$$(x + 5)(x - 3) = (\underline{\quad} + 5)(\underline{\quad} - 3)$$
$$= (\underline{\quad})(\underline{\quad})$$
$$= \underline{\quad}$$
$$-15 \underline{\quad} 0.$$

0; 0

5; -3

-15

<

10. Thus the points of Region B (*do/do not*) belong to the solution set.

do

11. Let us select x = 6 from Region C. This gives

$$(x + 5)(x - 3) = (\underline{\quad} + 5)(\underline{\quad} - 3)$$
$$= (\underline{\quad})(\underline{\quad})$$
$$= \underline{\quad}$$
$$33 \underline{\quad} 0.$$

6; 6

11; 3

33

≮

12. Hence the points of Region C (*do/do not*) belong to the solution set.

do not

13. Based on the work above, we can say that the solution set of $(x + 5)(x - 3) < 0$ is

_____.

(-5, 3)

14. The graph of the solution set is

Solve the inequalities in Frames 15–20.

15. $x^2 - x - 20 < 0$

First we solve _____ = 0. We find that
x = ___ or x = ___, which divide the number
line into three regions, as shown.

Do the points of Region A belong to the solution
set? (*yes/no*)

Of Region B? (*yes/no*)

Of Region C? (*yes/no*)

The solution set is _____.

16. $x^2 - x - 6 \geq 0$

We first solve _____ = 0, finding that
x = ___ or x = ___, which divide the number
line into three regions, as shown.

Do the points of Region A belong to the solution
set? (*yes/no*)

Of Region B? (*yes/no*)

Of Region C? (*yes/no*)

The solution set is _____.

$x^2 - x - 20$	
$-4;\ 5$	
$-4;\ 5$	
no	
yes	
no	
$(-4,\ 5)$	
$x^2 - x - 6$	
$-2;\ 3$	
$-2;\ 3$	
yes	
no	
yes	
$(-\infty,\ -2] \cup [3,\ \infty)$	

17. $6r^2 - r - 2 > 0$

    The solutions of the corresponding quadratic

    equations are r = _____ or r = _____.

    Solution set: _____

    $-\frac{1}{2}, \frac{2}{3}$

    $(-\infty, -\frac{1}{2}) \cup (\frac{2}{3}, \infty)$

18. $8y^2 - 4y < 0$

    First solve _____ = 0, getting y = ___

    or y = _____. The solution set is

    _____.

    $8y^2 - 4y; 0$

    1/2

    (0, 1/2)

19. $(4k - 1)^2 \le -1$

    Since any square is _____ than or equal

    to 0, it is not possible for $(4k - 1)^2$ to be

    _____. The solution set here is ___.

    greater

    negative; ∅

20. A _____ or _____ inequality

    such as $(3p - 1)(p + 2)(p - 4) > 0$ may be solved

    using the method we use for solving a quadratic

    inequality.

    cubic;
    third-degree

21. $(3p - 1)(p + 2)(p - 4) > 0$

    Here we use the numbers _____, _____ and ___

    to divide the number line into ___ regions, and

    then try one point from each region. Doing that,

    we find the solution set to be

    _____.

    1/3; -2; 4
    four

    $(-2, 1/3) \cup (4, \infty)$

22. To solve $\frac{1}{x + 1} < 2$, first write the corresponding

    _____ and solve it. _____ = ____

    Multiply by the common _____. 1 = _____

    equation; $\frac{1}{x + 1}$; 2

    denominator;
    2(x + 1)

Solve the resulting equation. x = _____

$-\dfrac{1}{2}$

Now find the number that makes the _____

denominator

zero.

_____ = 0

x + 1

x = _____

−1

Use −1/2 and −1 to divide the number line into

three _____ and test a point from each

regions

region. Doing this gives the solution set

_____.

$(-\infty, -1)\cup(-\dfrac{1}{2}, \infty)$

23.   Solve $\dfrac{y - 1}{y + 2} \geq 4$.

Write the corresponding equation and solve it.

_____ = _____

$\dfrac{y - 1}{y + 2}; 4$

y = _____

−3

Now, find the number that makes the _____

denominator

zero.

y = _____

−2

Use these numbers to divide a number line into

three regions, giving

_____.

$-3 \leq y \leq -2$

However, because of the = portion of ≥ in the

original inequality, it is necessary to check

−3 and −2; since −2 makes the denominator equal

_____, it cannot be part of the solution. The

0

solution set is

_____.

[−3, −2)

Solve each inequality in Frames 24–27.

24. $2 < \dfrac{1}{x - 1}$

Solution set: _____

$(1, \dfrac{3}{2})$

25. $-1 > \dfrac{2}{x + 2}$

Solution set: _____ $(-4, -2)$

26. $\dfrac{2}{p + 1} \le 1$

Solution set: _____ $(-\infty, -1) \cup [1, \infty)$

27. $\dfrac{3k}{k - 4} \le -2$

Solution set: _____ $[\dfrac{8}{5}, 4)$

Chapter 6 Test

The answers for these questions are at the back of this Study Guide.

Solve each equation.

1. $2m^2 + 7m = 4$ 1. _____

2. $x = \sqrt{\dfrac{3 - 7x}{6}}$ 2. _____

3. $\dfrac{18}{m^2} = 2 - \dfrac{9}{m}$ 3. _____

4. On a 24-mile walk, Tom needed 2 hours less time
    than Fred. If Tom's speed was 2 miles per hour
    more than Fred's, find Fred's speed.   4. _____

Solve each of the following by completing the square.

5. $x^2 - 4x - 2 = 0$ 5. _____

6. $3x^2 - 9x + 1 = 0$ 6. _____

Solve each of the following by the quadratic formula.

7. $4x^2 - 12x + 9 = 0$ 7. _____

8. $2x^2 - 3x + 1 = 0$ 8. _____

9. $y^4 + 5y^2 - 36 = 0$ (Find all solutions,
                       not just the real ones.)   9. _____

10. $k = \sqrt{k + 20}$ 10. _____

11. $3 = \dfrac{4}{x^2} - \dfrac{11}{x}$ 11. _____

12. Use the discriminant to predict the number and type of solutions for the equation $6z^2 - 11z + 12 = 0$.

12. _____

13. Use the discriminant to determine if $9x^2 - 5x - 4 = 0$ can be factored.

13. _____

14. Without solving, give the sum and product of the solutions of the equation $6z^2 - 11z + 12 = 0$.

14. _____

15. Write the quadratic equation whose solution set is $\left\{-3, \frac{1}{2}\right\}$.

15. _____

Solve for the indicated variable.

16. $m = 2kz^2$, for z

16. _____

17. $A = zy\sqrt{r^2 + t^2}$, for t

17. _____

18. A right triangle has a longest side 2 meters longer than the medium side. The medium side is 1 meter less than twice the short side. Find the lengths of the three sides of the triangle.

18. _____

19. The length of a rectangle is 4 meters less than twice the width. The area of the rectangle is 70 square meters. Find the length and width of the rectangle.

19. _____

20. Find the time it takes for an object to move 15 feet in a straight line given the equation $s = 2t^2 - t$, where s is the distance in feet and t is the time in seconds.

20. _____

Solve the following.

21. $6x^2 - 7x - 5 \leq 0$ 21. _____

22. $2x^2 - x - 6 > 0$ 22. _____

23. $(x + 3)(x + 1)(x - 4) \leq 0$ 23. _____

24. $\dfrac{2}{x - 3} \leq -1$ 24. _____

CHAPTER 7   THE STRAIGHT LINE

7.1   The Rectangular Coordinate System

1 Plot ordered pairs. (Frames 1-2)

2 Find ordered pairs which satisfy a given equation. (Frames 3-8)

3 Graph lines. (Frames 9-13)

4 Find x- and y-intercepts. (Frames 14-16)

5 Recognize equations of vertical and horizontal lines. (Frames 17-20)

6 Use the distance formula. (Frames 21-30)

---

1. (2, -3) is called an _____ pair. 2 is called the _____ component, and -3 is called the _____ component.

ordered
first
second

2. To graph ordered pairs we use the perpendicular crossed number lines of a _____ system. The horizontal line is called the _____, while the vertical line is called the _____. To locate a point such as (-3, 5), go 3 units to the _____ on the _____, and then _____ units up parallel to the y-axis.
Plot (-3, 5) on the graph at the right.
Also plot (2, -1), (3, -4), (2, 0), (-4, 0), (1, 6), (0, -3), and (-2, -5)

coordinate
x-axis
y-axis

left; x-axis; 5

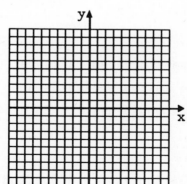

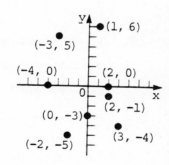

3. An equation with two _____ will have solutions which are _____.

variables
ordered pairs

4. For example, many ordered pairs are solutions
    for the equation $4x - 5y = 20$. If $x = 0$, then
    we get

$$4(\underline{}) - 5y = \underline{}$$
$$-5y = \underline{}$$
$$y = \underline{}.$$

    If $x = 0$, then $y = \underline{}$. This can be written
    as the ordered pair $\underline{}$.

0; 20
20
-4
-4
(0, -4)

Complete the ordered pairs of Frames 5–8 for the equation $4x - 5y = 20$.

5. $(\underline{}, 0)$
    Let $y = \underline{}$; then $x = \underline{}$, giving the ordered
    pair $\underline{}$.

    0; 5
    (5, 0)

6. $(-5, \underline{})$ −8

7. $(\underline{}, 2)$ 15/2

8. $(\underline{}, 6)$ 25/2

9. The equation $4x - 5y = 20$ is an example of a
    $\underline{}$ equation in two $\underline{}$.

    first−degree;
    variables

10. If we found many different ordered pairs for
    a first-degree equation in two variables and
    graphed them, the points would all lie on a
    $\underline{}$ line.

    straight

11. To graph the linear equation $2x + y = 6$, find some
    ordered pairs that satisfy the equation. For ex-
    ample, if $x = 0$, then $y = \underline{}$, so that $(0, \underline{})$
    belongs to the graph.

    6; 6

Complete each of the following ordered pairs: (3, ___), (-1, ___), and (4, ___). Graph these ordered pairs on the graph at the right and draw a line through them. These points lie on a _____ line.

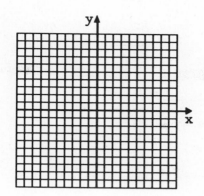

0

8

-2

straight

12. Graph $3x - 4y = 12$ by completing the following ordered pairs.

(8, ___), (-4, ___), (0, ___), (2, ___), (4, ___)

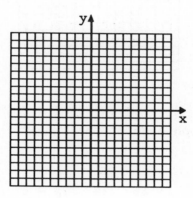

3; -6

-3; $-\frac{3}{2}$

0

13. An equation written in the form $Ax + By = C$ is called an equation in _____ form.

standard

14. The point where the graph crosses the x-axis is called the _____ of the graph. The point where the graph crosses the y-axis is called the _____. To find the x-intercept of

$$5x + 4y = 20,$$

substitute 0 for _____.

$$5x + 4(\_\_) = 20$$

x-intercept

y-intercept

y

0

or

$$x = \underline{\hspace{2em}},$$ 4

which leads to the _____, _____. To find   x-intercept; (4, 0)

the y-intercept, let _____ = 0.   x

$$5(\underline{\hspace{1.5em}}) + 4y = 20$$ 0

or

$$y = \underline{\hspace{2em}},$$ 5

which leads to the y-intercept, _____.   (0, 5)

Graph the intercepts
and draw a line
through them. This
line is the graph of
5x + 4y = 20.

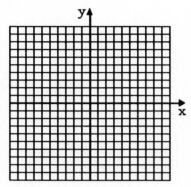

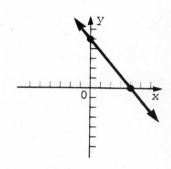

**Find the intercepts and graph each of the linear
equations of Frames 15–16.**

15. 2x + y = 4   **16.** 3x − 2y = 6

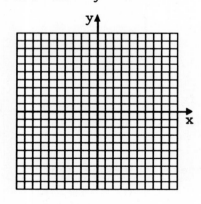   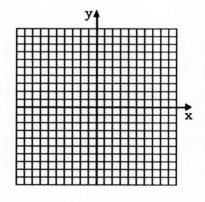

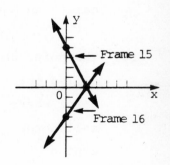

x-intercept: _____   x-intercept: _____   (2, 0); (2, 0)

y-intercept: _____   y-intercept: _____   (0, 4); (0, −3)

17. In the equation x = -2, the value of x is always
     _____, so that it is not possible for x to equal
     0. For this reason, the graph of x = -2 has no
     (x/y) intercept. This means that the line is
     (*horizontal/vertical*).

     -2

     y
     vertical

18. In similar manner, the graph of y = 3 has no
     (x/y) intercepts, and is (*horizontal/vertical*).

     x; horizontal

19. Graph x = -2.   20. Graph y = 3.

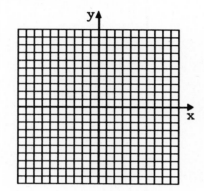   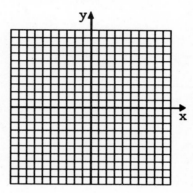

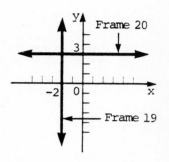

21. The distance between the points (x_1, y_1) and
     (x_2, y_2) is given by

$$d = \sqrt{(\underline{\hspace{1cm}})^2 + (\underline{\hspace{1cm}})^2}.$$

     This result is called the _____ formula.

     $x_2 - x_1$; $y_2 - y_1$

     distance

22. The distance between (5, -1) and (-3, 4) can be
     found by letting $y_2 = 4$, so that

     $y_1 =$ _____, $x_2 =$ _____, and $x_1 =$ _____.

     -1; -3; 5

     By the distance formula, the distance between the
     points is

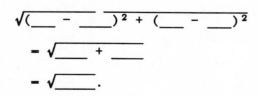

     -3; 5; 4; -1

     64; 25

     89

Find the distance between the pairs of points listed in Frames 23—26.

23.   (0, 1) and (2, 3)

$$\sqrt{(\underline{} - \underline{})^2 + (\underline{} - \underline{})^2}$$

$$= \sqrt{\underline{} + \underline{}}$$

$$= \sqrt{\underline{}}$$

$$= \underline{}$$

2; 0; 3; 1

4; 4

8

$2\sqrt{2}$

24.   (−1, 3) and (−5, 2)

$$\sqrt{(\underline{} - \underline{})^2 + (\underline{} - \underline{})^2}$$

$$= \sqrt{\underline{} + \underline{}}$$

$$= \sqrt{\underline{}}$$

−5; −1; 2; 3

16; 1

17

25.   (−3, −1) and (2, −4)   _____

$\sqrt{34}$

26.   (3, 0) and (7, 3)   _____

5

27.   To see whether or not the points (2, −1), (5, 8), and (−2, −13) lie on the same straight line, we can use the _____ formula. The distance between (2, −1) and (5, 8) is _____. The distance between (2, −1) and (−2, −13) is

_____. The distance between (5, 8) and (−2, −13) is _____. Since

_____ + _____ = _____,

the points (*do/do not*) lie on the same line.

distance

$\sqrt{90} = 3\sqrt{10}$

$\sqrt{160} = 4\sqrt{10}$

$\sqrt{490} = 7\sqrt{10}$

$3\sqrt{10}$; $4\sqrt{10}$; $7\sqrt{10}$

do

Determine whether or not the points of Frames 28-30 lie on the same lines.

28. A = (1, 5), B = (-2, 14), C = (3, -1)

distance from A to C = _____ | $\sqrt{40} = 2\sqrt{10}$

distance from A to B = _____ | $\sqrt{90} = 3\sqrt{10}$

distance from B to C = _____ | $\sqrt{250} = 5\sqrt{10}$

Therefore, the points (*do/do not*) lie on the same line. | do

29. Do (2, 9), (4, 18), (0, -1) lie on the same line? (*yes/no*) | no

30. Do (1, -7), (2, -15), (-1, 9) lie on the same line? (*yes/no*) | yes

7.2 The Slope of a Line

[1] Find the slope of a line given two points on the line. (Frames 1-6)

[2] Find the slope of a line given the equation of the line. (Frames 7-9)

[3] Graph a line given its slope and a point on the line. (Frames 10-12)

[4] Use slope to decide whether two lines are parallel or perpendicular. (Frames 13-20)

[5] Solve problems involving average rate of change. (Frames 21-23)

1. The slope of a line is a measure of the _____ of the line. The slope of the line through (x_1, y_1) and (x_2, y_2) is given by | steepness

$$m = \underline{\qquad\qquad}.$$ | $\dfrac{y_2 - y_1}{x_2 - x_1}$

2. To find the slope of the line through (−2, 4) and
   (6, −8), let $x_1 = -2$, so that

$$y_1 = \underline{\hspace{1cm}}, \quad x_2 = \underline{\hspace{1cm}}, \quad \text{and } y_2 = \underline{\hspace{1cm}}.$$

   4; 6; −8

   The slope is thus

$$m = \underline{\hspace{3cm}}$$

$$\frac{-8 - 4}{6 - (-2)}$$

$$= \underline{\hspace{2cm}}.$$

   −3/2

**Find the slopes of the lines through the pairs of
points in Frames 3–6.**

3. (2, 0), (−3, 5)

$$m = \underline{\hspace{3cm}}$$

$$\frac{5 - 0}{-3 - 2}$$

$$= \underline{\hspace{2cm}}$$

   −1

4. (−5, 1), (2, 6)

$$m = \underline{\hspace{3cm}}$$

$$\frac{5}{7}$$

5. (11, −2), (−8, 7)

$$m = \underline{\hspace{3cm}}$$

$$-\frac{9}{19}$$

6. (4, 5), (−1, 5)

$$m = \underline{\hspace{3cm}}$$

   0

7. The slope of $3x + 4y = 12$ can be found by obtaining two points that the line goes through. If
   $x = 0$, then $y = \underline{\hspace{1cm}}$, so that the line goes

   3

   through (0, ___). Another point the line goes

   3

   through is (___, 0). Use these points to find

   4

   that the slope is $m = \underline{\hspace{1cm}}$.

   −3/4

8. The slope of the line $y = 3$ can be found if we
   know two points of the line. For example,
   (5, ___) and (−2, ___) lie on the line. Find

   3; 3

   the slope.

$$m = \frac{3 - \underline{\hspace{1cm}}}{-2 - \underline{\hspace{1cm}}} = \underline{\hspace{1cm}}$$

$$\frac{3 - 3}{-2 - 5} = 0$$

   The graph of $y = 3$ is a _____ line. The

   horizontal

   slope of any horizontal line is 0.

9. The line x = -4 has a graph which is a _____
    line. Two points on the line are (___, 2) and
    (___, 5). The slope is

    $$m = \underline{\hspace{2cm}} = \underline{\hspace{1.5cm}}.$$

    Since division by _____ is not permitted, the
    slope is _____. All vertical lines have
    _____ slope.

vertical	
-4	
-4	
$\dfrac{2-5}{-4-(-4)} = \dfrac{-3}{0}$	
0	
undefined	
undefined	

10. If a line has slope 2/3 and goes through (0, -4),
    we can find its graph. First plot the point
    (___, -4). From this
    point, go to the
    right _____ units and
    then up _____ units,
    to get another point
    of the graph. We can
    then draw a line
    through these two
    points to complete
    the graph. Sketch
    the graph at the
    right.

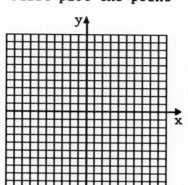

0

3

2

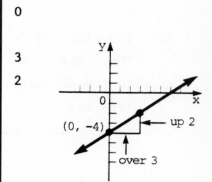

Graph the lines in Frames 11 and 12.

11. m = -3/4,
    through (-2, 3)

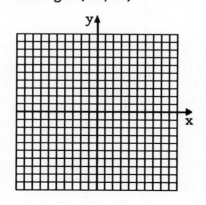

12. m = -3
    through (3, -2)

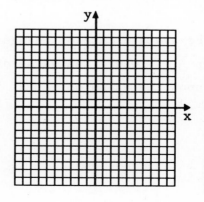

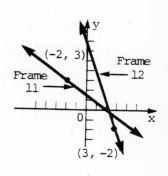

13. Two different lines with the same slope are
     _____.   parallel

14. Are 4x + 3y = 8 and 6y = 7 - 8x parallel? To
     find out, get the slope of each line. The slope
     of 4x + 3y = 8 is _____, and the slope of   -4/3
     6y = 7 - 8x is _____. Are the slopes equal?   -4/3
     (*yes/no*) Are the lines parallel? (*yes/no*)   yes; yes

15. Two different nonvertical lines are perpendicular
     if the product of their slopes is _____.   -1

16. The slope of 7x - 4y = 11 is _____.   7/4
     The slope of 4x + 7y = 3 is _____.   -4/7
     Is the product of these slopes equal to -1?
     (*yes/no*)   yes
     Are the lines perpendicular? (*yes/no*)   yes

**Write parallel, perpendicular, or neither for each of
the pairs of lines in Frames 17-20.**

17. x + y = 8,   2x = 8 - 2y   _____   parallel

18. 3x + y = 4,   6 + 3y = x   _____   perpendicular

19. 5x + 2y = 0,   4y = 7 - 10x   _____   parallel

20. 3x - 4y + 2,   3x + 4y = 6   _____   neither

21. The slope gives the average rate of change of ____   y
     per unit change in ____, where the value of y is   x
     _____ upon the value of x.   dependent

In Frames 22-23, find the average rate of change of y per unit change in x.

22. When Lins Loc was 3 years old she watched about 7 hours of television each week. Now that she is 18 years old, she watches about 17 hours of television each week. What is the average rate of change (in hours per week) in her television watching habit over that period of time?

At 3 years of age she watched about _____ hours per week, so (x_1, y_1) = _____ .

7

(3, 7)

At 18 years of age she watches about _____ hours per week, so (x_2, y_2) = _____ .

17

(18, 17)

Average rate of change of y per unit change in x $= \dfrac{y_2 - y_1}{\underline{\hspace{1cm}}} = \underline{\hspace{1cm}} = \underline{\hspace{1cm}} = \underline{\hspace{1cm}}$

17 - 7; 10; 2

$x_2 - y_2$; 18 - 3;

15; 3

Thus, she increased her television watching an average of _____ hours (or _____ minutes) per week each year.

2/3; 40

23. During the first week of October the price of oranges was $.99 per pound. Because of an early freeze in Florida the price was $1.47 per pound 6 weeks later. What was the average price increase each week during this time?

$.08 per pound

7.3 Linear Equations

[1] Write the equation of a line given its slope and a point on the line. (Frames 1–6)

[2] Write the equation of a line given two point on the line.   (Frames 7–10)

[3] Write the equation of a line given its slope and y-intercept. (Frames 11–14)

[4] Find the slope and y-intercept of a line given its equation. (Frames 15–18)

[5] Write the equation of a line parallel or perpendicular to a given line. (Frames 19–28)

[6] Find an equation relating two unknowns in a linear relationship. (Frames 29–30.)

---

1. A line having slope m and going through the point (x_1, y_1) has equation

$$y - \underline{\quad} = \underline{\quad} (x - \underline{\quad}).$$

This equation is called the \underline{\qquad} – \underline{\qquad} form of the equation of a line.

y_1; m; x_1

point; slope

2. The equation of the line through $(-4, 6)$ with slope $-3/2$ is given by

$$y - \underline{\quad} = \underline{\quad} (x - \underline{\quad}),$$

or, upon simplifying,

$$\underline{\qquad\qquad}.$$

6; $-3/2$; -4

$3x + 2y = 0$

Find the equations of the lines in Frames 3–6.

3. $m = -1$, through $(-3, 4)$

$$y - \underline{\quad} = \underline{\quad} (x - \underline{\quad})$$
$$y = \underline{\qquad}$$
$$\underline{\qquad}$$

4; -1; -3

$-x + 1$

$x + y = 1$

4. m = -4/5, through (-7, -3)

$$y - \underline{\quad} = \underline{\quad} (x - \underline{\quad})$$

$$y + 3 = -\frac{4}{5}(\underline{\quad})$$

$$\underline{\quad} = -4(x + 7)$$

$$5y + 15 = \underline{\quad}$$

$$\underline{\quad} = -43$$

$-3; -\frac{4}{5}; -7$
$x + 7$
$5(y + 3)$
$-4x - 28$
$4x + 5y$

5. m = 0, going through (5, 7)

$$y - \underline{\quad} = \underline{\quad} (x - \underline{\quad})$$

$$y = \underline{\quad}$$

7; 0; 5

7

6. The line with undefined slope through (-5, 2)

$$\underline{\quad}$$

x = -5

7. To find the equation of a line through the points (-1, 7) and (9, -3), first find the _____ of the line.

slope

$$m = \underline{\quad}$$

-1

Now use the _____ form. We can use either (-1, 7) or _____. Let us use (-1, 7).

point-slope

(9, -3)

$$y - \underline{\quad} = \underline{\quad} [x - (-1)]$$

$$y - 7 = -1(\underline{\quad})$$

$$y - 7 = \underline{\quad}$$

$$\underline{\quad} = 6$$

7; -1

x + 1

-x - 1

x + y

Find the equations of the lines in Frames 8–10.

8. Through (-2, 1) and (6, -3)
First find the slope.

$$m = \underline{\quad}$$

$-\frac{1}{2}$

The equation of the line is _____.

x + 2y = 0

9. Through (2, 5) and (−1, −1)

    _____ $2x - y = -1$

10. Through (−3, 5) and (8, 9)

    _____ $4x - 11y = -67$

11. An equation solved for y has the form $y = mx + b$,
    where the slope is _____, and the y−intercept is   m
    _____.   (0, b)

12. $y = mx + b$ is called the slope−_____ form   intercept
    of the equation of a line.

Write the equations of the lines in Frames 13–14.

13. Slope: −1/2, y−intercept: (0, 1)
    By the slope−intercept form, the equation is

    $y =$ _____. $-\frac{1}{2}x + 1$

    Multiply both sides by ___ to clear the equation   2
    of fractions.

    _____ $2y = -x + 2$
    or   _____ = 2 $x + 2y$

14. Slope: −4/5, y−intercept: $\left(0, \frac{2}{3}\right)$

    _____ $12x + 15y = 10$

15. The slope of $3x + 4y = 12$ can be found by solving
    the equation for _____. Subtract _____ from both   y; 3x
    sides to get

    $4y =$ _____. $-3x + 12$

    Divide both sides by _____.   4

    $y =$ _____ $-\frac{3}{4}x + 3$

The slope is given by the coefficient of _____, | x
which is _____. The number 3 at the right in- | $-3/4$
dicates the _____ which is _____. | y-intercept;
| (0, 3)

Find the slopes and y-intercepts of the lines in Frames 16–18.

16. $3x - y = 5$

    Solve for y.

    $$y = \underline{\hspace{2cm}}$$ | $3x - 5$
    The slope is _____. | 3
    The y-intercept is _____. | (0, -5)

17. $5x - 4y = 7$

    slope = _____ | $\dfrac{5}{4}$
    y-intercept = _____ | $\left(0, -\dfrac{7}{4}\right)$

18. $3x = 2y$

    slope = _____ | $\dfrac{3}{2}$
    y-intercept = _____ | (0, 0)

19. Two lines are parallel if they have the _____ slope. | same

20. The slope of $2x + 3y = 6$ is _____, while the | $-2/3$
    slope of $4x = 2 - 6y$ is _____. These two lines | $-2/3$
    (*are/are not*) parallel. | are

21. The lines $3x - 5y = 8$ and $6x = 4 + 10y$ |
    (*are/are not*) parallel. | are

22. Two lines are perpendicular if their slopes |
    are _____ _____. | negative
    | reciprocals

23. The lines $3x - y = 6$ and $x + 3y = 12$ (*are/are not*)
    perpendicular.

    are

24. The lines $4x - 1 + y$ and $x - 3 + y$ (*are/are not*)
    perpendicular.

    are not

Write the equations for the lines in Frames 25–28.

25. Parallel to $2x + y = 7$, going through $(-1, 5)$
    The slope of the desired line is the _____ as the
    slope of $2x + y = 7$, which is ____. Hence we can
    use the _____-_____ form to find the
    equation, which is _____.

    same
    -2
    point; slope
    $2x + y = 3$

26. Parallel to $3x - 2y = 4$, going through $(0, 1)$

    _____

    $3x - 2y = -2$

27. Perpendicular to $x + y = 4$, going through $(1, -2)$
    The line $x + y = 4$ has slope _____, so that the
    desired line has slope _____. Use the _____-
    _____ form to find the equation, which is
    _____.

    -1
    1; point
    slope
    $-x + y = -3$ or
    $x - y = 3$

28. Perpendicular to $3x - y = 2$, going through $(0, -4)$

    _____

    $x + 3y = -12$

29. A team of 15 telephone solicitors can make 450
    calls in an hour. 20 solicitors can make 800
    calls in an hour. Write an equation that gives
    the number of calls made, y, by the solicitors,
    x. Predict the number of calls that can be made
    by 18 solicitors.

    The equation of the line contains the points
    (____, 450) and (20, _____). The slope of
    the line is

    15; 800

$$\frac{\text{---} - 450}{\text{---}} = \frac{\text{---}}{\text{---}} = \text{---}.$$

	800; 350; 70
	20 - 15; 5

Using the point-slope form with the point

(___, 450) and the slope _____, we can write   15; 70

$$y - \text{---} = \text{---}(x - \text{---})$$ 450; 70; 15

Simplifying gives y = _____.   70x - 600

When x = 18, y = _____, so 18 solicitors can make   660

_____ telephone calls.   660

30. **On a typical summer day 3 Coast Guard Auxiliary
boats can respond to 25 calls for help. If 7
boats are available, 85 calls can be answered.
Write an equation that gives the number of calls
answered, y, by the boats, x. Estimate the number of calls for help that 5 boats could handle.**

$$y = \text{_____}$$ 15x - 20

Five Coast Guard Auxiliary boats could handle

_____ calls for help.   55

7.4 Linear Inequalities

[1] Graph linear inequalities. (Frames 1-4)

[2] Graph the intersection of two linear inequalities. (Frames 5-6)

[3] Graph the union of two linear inequalities. (Frames 7-8)

[4] Graph first-degree absolute value inequalities. (Frames 9-12)

1. An expression of the form Ax + By ≥ C where A, B,
and C are real numbers and A and B are not both 0,
is called a _____ inequality.   linear

2. To graph the linear inequality x - 2y ≤ 4, first
graph the boundary, _____. Graph this line   x - 2y = 4
on the axes below.

The graph of the
linear inequality
will include the
region on one side
or the other of the
line. To find out
which, select any
point not on the

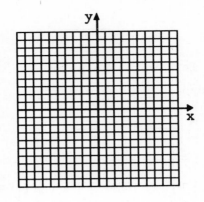

_____, and try it in the original inequality.
Here we can try (0, 0). We have 0 − 2 • 0 ≤ 4,
a (*true/false*) statement. Hence the graph
includes the region that contains (___, ___).
Shade that region on the graph.

line

true

0, 0

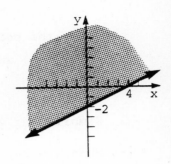

3.   To graph 2x + 3y < 6, we first graph _____.

2x + 3y = 6

Make this line dashed,
since the points of
the line itself do not
belong to the linear
inequality. Graph
this line on the given
axes. The point (0, 0)
(*does/does not*) satisfy

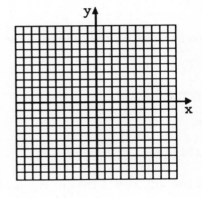

the inequality, so that we (*do/do not*) shade the
side of the graph that includes (0, 0). Complete
the graph.

does

do

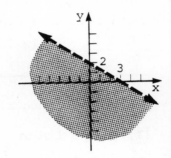

4. Graph $3x - y > 2$.

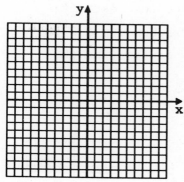

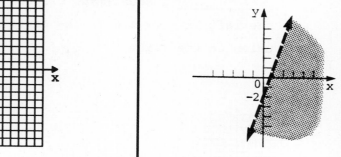

5. To graph the intersection of $3x + y \leq 4$ and
    $y \geq 2$, graph _____ and _____ separately,
    and then find the _____ of the two
    graphs. Sketch $3x + y \leq 4$ on the graph at the
    left below, $y \geq 2$ on the graph in the center,
    and the intersection at the right.

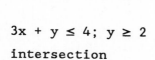

$3x + y \leq 4$; $y \geq 2$
intersection

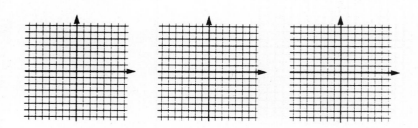

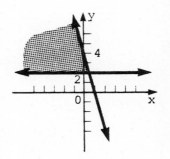

6. Graph the intersection of $x - 3y \leq 6$ and $x \geq 2$.

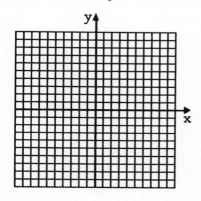

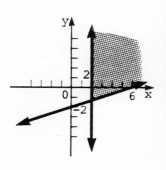

7. Graph the union of the x + y ≥ 2 and x − 2y ≤ 6.
    The union includes all points that satisfy
    _____ inequality. Graph x + y ≥ 2 on   either
    the left, x − 2y ≤ 6 in the center, and the
    union on the right.

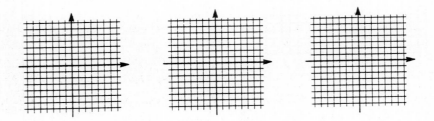

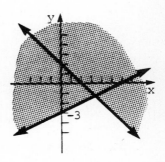

8. Graph the union of x > −2 and 2x − 5y < 10.

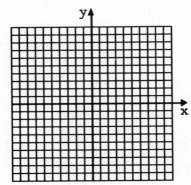

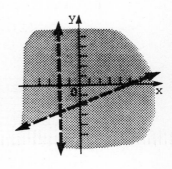

9. To graph |x| ≤ 2, use the equation of the bound-
    ary, _____, to determine x = _____ or |x| = 2; 2
    x = _____. Complete the graph.   −2

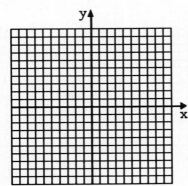

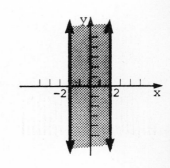

10. To graph $|y - 1| > 2$, first write

 _____ - _____ or _____ - _____,

$y - 1 = 2;$
$y - 1 = -2$

which leads to

 y - _____ or y - _____.

Complete the graph.

3; -1

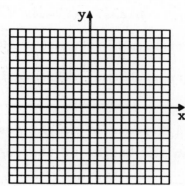

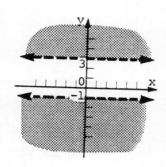

Graph each of the inequalities in Frames 11–12.

11. $|x + 3| < 2$

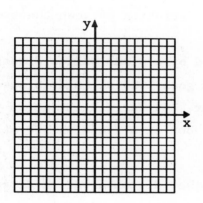

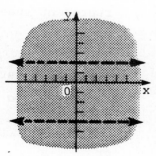

12. $|y + 1| \neq 3$

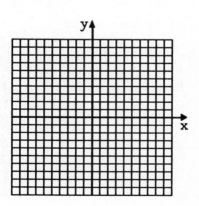

All points belong
to graph except
those on the
horizontal lines
$y = 2$ and $y = -4$.

7.5 Variation

[1] Write an equation expressing direct variation. (Frame 1)

[2] Find the constant of variation and solve direct variation problems. (Frames 2-3)

[3] Solve inverse variation problems. (Frames 4-7)

[4] Solve joint variation problems. (Frames 8-10)

[5] Solve combined variation problems. (Frames 11-14)

---

1. Suppose a puppy gains 1/2 pound of weight for each week of its life. Then the weight of the puppy is said to vary _____ as its age.

    | directly

    If y represents the weight of the puppy and x represents his age, then the equation _____ indicates the direct variation between the two.

    | $y = \frac{1}{2}x$

2. The distance traveled by a plane varies directly as its speed. If it goes 1000 miles at a speed of 400 miles per hour, how far will it go in the same amount of time, but at a speed of 500 miles per hour?

    To express the fact that distance varies directly as speed, we can let d represent distance and s represent speed, so that we can write

    $$d = \underline{\hspace{1cm}},$$

    | ks

    for some constant k. Replace d with _____ and s with _____ to get

    | 1000
    | 400

    $$\underline{\hspace{1.5cm}} = \underline{\hspace{1.5cm}} \cdot k,$$

    | 1000; 400

    or

    $$k = \underline{\hspace{1cm}}.$$

    | $\frac{5}{2}$

    Hence the relationship between distance and speed is given by

    $$d = \underline{\hspace{1cm}}.$$

    | $\frac{5}{2}s$

If s = 500 mph, we have

$$d = \left(\frac{5}{2}\right)(\underline{})$$ 500

$$d = \underline{} \text{ miles.}$$ 1250

3. **The distance a body falls from rest varies directly as the square of the time it falls. If an object falls 144 feet in 3 seconds, how far will it fall in 6 seconds?**

Here

$$\underline{} = \underline{}.$$ $d = kt^2$

If t = ____, then d = _____. Hence   3; 144

$$\underline{} = \underline{}$$ 144; $k \cdot 3^2$

or

$$k = \underline{}.$$ 16

Finally,

$$d = \underline{}.$$ $16t^2$

If t = 6, we have d = $16 \cdot 6^2$ = _____ feet.   576

4. **If x increases while y decreases, then x is said to vary _____ as y.**   inversely

5. **Suppose x varies inversely as y^2, and x = 18 when y = 2. Find x when y = 4.**

Start with

$$x = \underline{}$$ $\dfrac{k}{y^2}$

for some real number k. We know x = 18 when
y = ____, so that   2

$$\underline{} = \frac{k}{\underline{}}$$ 18; 2^2 or 4

from which   k = _____,   72

or

$$x = \frac{}{y^2}.$$ 72

When y = 4,

$$x = \frac{72}{(\underline{\quad})^2}$$

4

$$x = \underline{\quad}.$$

$\dfrac{9}{2}$

6. Suppose x varies inversely as y. If x = 12 when
    y = 6, find x if y = 4.

    We can write

    $$x = \underline{\quad}.$$

    $\dfrac{k}{y}$

    Next, find k: k = \underline{\quad}.   72

    If y = 4, then   x = \underline{\quad}.   18

7. The current in a simple electrical circuit is in-
    versely proportional to the resistance. If the
    current is 30 ohms when the resistance is 7.5 ohms,
    find the current when the resistance is 10 ohms.

    Let c represent the \underline{\quad}, and r the \underline{\quad}.   current; resistance

    The appropriate formula is \underline{\quad}.   c = k/r

    Find the value of the constant of variation.

    k = \underline{\quad}.   225

    Now, substitute k in the original formula with the
    new value for the resistance to find the new value
    for the \underline{\quad}.   current

    $$c = \frac{\underline{\quad}}{\underline{\quad}} = \underline{\quad}.$$ The new current is \underline{\quad}   225; 22.5; 22.5;
    ohms.   10

8. The situation where one variable varies as the
    product of several others is called \underline{\quad}   joint
    variation.

9. Suppose x varies jointly as y, z, and w. If
    x = 18 when y = 3, z = 2, and w = 4, find x
    if y = 4, z = 3, and w = 8. Here

    $$x = \underline{\quad} \cdot \underline{\quad} \cdot \underline{\quad} \cdot \underline{\quad}$$ k; y; z; w

for some real number k. If we substitute in the given values, we have

_____ = _____ · _____ · _____ · _____ ,

18; k; 3; 2; 4

or

k = _____ .

3/4

If y = 4, z = 3, and w = 8,

x = _____ · _____ · _____ · _____

3/4; 4; 3; 8

x = _____ .

72

10. The force of the wind blowing on a vertical surface varies jointly as the area of the surface and the square of the velocity. If a wind of 60 miles per hour exerts a force of 75 pounds on a surface of 3/4 square foot, how much force will a wind of 50 miles per hour place on a surface of a square feet?

Let F represent the force, A the surface area and v the velocity.

Then, F = _____ .

kAv^2

Solve for the constant of variation by substituting the original values.

_____ = k(___)(___), so k = _____ .

75; 3/4; 60^2; .0277$\overline{7}$

Now, use this value of k in the formula with the new values for area and velocity.

F = (___)(___)(___). The wind will exert a force of about _____ pounds.

.0277$\overline{7}$; 3; 50^2; 208.3

11. Suppose x varies jointly as y and the square of z and inversely as w. If x = 36 when y = 2, z = 3, and w = 4, find x if y = 3, z = 2, and w = 6.

Here

$$x = \dfrac{\underline{\qquad} \cdot \underline{\qquad}}{\underline{\qquad}}$$

for some real number k. If we substitute in the known values,

$$\underline{\qquad} = \dfrac{\underline{\qquad} \cdot \underline{\qquad}}{\underline{\qquad}}$$

or

$$k = \underline{\qquad}.$$

If y = 3, z = 2, and w = 6, x = _____.

12. Suppose x varies directly as the square of y and inversely as the product of z and w. If x = 10 when y = 5, z = 2, and w = 10, find x if y = 10, z = 5, and w = 2.

$$x = \underline{\qquad}$$

13. Suppose m varies directly as the square of r and the cube of s, and inversely as the square of t. If m = 20/9 when r = 1, s = 2, and t = 6, find m if r = 3, s = 1, and t = 5.

$$m = \underline{\qquad}$$

14. The force needed to keep a car from skidding on a curve varies inversely as the radius of the curve and jointly as the weight of the car and the square of the speed. If 4000 pounds of force keep a 2500-pound car from skidding on a curve of radius 600 feet at 40 miles per hour, what force would keep a 2000-pound car from skidding on a curve of radius 800 feet at 60 miles per hour?

Let F represent the _____, w represent the _____, s represent the _____ and r represent the _____.

---

k; y; z²
w

36; k; 2; 3²
4

8

16

80

$\dfrac{18}{5}$

force
weight; speed
radius

Then, the appropriate formula is _____.	$F = (kws^2)/r$
Substitute the original values and solve for the constant of variation. k = _____	.6
Now, use this value in the formula along with the new values for weight, speed and radius to determine the new force. F = _____	5400
Thus, it would take _____ pounds of force to keep the car from skidding.	5400

7.6   Introduction to Functions; Linear Functions

1 Define and identify relations and functions. (Frames 1–12)

2 Find the domain and range of a relation. (Frames 13–20)

3 Use the vertical line test. (Frames 21–24)

4 Use f(x) notation. (Frames 25–34)

5 Identify and graph linear functions. (Frames 35–40)

1. A set of ordered pairs is called a _____.	relation
2. Thus $\{(2, -1), (-3, 1), (0, 5)\}$ is an example of a _____.	relation
3. If the value of y depends on the value of x, then y is the _____ variable and x is the _____ variable.	dependent independent
4. The cost of gasoline depends on the number of miles driven. In this example, cost is the _____ variable, and the number of miles is the _____ variable.	dependent independent
5. A relation defined by the ordered pairs (x, y) in which for each value of x there is exactly one value of y is called a _____.	function

6. $\{(-2, 1), (-2, 2)\}$ *(is/is not)* a function, since
    to the x-value _____ there corresponds two
    y-values: _____ and _____ .

is not
-2
1; 2

**Identify any of the relations of Frames 7–12 which
represent functions.**

7. $x = y^2$ If $x = 16$, then $y =$ ____ or $y =$ ____ .
    This relation *(is/is not)* a function.

4; -4
is not

8. $y = 3x + 2$ For each x, there is _____ one
    value of y, so that the relation *(is/is not)* a
    function.

exactly
is

9. $y = x^2$ For each x, there is _____ one y,
    so that the relation *(is/is not)* a function.

exactly
is

10. $x = 2$ If $x = 2$, then y can take on ____ value,
    so that the relation *(is/is not)* a function.

any
is not

11. $y = -5$ For any value of x, there is _____
    one value of y; that is, $y =$ ____ . This relation
    (is/is not) a function.

exactly
-5
is

12. $y < 2x + 1$ For each value of x, there are
    _____ many values of y, so that this
    relation *(is/is not)* a function.

infinitely
is not

13. The set of first components of the ordered pairs
    of a relation is called the _____ of the
    relation. The domain of the relation in Frame 2
    above is $\{$_____$\}$.

domain
2, -3, 0

14. The set of second components of the ordered pairs
    of a relation is called the _____ of the re-
    lation. The range of the relation in Frame 2
    above is $\{$_____$\}$.

range
-1, 1, 5

15. To find the domain and range of the relation

$$y = x^2,$$

note that x can take on any value, so that the
domain of the relation is _____. $(-\infty, \infty)$
The square of a real number is never _____,   negative
so that the range is _____. $[0, \infty)$

**Find the domain and range of each of the relations in
Frames 16–18.**

16. $3x + 2y = 6$

       domain: _____ $(-\infty, \infty)$

        range: _____ $(-\infty, \infty)$

17. $x = 3 + 0y$

       domain: _____ $\{3\}$

        range: _____ $(-\infty, \infty)$

18. $4x - y \leq 6$

       domain: _____ $(-\infty, \infty)$

        range: _____ $(-\infty, \infty)$

**Use the graph to give the domain and range in
Frames 19–20.**

19.   20.

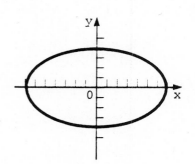   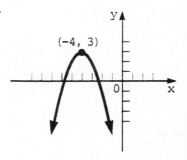

(-4, 3)

Frame 19:

$[-7, 7]$

$[-4, 4]$

Frame 20:

    domain: _____   domain: _____ $(-\infty, \infty)$

     range: _____   range: _____ $(-\infty, 3]$

21. The graph below is the graph of a relation. No
    vertical line cuts the graph in more than one
    point. Hence the graph (*is/is not*) the graph of
    a function.

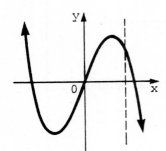

is

Identify any of the graphs in Frames 22—24 which are
the graphs of functions.

22.

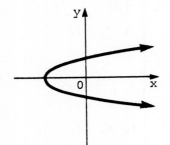

(*is/is not*)

23.

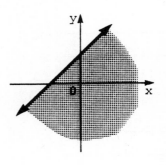

(*is/is not*)

is not; is not

24.

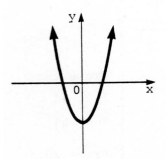

(*is/is not*)

is

25. To emphasize that y is the dependent variable and
     x the _____ variable, it is common to write   independent

                    y = _____   f(x)

     to show that y is a _____ of x.   function

26. For example, if f(x) = 5x − 7, then f(2) is the
     result when x is replaced with _____. Find f(2).   2

                 f(2) = 5(____) − 7   2

                      = _____   3

**In Frames 27–30, let f(x) = −x² + 5x − 9, and find the
indicated values.**

27. f(2) = −(____)² + 5(____) − 9   2; 2

          = _____ + 10 − 9   −4

          = _____   −3

28. f(−3) = _____   −33

29. f(5) = _____   −9

30. f(0) = _____   −9

31. If a function is given with both x and y, write
     the function with f(x) notation by first solving
     for _____. For example, to write 2x − 7y = 3   y
     with f(x) notation, first solve for y.

                 −7y = _____   −2x + 3

                   y = _____ $\frac{2}{7}x - \frac{3}{7}$

     Now replace y with _____.   f(x)

                 f(x) = _____ $\frac{2}{7}x - \frac{3}{7}$

In Frames 32-34, find an expression for f(x), and then find the indicated values.

32. x - 4y = 9, find f(2) and f(-1).

$$f(x) = _____$$

$$f(2) = \_\_\_\_\_, \quad f(-1) = \_\_\_\_\_$$

$\frac{1}{4}x - \frac{9}{4}$

-7/4; -5/2

33. y - x² = 3, find f(-2) and f(p).

$$f(x) = _____$$
$$f(-2) = \_\_\_\_\_$$
$$f(p) = _____$$

$x^2 + 3$

7

$p^2 + 3$

(To find f(p), replace each ____ with ____.)

x; p

34. y + x³ = 1, find f(2) and f(r).

$$f(x) = _____$$
$$f(2) = \_\_\_\_\_$$
$$f(r) = _____$$

$1 - x^3$

-7

$1 - r^3$

35. Straight lines are graphs of _____ functions. A linear function is a function that can be written in the form

$$f(x) = _____,$$

for real numbers m and b.

linear

mx + b

36. As we saw in Section 7.3, the value of ____ is the slope of the line, while (0, b) is the _____.

m

y-intercept

37. To write 2x - 5y = 4 as a linear function, we must solve for ____. First, subtract ____ from each side.

$$-5y = _____$$

Then divide both sides by ____.

$$y = _____$$

y; 2x

-2x + 4

-5

$\frac{2}{5}x - \frac{4}{5}$

Replace y with _____ to get

$$\text{_____} = \frac{2}{5}x - \frac{4}{5}.$$

f(x)

f(x)

38. The slope of the line in Frame 37 is _____ and the
y-intercept is _____. Graph the line.

2/5

(0, -4/5)

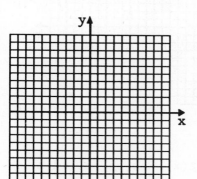

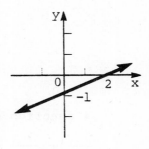

39. Write x - 2y = 0 in the form f(x) = mx + b and
graph the line.

$$f(x) = \text{_____}$$

The slope is _____ and the y-intercept is _____.

$\frac{1}{2}x$

1/2; (0, 0)

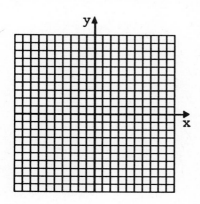

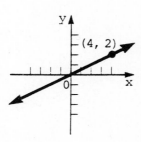

40. Write y + 4 = 0 in the form f(x) = mx + b.

$$f(x) = \text{_____}$$

-4

The slope is _____, and the y-intercept is _____.   0; (0, -4)
Graph the line.

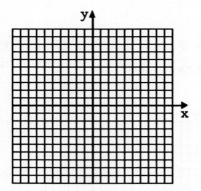

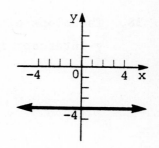

A function of the form f(x) = k is called a

_____ function.   constant

Chapter 7 Test

The answers for these questions are at the back of this Study Guide.

1. Find the slope of the line through (-1, 7) and
   (8, -3).   1. _____

For each line, find the slope and the x- and y-intercepts.

2. 4x + 7y = 8   2. _____

3. 2x - 5y = 1   3. _____

4. x = 5   4. _____

5. y + 2 = 0   5. _____

Find an equation for each of the lines in questions 6-11.

6. m = 3/2, goes through (-1, 3)   6. _____

7. m = -3, y-intercept (0, 6)   7. _____

8. slope undefined, goes through (8, -2)   8. _____

9. through (-1, 2) and (1, 5)   9. _____

10. through (-8, 9) and (4, 1)   10. _____

11. parallel to 2x + y = 3, through (-3, 4)   11. _____

12. Find the distance between (-5, 8) and (3, -2).   12. _____

13. The city roads can be cleared by 8 snowplows
     in three hours. If 12 snowplows are used, the
     roads can be cleared in 2 hours. Write an equa-
     tion that gives the number of hours, y, needed
     to clear the roads, by the number of snowplows,
     x. How long would it take 11 snowplows to
     clear the roads?   13. _____

                                                            _____

**Use slope to determine if the following pairs of lines are parallel,
perpendicular or neither.**

14. $3y - 12x = 6$ and $4x + y = -3$ 14. _____

15. $2x - y = -3$ and $x + 2y = 8$ 15. _____

16. During a recent thunderstorm, .1 inch of rain
     fell the first hour. After 4 hours, 1.3 inches
     of rain had fallen. What was the average rate
     of change in rainfall each hour?   16. _____

Graph each of the following.

17. $3x + 2y = 6$ 18. $3y + 4x = 12$ 19. $x = 6$

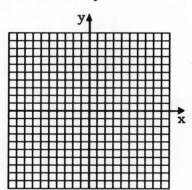
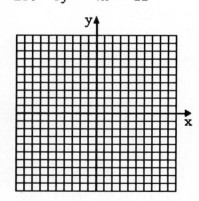
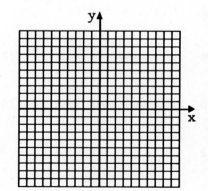

20. 3x + 2y = 0

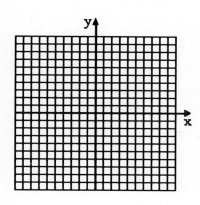

21. The line with
     slope 2, through
     (-1, 4)

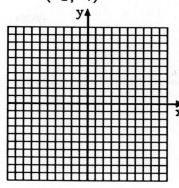

22. The line with
     slope 0, through
     (3, -2)

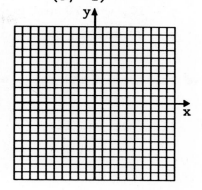

23. 5x - 2y ≤ 10

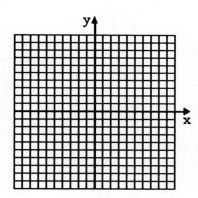

24. Find the union
     of 4x - 2y > 6
     and x + y < 1.

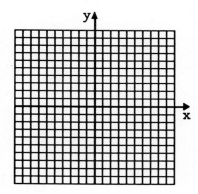

25. Find the inter-
     section of
     y - 5x > -5
     and x ≥ 0.

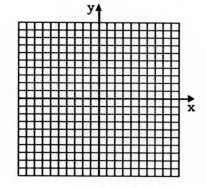

26. Suppose x varies directly as y and inversely
     as z. If x = 12 when y = 8 and z = 5, find x
     if y = 10 and z = 15.   26. _____

27. The distance traveled by a freely falling
     body varies directly as the square of the
     number of seconds of fall. A body falls
     256 feet in 4 seconds. Find the distance
     it would fall in 10 seconds.   27. _____

28. Suppose x varies inversely as y and jointly
    as w and the square of z. If x = 32 when
    y = 10, w = 10 and z = 4, find x when y = 32,
    w = 20 and z = 8.   28. _____

Give the domain and range of each relation. Identify any functions.

29. $\{(2, 5), (3, 11), (5, 17), (9, 11)\}$ 29. _____

30. $x = y^2 - 4$ 30. _____

31.   32.

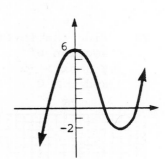   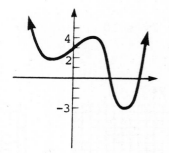

33. Graph $|x - 1| \leq 3$.   34. Graph $f(x) = 2x - 3$.

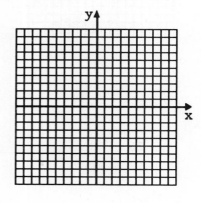   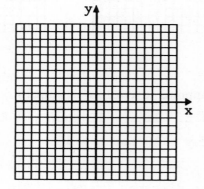

35. Given $3y + 2x = 6$, find an expression for $f(x)$.
    Then find $f(2)$ and $f(j)$.   35. _____

                                                            _____

                                                            _____

CHAPTER 8   GRAPHING RELATIONS AND FUNCTIONS

8.1   The Algebra of Functions

1 Form a new function by adding, subtracting, multiplying, or dividing functions. (Frames 1–8)

2 Determine the domains of functions formed as in Objective 1. (Frames 9–13)

3 Form composite functions. (Frames 14–24)

---

1. Given two functions f and g, we can use the

   _____ on functions to get four other

   functions. By definition,

   $$(f + g)(x) = \underline{\hspace{1cm}} + \underline{\hspace{1cm}}$$

   $$(f - g)(x) = \underline{\hspace{2cm}}$$

   $$(fg)(x) = \underline{\hspace{2cm}}$$

   $$\left(\frac{f}{g}\right)(x) = \underline{\hspace{1.5cm}}, \text{ if } g(x) \neq 0.$$

   | |
   | operations |
   | $f(x)$; $g(x)$ |
   | $f(x) - g(x)$ |
   | $f(x) \cdot g(x)$ |
   | $\dfrac{f(x)}{g(x)}$ |

In Frames 2–8, let $f(x) = 2 + 3x$, and $g(x) = x^2 + x$.
Find the indicated values.

2. $(f + g)(2)$

   Find f(2) and _____.

   $f(2) = \underline{\hspace{1cm}}$ and $g(2) = \underline{\hspace{1cm}}$

   Then $(f + g)(2) = f(2) + g(2)$

   $= \underline{\hspace{1cm}} + \underline{\hspace{1cm}}$

   $= \underline{\hspace{1cm}}.$

   | |
   | $g(2)$ |
   | 8; 6 |
   | |
   | 8; 6 |
   | 14 |

3. $\left(\dfrac{f}{g}\right)(2) = \underline{\hspace{2cm}}$

   $\dfrac{8}{6} = \dfrac{4}{3}$

4. $(fg)(-1) = f(-1) \cdot \underline{\hspace{1.5cm}}$

   $= (\underline{\hspace{1cm}}) \cdot (\underline{\hspace{1cm}})$

   $= \underline{\hspace{1cm}}$

   | |
   | $g(-1)$ |
   | -1; 0 |
   | 0 |

5. $(f - g)(4) = \underline{\hspace{1cm}}$

   -6

6. $\left(\dfrac{f}{g}\right)(-2) = \underline{\hspace{1cm}}$

   -2

7. $\left(\dfrac{f}{g}\right)(3r) =$ _____

$\dfrac{2 + 9r}{9r^2 + 3r}$

8. $(fg)(m - 1) =$ _____

$(3m - 1)(m^2 - m)$, or

$3m^3 - 4m^2 + m$

9. For functions f and g, the domains of f + g,
    f − g, and fg include all real numbers in the
    _____ of the domains of f and g, while

intersection

    the domain of f/g includes those real numbers
    in the _____ of the domains of f and g

intersection

    for which $g(x) \neq$ _____ .

0

In Frames 10–13, let $f(x) = \sqrt{5 + 2x}$ and $g(x) = x^2 - 4$.

10. The domain of f(x) includes just those real
    numbers that make $5 + 2x \geq$ _____ ; the domain

0

    of f is _____ .

$[-\dfrac{5}{2}, \infty)$

11. The domain of g(x) is _____ , or

the set of all
real numbers

    written as an interval, _____ .

$(-\infty, \infty)$

12. The domain of f + g, f − g, and fg is _____ .

$[-\dfrac{5}{2}, \infty)$

13. The domain of f/g is _____ because

$[-\dfrac{5}{2}, -2) \cup (-2, 2) \cup (2, \infty)$

    the values _____ and _____ make g(x) = 0.

−2, 2

14. The symbol $(f \circ g)(x)$ denotes the _____

composite function, or
composition

    of f and g.

            $(f \circ g)(x) =$ _____

f[g(x)]

In Frames 15–18 let $f(x) = 7 + 3x$ and $g(x) = x^2 - 2$.
Find the indicated values.

15. $(f \circ g)(1)$

    First find _____ .

g(1)

        $g(1) = ($___$)^2 - 2 =$ _____

1; −1

    Now,

        $f($___$) = 7 + 3($___$) =$ _____ .

−1; −1; 4

16. $(g \circ f)(1)$

First, $f(1) =$ _____ .

Then, $g(\_\_\_) =$ _____ .

10

10; 98

17. $(f \circ f)(2) = f(\_\_\_) =$ _____

13; 46

18. $(g \circ f)(0) =$ _____

47

In Frames 19–20, let $f(x) = 3x + 2$ and $g(x) = x^2 - 5$. Find the indicated functions.

19. $(g \circ f)(x) = g[\_\_\_\_\_]$

= $g(_____)$

= $(_____)^2 - 5$

= _____ $- 5$

= _____ .

f(x)

3x + 2

3x + 2

$9x^2 + 12x + 4$

$9x^2 + 12x - 1$

20. $(f \circ g)(x) = f[\_\_\_\_\_]$

= $f(_____)$

= $3(_____) + 2$

= _____ $+ 2$

= _____ .

g(x)

$x^2 - 5$

$x^2 - 5$

$3x^2 - 15$

$3x^2 - 13$

21. As shown in Frames 19 and 20, $(g \circ f)(x)$ (*is/is not*) always equal to $(f \circ g)(x)$.

is not

In Frames 22–24, h(x) is given. Find functions f and g such that $h(x) = (f \circ g)(x)$. Many such pairs of functions exist so other correct answers are possible.

22. $h(x) = (5x - 1)^2$

f(x) = _____   g(x) = _____

x^2; 5x - 1

23. $h(x) = \sqrt{x + 3} + x + 3$

f(x) = _____   g(x) = _____

$\sqrt{x} + x$; x + 3

24. $h(x) = \dfrac{1 + 2x}{2x + 3}$

f(x) = _____   g(x) = _____

$\dfrac{1 + x}{x + 3}$; 2x

8.2 Quadratic Functions; Parabolas

1. Graph $f(x) = x^2$. (Frames 1 and 2)

2. Graph $f(x) = ax^2$. (Frame 3)

3. Graph translations of $f(x) = ax^2$. (Frames 4 and 5)

4. Graph $f(x) = a(x - h)^2 + k$. (Frames 6–10)

5. Use the geometric definition of a parabola. (Frames 11–15)

---

1. A function f is a quadratic function if
    $f(x) =$ _____, where a, b, and c are
    real numbers with a $\neq$ ____.

    $ax^2 + bx + c$

    0

2. To sketch the graph of $y = x^2$, first complete
    a table of values, plot the ordered pairs, and
    then draw the graph.

x	y
-3	____
-2	____
-1	____
0	____
1	____
2	____
3	____

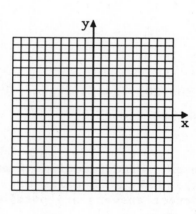

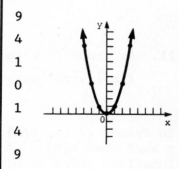

9
4
1
0
1
4
9

This graph is called a _____. The bottom
point on this graph is called the _____.
The vertex here is (____, ____).

parabola
vertex
0; 0

3. To sketch the graph of $g(x) = -2x^2$, first com-
    plete the table of values, and then draw the
    graph.

x	y
-2	___
-1	___
0	___
1	___
2	___

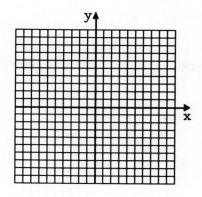

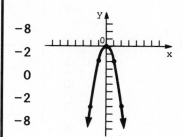

-8	
-2	
0	
-2	
-8	

This parabola opens _____ because
the coefficient of x² is _____.
The absolute value of the coefficient, or
_____, will cause the resulting value of
g(x) to be greater than for f(x) — x², making
the parabola of g(x) _____ than the
graph of f(x) = x². The vertex here is
(_____, _____).

downward
negative

2

narrower

0; 0

4. Sketch the graph of y = x² + 3.

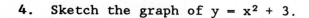

x	y
-2	___
-1	___
0	___
1	___
2	___

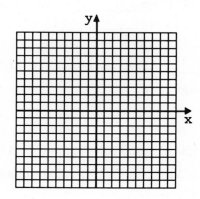

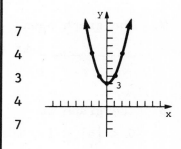

7	
4	
3	
4	
7	

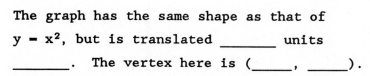

The graph has the same shape as that of
y — x², but is translated _____ units
_____. The vertex here is (_____, _____).

3

up; 0; 3

5. Sketch the graph of $y = (x - 3)^2$.

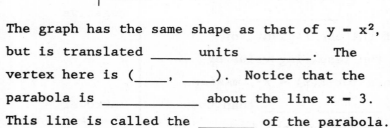

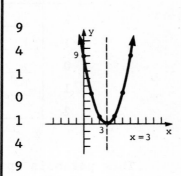

x	y
0	____
1	____
2	____
3	____
4	____
5	____
6	____

The graph has the same shape as that of $y = x^2$,
but is translated _____ units _____. The
vertex here is (____, ____). Notice that the
parabola is _____ about the line x = 3.
This line is called the _____ of the parabola.

3; right
3; 0
symmetrical
axis

6. In general, the graph of $f(x) = a(x - h)^2 + k$,
where $a \neq 0$, is a _____ with _____
(h, k), and the vertical line _____ as axis.

parabola; vertex
x = h

7. The graph opens _____ if a is positive and
_____ if a is negative.

upward
downward

8. The graph is _____ than that of $f(x) = x^2$ if
$0 < |a| < 1$. The graph is _____ than that
of $f(x) = x^2$ if $|a| > 1$.

wider
narrower

9. Graph $f(x) = 2(x + 1)^2 - 3$.
The parabola opens _____. It is
(*wider/narrower*) than the graph of $f(x) = x^2$.
It is translated _____ unit(s) down and _____
unit(s) left. The vertex is _____. The
axis is _____.

upward
narrower
3; 1
(-1, -3)
x = -1

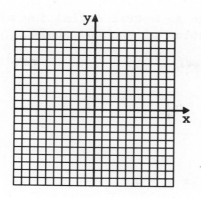

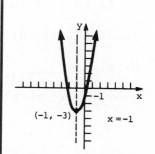

10. Graph $f(x) = -(x - 2)^2 + 2$.

The parabola opens _____. It is translated
_____ units up and _____ units right. The vertex
is _____.

downward

2; 2

(2, 2)

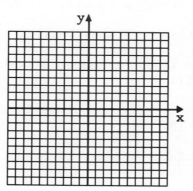

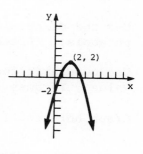

11. Geometrically, a parabola is the set of all
points in a plane that are equally distant from
a fixed point, the _____, and a fixed line,
the _____. The _____ is the line
through the _____ and perpendicular to the
_____.

focus

directrix; axis

focus

directrix

12. The _____ is the point on the axis that is
equally distant from the focus and the directrix.

vertex

13. A parabola with the point (0, p) as focus and
the line y = -p as directrix has the equation
$x^2 =$ _____ or y = _____.

$4py$; $\frac{1}{4p}x^2$

In **Frames 14 and 15**, use the geometric definition of
a parabola to find the equation of each parabola.

14.   focus (0, 5), directrix y = -5

Here, p = ____, so y = $\frac{1}{4p}x^2$ becomes

y = ____x² or x² = ____.

| 5 |
| $\frac{1}{20}$; 20y |

15.   focus (0, -3), directrix y = 3

Here p = ____, so y = ____ or x² = ____.

$-3; -\frac{1}{12}x^2; -12y$

8.3   More About Parabolas and Their Applications

1. Find the vertex of a vertical parabola. (Frames 1-7)

2. Graph a quadratic function. (Frames 8-12)

3. Use the discriminant to find the number of x-intercepts of a vertical
parabola. (Frames 13-17)

4. Use quadratic functions to solve problems involving maximum or minimum
value. (Frames 18 and 19)

5. Graph horizontal parabolas. (Frames 20-24)

---

1.   Find the vertex of the graph of f(x) = x² - 6x + 8.
Begin by completing the _____. Simplify the
notation by letting f(x) = ____.

                y = _____

Get the constant term on the left side.

        y - ____ = _____

Half of ____ is ____; (___)² = ____. Add ____
to both sides.

        y + ____ = _____

Factoring on the right gives

        y + 1 = (_____)².

Finally,

            y = _____.

The vertex is _____, as shown in Section 8.2.

| square |
| y |
| x² - 6x + 8 |
| 8; x² - 6x |
| -6; -3; -3; 9; 9 |
| 1; x² - 6x + 9 |
| x - 3 |
| (x - 3)² - 1 |
| (3, -1) |

2. Find the vertex of the graph of $f(x) = 2x^2 + 12x + 5$.

Start with y = _____.

Divide both sides by ____.

$$\frac{y}{\rule{1cm}{0.4pt}} = \rule{3cm}{0.4pt}$$

Move the constant term to the left.

$$\frac{y}{2} - \rule{1.5cm}{0.4pt} = x^2 + 6x$$

Half of ____ is ____; $(\rule{0.6cm}{0.4pt})^2 = \rule{0.8cm}{0.4pt}$. Add ____ to both sides.

$$\frac{y}{2} + \rule{1cm}{0.4pt} = \rule{3cm}{0.4pt}$$

Factor on the right side and subtract _____ from both sides.

$$\frac{y}{2} = \rule{2cm}{0.4pt} - \frac{13}{2}$$

Finally, multiply by ____. y = _____

The vertex is _____.

$2x^2 + 12x + 5$
2
$2;\ x^2 + 6x + \dfrac{5}{2}$
$\dfrac{5}{2}$
6; 3; 3; 9; 9
$\dfrac{13}{2};\ x^2 + 6x + 9$ $\dfrac{13}{2}$
$(x + 3)^2$
$2;\ 2(x + 3)^2 - 13$ $(-3,\ 13)$

3. By completing the square for $y = ax^2 + bx + c$, we find that $y = a[x - (\rule{1.5cm}{0.4pt})]^2 + \rule{2cm}{0.4pt}$.

$$\frac{-b}{2a};\ \frac{4ac - b^2}{4a}$$

4. From Section 8.2, when $y = a(x - h)^2 + k$, the vertex of the parabola is _____. Thus, the vertex of the parabola $y = ax^2 + bx + c$ is _____. The expression for the y-coordinate is found by replacing x with _____. Thus, the vertex may be written $\left[\dfrac{-b}{2a},\ f\left(\rule{1cm}{0.4pt}\right)\right]$.

$(h,\ k)$

$\left(\dfrac{-b}{2a},\ \dfrac{4ac - b^2}{4a}\right)$

$\dfrac{-b}{2a}$

$\dfrac{-b}{2a}$

In Frames 5–7, use the vertex formula to find the vertex of the graph of the parabola.

5. $f(x) = x^2 + 6x - 1$

a = ____   b = ____   c = ____

$$\frac{-b}{2a} = \frac{\rule{1cm}{0.4pt}}{2} = \rule{1cm}{0.4pt}$$

$f(\rule{0.8cm}{0.4pt}) = (\rule{0.8cm}{0.4pt})^2 + 6(\rule{0.8cm}{0.4pt}) - 1 = \rule{1cm}{0.4pt}$

The vertex is _____.

1; 6; −1

−6; −3

−3; −3; −3; −10

(−3, −10)

6. $f(x) = 2x^2 - 4x + 3$

$\dfrac{-b}{2a} = $ _____   → 1

$f\left(\dfrac{-b}{2a}\right) = f(\text{___}) = \text{___}$ → 1; 1

The vertex is _____.   → (1, 1)

7. $f(x) = -2x^2 - 3x + 5$

$\dfrac{-b}{2a} = $ _____   → $-\dfrac{3}{4}$

$f(\text{___}) = -2(\text{___})^2 - 3(\text{___}) + 5$ → $-\dfrac{3}{4}$; $-\dfrac{3}{4}$; $-\dfrac{3}{4}$

$= \text{___} + \text{___} + \text{___}$ → $-\dfrac{9}{8}$; $\dfrac{9}{4}$; 5

$= \text{___}$ → $\dfrac{49}{8}$

The vertex is _____.   → $\left(-\dfrac{3}{4}, \dfrac{49}{8}\right)$

8. The vertex, along with the y-_____ and x-_____, is useful in graphing parabolas.   → intercept; intercept

9. Find the y-intercept of $f(x) = y$ by evaluating _____.   → $f(0)$

Find the x-intercept by solving _____.   → $f(x) = 0$

10. Use the formula or complete the square to find the _____.   → vertex

Find the following information for each parabola in Frames 11 and 12. Then graph the parabola.

11. Graph $y = x^2 + 4x$.

The y-intercept is _____.   → (0, 0)

The x-intercepts are _____ and _____.   → (0, 0); (-4, 0)

The vertex is _____.   → (-2, -4)

The parabola opens _____.   → upward

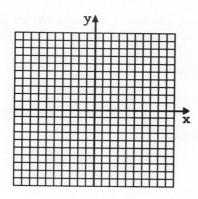

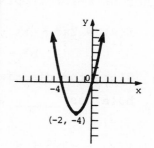

12. Graph $y = -2x^2 - 4x + 3$.

The y-intercept is _____ .   (0, 3)

The x-intercepts are _____ and _____ . $\left(-1 + \dfrac{\sqrt{10}}{2}, 0\right)$;

$\left(-1 - \dfrac{\sqrt{10}}{2}, 0\right)$

The vertex is _____ .   (-1, 5)

The parabola opens _____ .   downward

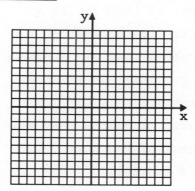

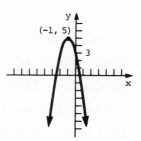

13. For the parabola $y = ax^2 + bx + c$, the discriminant of the quadratic equation $ax^2 + bx + c = 0$
can be used to decide how many _____   x-intercepts
the graph of the parabola has.

14. If the discriminant is _____ , the parabola   positive
will have two x-intercepts. If the discrimi-
nant is _____ , the parabola will have one   zero
x-intercept. If the discriminant is _____ ,   negative
the parabola will have no x-intercepts.

In Frames 15–17, use the discriminant to determine the
number of x-intercepts of the graph of each parabola.

15. $f(x) = 2x^2 - 6x + 3$

$$b^2 - 4ac = \underline{\hspace{1cm}}$$

Since the discriminant is _____, the parabola has _____ x-intercept(s).

	12
	positive
	two

16. $f(x) = -3x^2 + x - 5$

$$b^2 - 4ac = \underline{\hspace{1cm}}$$

Since the discriminant is _____, the parabola has _____ x-intercept(s).

	-59
	negative
	no

17. $f(x) = 2x^2 - 8x + 8$

$$b^2 - 4ac = \underline{\hspace{1cm}}$$

Since the discriminant is ____, the parabola
has _____ x-intercept(s).

	0
	0
	no

18.   The number, in thousands, of ladybugs in a
certain area of North Carolina depends on the
number of quarts of ladybugs that are stocked.
If the number of ladybugs, T(x), is given by
$T(x) = 8x - x^2$, where x represents the number
of quarts, find the number of quarts that will
produce the maximum number of ladybugs. What
is the maximum number of ladybugs?

First, we recognize that the graph of the
parabola $T(x) = 8x - x^2$ opens _____. Thus,
the vertex will be the _____ point of the
parabola $T(x) = 8x - x^2$.

	downward
	highest

Complete the square.

$$T(x) = -(\underline{\hspace{1.5cm}}) = -(x - 4)^2 + 16.$$

The vertex is _____. The number of quarts
that will produce the maximum number of ladybugs
is the x-value of the vertex. Thus _____ quarts
will produce the maximum number of ladybugs.
The maximum number in thousands of ladybugs is

_____.

	$x^2 - 8x$
	(4, 16)
	4
	16

19. The profit of a 500-room hotel during a 3-day convention depends on the number of unoccupied rooms. If the profit, P(x), is given by $P(x) = 10,000 + 100x - x^2$, where x is the number of unoccupied rooms, find the maximum profit and the number of unoccupied rooms which produce maximum profit.
_____

$12,500 with 50 unoccupied rooms

20. The graph of $y = x^2$ is a _____ opening _____. Exchanging x and y gives _____, a parabola opening to the _____.

parabola upward; $x = y^2$ right

21. The vertex of $x = -y^2$ is _____. Graph this parabola on the grid below.

(0, 0)

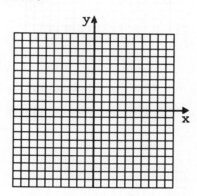

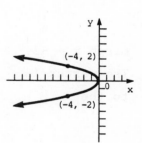

22. Graph $x = (y + 2)^2 - 3$ on the grid below. The vertex is _____. (Remember to reverse the ordered pair.)

(-3, -2)

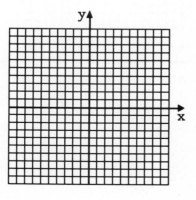

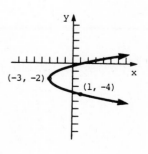

23. Graph $x = 2y^2 + 4y + 2$.

    Complete the square on $2y^2 + 4y$.

    $$\frac{x}{2} = \underline{\hspace{2cm}}$$ $y^2 + 2y + 1$

    $$x = 2(y + 1)^2$$

    The vertex is (____, ____). (Remember to reverse   0; −1
    the ordered pair.) The parabola opens to the
    _____ because the 2 is positive. Sketch the   right
    graph.

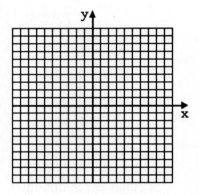

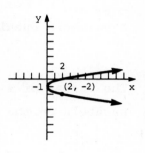

24. Graph $x = -2y^2 + 4y + 3$ on the following grid.
    The vertex is _____, and the parabola opens   (5, 1)
    to the _____.   left

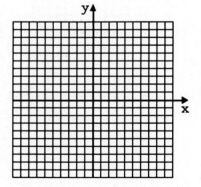

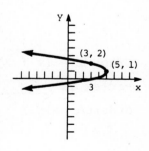

8.4 Symmetry; Increasing/Decreasing Functions

1️⃣ Test for symmetry with respect to an axis. (Frames 1–7)

2️⃣ Test for symmetry with respect to the origin. (Frames 8–10)

3️⃣ Decide if a function is increasing or decreasing on an interval. (Frames 11–13)

---

1. A graph is symmetric with respect to the y-axis if the replacement of ____ with ____ results in an equivalent equation.	x; -x
2. A graph is symmetric with respect to the x-axis if the replacement of ____ with ____ results in in a equivalent equation.	y; -y

In Frames 3–5, test for symmetry with respect to the x-axis or y-axis.

3. $y = x^4 - 1$ Replace x with _____. $y = (\underline{})^4 - 1$ $= \underline{}$ The graph is symmetric with respect to the _____.	-x -x $x^4 - 1$ y-axis
4. $x + y = 3$ Replace x with -x: _____ Replace y with -y: _____ Neither equation is equivalent to the original. This graph is not symmetric with respect to either axis.	 $-x + y = 3$ $x + (-y) = 3$
5. $y^2 + x = 4$ Replace x with -x: _____ $= 4$ _____ $= 4$ *Simplify*	 $y^2 + (-x)$ $y^2 - x$

Replace y with −y: _____ − 4 $(-y)^2 + x$

                       _____ − 4 *Simplify* $y^2 + x$

This graph is symmetric with respect to the

_____.   x−axis

6.

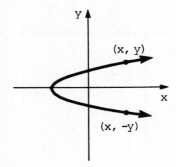

This graph is symmetric with respect to the

_____.   x−axis

7.

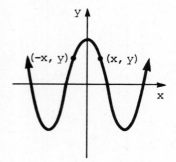

This graph is symmetric with respect to the

_____.   y−axis

8. A graph is symmetric with respect to the _____   origin
    if the replacement of both x with _____ and y   −x
    with _____ results in an equivalent equation.   −y

9. Test $x^2 + 3y^2 = 5$ for symmetry with respect to
    the origin.

         Replace x with −x and y with −y. The graph
    (*is/is not*) symmetric with respect to the origin.   is

10.

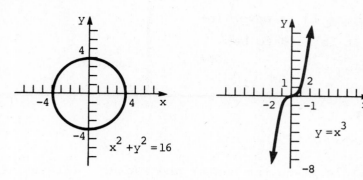

Both of these graphs are symmetric with respect
to the _____.

origin

11. A function is said to be increasing on an interval if the graph goes _____ from left to right,   up
and decreasing if the graph goes _____ from   down
left to right. If a function neither increases
nor decreases on an interval, it is said to be
_____ on that interval.   constant

12.

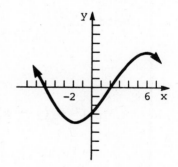

The function graphed above increases on the domain
interval _____. It is decreasing on the in-   (−2, 6)
tervals _____ and _____.   (−∞, −2); (6, ∞)

13.

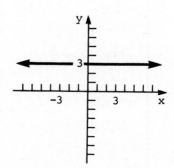

The function graphed in Frame 13 is never increasing or decreasing. It is said to be _____ on its domain.	constant

8.5   The Circle and the Ellipse

1. Find the equation of a circle given the center and radius. (Frames 1-4)

2. Determine the center and radius of a circle given its equation. (Frames 5-10)

3. Find the equation of a circle given information other than the center and radius. (Frames 11 and 12)

4. Recognize the equation of an ellipse. (Frames 13 and 14)

5. Graph ellipses. (Frames 15-18)

1. A _____ is the set of all points in a plane which lie a fixed distance from a fixed point. The fixed distance is called the _____, and the fixed point is called the _____.	circle radius center
2. An equation of the circle with radius r and center at (h, k) is $\rule{3cm}{0.4pt} = r^2.$	$(x - h)^2 + (y - k)^2$

Find an equation of each circle in Frames 3 and 4.

3. Center (-1, 7), radius 2 $\rule{3cm}{0.4pt}$	$(x + 1)^2 + (y - 7)^2 = 4$
4. Center at the origin, radius $\frac{1}{3}$ $\rule{3cm}{0.4pt}$	$x^2 + y^2 = \frac{1}{9}$
5. The equation $x^2 + y^2 = 25$ represents a _____ of radius _____, with center at (_____, _____). Graph the circle.	circle 5; 0; 0

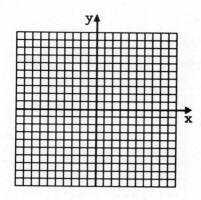

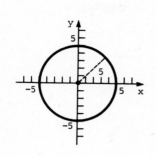

6. The equation $(x - 5)^2 + (y + 3)^2 = 9$ represents
a circle with center at (____, ____) and radius
_____. Graph the circle.

5; −3

3

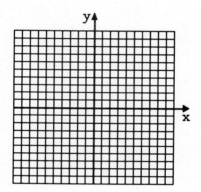

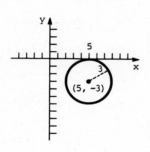

Graph the circles of Frames 7 and 8.

7. $(x + 2)^2 + (y - 1)^2 = 16$
center (____, ____) radius ____

−2; 1; 4

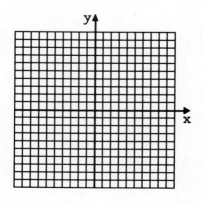

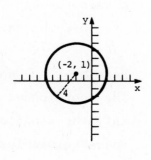

8. $(x - 1)^2 + (y - 3)^2 = 4$

center (____, ____) radius ____   1; 3; 2

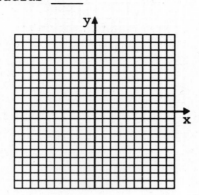

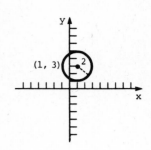

9. To find the center and radius of

$$x^2 + 6x + y^2 - 4y = -4,$$

we must _____ the square. Work as follows.   complete

(Remember: Take _____ the coefficient of x   half

or y and square.)

$$(x^2 + 6x + \underline{\quad}) + (y^2 - 4y + \underline{\quad})$$ 9; 4

$$= -4 + \underline{\quad} + \underline{\quad}$$ 9; 4

Now factor.

$$(\underline{\hspace{2cm}})^2 + (\underline{\hspace{2cm}})^2 = \underline{\quad}$$ x + 3; y - 2; 9

The center is _____, and the radius is ____.   (-3, 2); 3

10. Find the center and radius of

$$x^2 + y^2 - 5x + 8y + 2 = 0.$$

center: _____ radius: _____ $(\frac{5}{2}, -4)$; $\frac{9}{2}$

11. Find the equation of a circle that satisfies

the conditions that its center is (-1, -3) and

passes through the point (2, 1).

Use the equation

$$(x - h)^2 + (y - k)^2 = r^2.$$

Here h = ____ and k = ____ and r must be found.   -1; -3

The coordinates x = ____ and y = ____ must sat-   2; 1

isfy the equation of the circle. We substitute

these values into the equation given above.

$$[2 - (-1)]^2 + [1 - (-3)]^2 = r^2$$

$$\underline{\hspace{2cm}}^2 + \underline{\hspace{2cm}}^2 = r^2$$ $\quad$ 3; 4

$$25 = r^2$$

$$r = \underline{\hspace{1cm}}$$ $\quad$ 5

and so the equation is $\underline{\hspace{4cm}}$. $\quad (x + 1)^2 + (y + 3)^2 = 25$

12. Find the equation of a circle which has endpoints
     with a diameter at (2, -4) and (-4, 6).

     The coordinates of the center of the circle
     are obtained by finding the midpoint of the
     diameter. The center has coordinates $\underline{\hspace{2cm}}$. $\quad$ (-1, 1)
     Find the radius by determining the distance be-
     tween one endpoint, say (2, -4), and $\underline{\hspace{2cm}}$. $\quad$ (-1, 1)

$$r = \sqrt{(2 - \underline{\hspace{0.6cm}})^2 + (-4 - \underline{\hspace{0.6cm}})^2}$$ $\quad$ -1; 1

$$= \sqrt{\underline{\hspace{0.6cm}}^2 + \underline{\hspace{0.6cm}}^2}$$ $\quad$ 3; -5

$$= \sqrt{9 + 25}$$

$$= \sqrt{34}$$

So the equation is $\underline{\hspace{4cm}}$. $\quad (x + 1)^2 + (y + 1)^2 = 34$

13. An $\underline{\hspace{2cm}}$ is the set of all points on a $\quad$ ellipse
     plane the sum of whose distance from two fixed
     points is constant. The two fixed points are
     called $\underline{\hspace{1.5cm}}$. $\quad$ foci

14. The ellipse whose x-intercepts are (a, 0) and
     (-a, 0) and whose y-intercepts are (0, b) and
     (0, -b) has an equation given by

     $\underline{\hspace{3cm}}$. $\quad \dfrac{x^2}{a^2} + \dfrac{y^2}{b^2} = 1$

15. $\dfrac{x^2}{16} + \dfrac{y^2}{4} = 1$ is an $\underline{\hspace{1.5cm}}$ having x-intercepts $\quad$ ellipse
     $\underline{\hspace{1cm}}$ and $\underline{\hspace{1cm}}$, and y-intercepts $\underline{\hspace{1cm}}$ and $\quad$ (4, 0); (-4, 0); (0, 2)
     $\underline{\hspace{1cm}}$. Graph the ellipse. $\quad$ (0, -2)

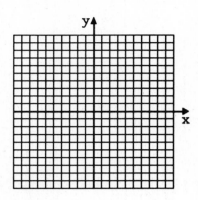

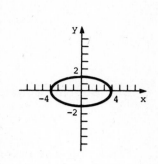

Graph the ellipses of Frames 16 and 17.

16. $\dfrac{x^2}{25} + \dfrac{y^2}{9} = 1$

        x-intercepts: _____ and _____   (5, 0); (-5, 0)

        y-intercepts: _____ and _____   (0, 3); (0, -3)

17. $16x^2 + 9y^2 = 144$

        x-intercepts: _____ and _____   (3, 0); (-3, 0)

        y-intercepts: _____ and _____   (0, 4); (0, -4)

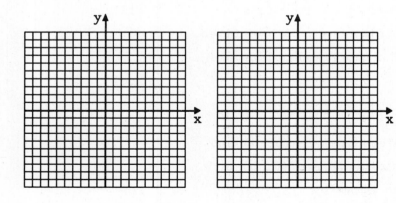

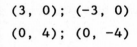

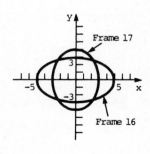

18. The graph of $\dfrac{(x-2)^2}{9} + \dfrac{(y+3)^2}{4} = 1$ is an ellipse with center at _____. Graph this ellipse.   (2, -3)

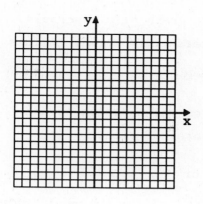

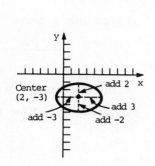

8.6 The Hyperbola; Square Root Functions

1️⃣ Recognize the equation of a hyperbola. (Frames 1–4)

2️⃣ Graph hyperbolas by using the asymptotes. (Frames 5–9)

3️⃣ Identify conic sections by their equations. (Frames 10–19)

4️⃣ Graph square root functions. (Frames 20–24)

---

1. A _____ is the set of all points in a plane such that the absolute value of the difference of the distances from two fixed points is constant.

    hyperbola

2. Thus $\frac{x^2}{9} - \frac{y^2}{16} = 1$ is a _____ having x-intercepts _____ and _____. The graph has _____ y-intercepts.

    hyperbola
    (3, 0); (-3, 0)
    no

3. A hyperbola with x-intercepts (a, 0) and (-a, 0) has an equation of the form

    $$\underline{\hspace{1cm}} - \underline{\hspace{1cm}} = 1.$$

    $\frac{x^2}{a^2}$; $\frac{y^2}{b^2}$

4. A hyperbola with y-intercepts (0, b) and (0, -b) has an equation of the form

    $$\underline{\hspace{1cm}} - \underline{\hspace{1cm}} = 1.$$

    $\frac{y^2}{b^2}$; $\frac{x^2}{a^2}$

5. The two branches of the graph of a hyperbola
    approach a pair of intersecting straight lines
    called _____.   asymptotes

6. Find the asymptotes from the equation of the
    hyperbola. The asymptotes of either

    $$\frac{x^2}{a^2} - \frac{y^2}{b^2} = 1, \quad \text{or} \quad \frac{y^2}{b^2} - \frac{x^2}{a^2} = 1$$

    are extended _____ of the rectangle with   diagonals
    corners at (a, b), (a, ___), (−a, b), (−a, ___).   −b; −b
    This rectangle is called the _____   fundamental
    rectangle.

**Draw the asymptotes and graph the hyperbolas in Frames
7 and 8.**

7. $\frac{x^2}{4} - \frac{y^2}{25} = 1$

    The asymptotes are extended diagonals of the
    rectangle with corners at (2, ____), (2, −5),   5
    (____, 5), and (−2, ____).   −2; −5

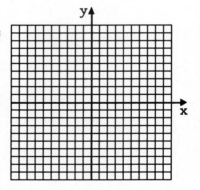

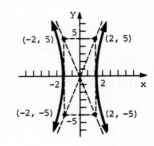

8. $y^2 - \frac{x^2}{9} = 1$

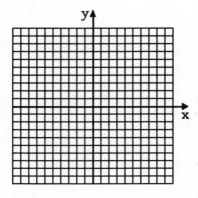

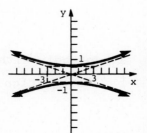

9. The hyperbola $\dfrac{(x + 1)^2}{4} - \dfrac{(y - 3)^2}{25} = 1$ has the same graph as _____ = 1, but is centered at _____ . Graph the hyperbola.

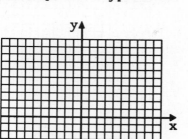

$\dfrac{x^2}{4} - \dfrac{y^2}{25}$

$(-1, 3)$

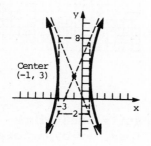

In Frames 10–13, first decide the type of graph indicated, and then sketch the graph.

10. $x^2 = 16 - y^2$

   type of graph:

   _____

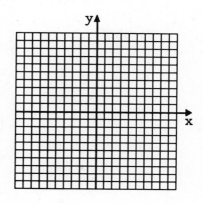

11. $\dfrac{x^2}{8} + \dfrac{y^2}{9} = 1$

   type of graph:

   _____

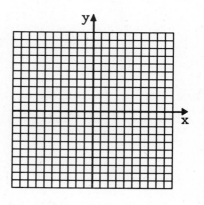

circle; ellipse

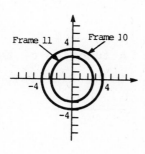

12. $x^2 - y^2 = 4$

   type of graph:

   _____

13. $y^2 - 4x^2 = 1$

   type of graph:

   _____

hyperbola; hyperbola

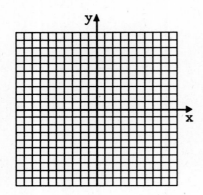 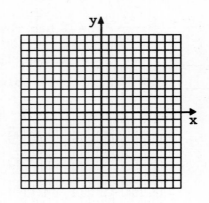 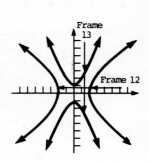

Identify each of the following as a *parabola*, *circle*, *ellipse*, or *hyperbola*.

14. $x^2 - 4y = 2$ _____   parabola

15. $16 - y^2 = x^2$ _____   circle

16. $12x^2 - y^2 = 6$ _____   hyperbola

17. $5x^2 + y^2 = 4$ _____   ellipse

18. $9x^2 = 16 + y^2$ _____   hyperbola

19. $y^2 = x - 1$ _____   parabola

20. A function of the form $f(x) = \sqrt{u}$ for an alge-
braic expression u, with $u \geq 0$, is called a
_____ _____ function.   square root

21. To graph $y = \sqrt{16 - x^2}$, square both sides to get
_____ or $x^2 + y^2 =$ _____, a circle $y^2 = 16 - x^2$; 16
with center at (0, 0) and radius ____. Since y   4
equals a square root, y must be _____,   nonnegative
so we graph only a _____. Draw the graph.   semicircle

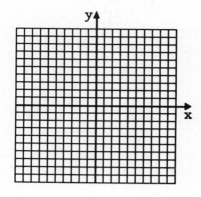

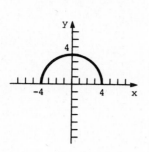

Draw the graphs in Frames 22—24.

22. $\dfrac{x}{2} = \sqrt{1 - \dfrac{y^2}{9}}$

       To graph, first square both sides:

$$\dfrac{x^2}{4} = \underline{\hspace{2cm}} \text{ or } \underline{\hspace{2cm}} = 1.$$

$1 - \dfrac{y^2}{9}; \ \dfrac{x^2}{4} + \dfrac{y^2}{9}$

This is an _____ centered at the origin

ellipse

with x—intercepts _____ and

(2, 0) and (−2, 0)

y—intercepts _____. However,

(0, 3) and (0, −3)

the given equation has a square root, so that x

must be _____. Graph only the portion

nonnegative

of the ellipse where x is not negative.

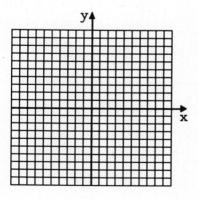

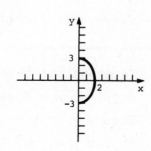

23. $y = \sqrt{1 + \dfrac{x^2}{25}}$

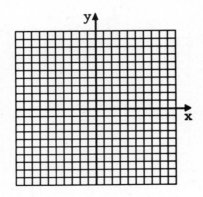

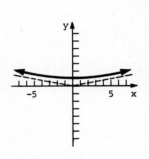

24. $x = -\sqrt{100 - y^2}$

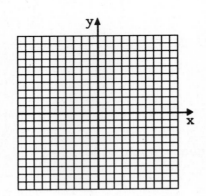

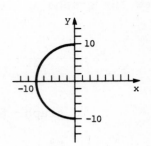

8.7   Other Useful Functions

[1]   Graph absolute value functions.   (Frames 1-3)

[2]   Graph functions defined piecewise.   (Frames 4-6)

[3]   Graph greatest integer functions.   (Frames 7 and 8)

[4]   Graph step functions.   (Frames 9 and 10)

1.   To graph the absolute value function $y = |x|$,
     consider $x \geq 0$ and $x < 0$ separately.
     When $x \geq 0$, $y = |x| = $ _____ , so we graph $y = $ _____ .   x; x
     When $x < 0$, $y = |x| = $ _____ , so we graph $y = $ _____ .   -x; -x
     Graph $y = |x|$ on the grid at the left.

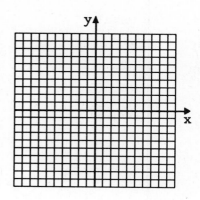
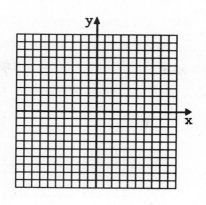
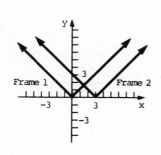

2. Graph y = |x - 3|.

    With the absolute value function y = |x|, we know
    that y ≥ ____. Changing to y = |x - 3| will have   0
    the effect of translating the "vertex" on the
    x-axis. The "vertex" of this graph is at _____.   (3, 0)
    Notice that absolute value graphs have the same
    basic shape but may be in different positions on
    the plane. Graph y = |x - 3| at the right above.

3. Graph y = |x + 1| - 2.

    The "vertex" of this graph is _____. The   (-1, -2)
    -2 has the effect of translating the graph ____   2
    units down. Note that some negative values for
    y are allowed, although |x + 1| is always non-
    negative. Graph y = |x + 1| - 2 below.

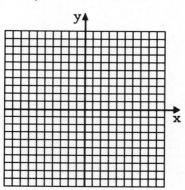

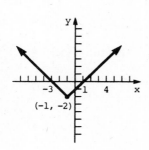

4. The function

    $$f(x) = \begin{cases} x \text{ if } x \geq 0 \\ -2x + 1 \text{ if } x < 0 \end{cases}$$

    is a function defined _____. Each part   piecewise

of the _____ must be graphed separately.   domain

First graph y = _____ when x ≥ 0. Then graph   x

y = _____ when x < 0. There is a solid   −2x + 1

endpoint at _____ and an open endpoint at   (0, 0)

_____ .   (0, 1)

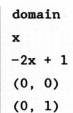

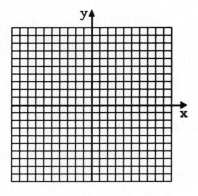

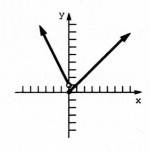

5. Graph $f(x) = \begin{cases} 3 \text{ if } x \geq -1 \\ x + 4 \text{ if } x < -1 \end{cases}$.

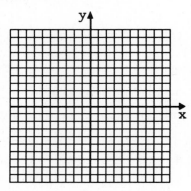

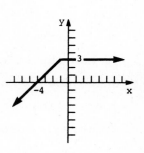

6. Graph $f(x) = \begin{cases} (x - 2)^2 \text{ if } x > 2 \\ -x \text{ if } x \leq 2 \end{cases}$.

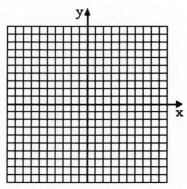

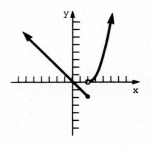

7. f(x) = ⟦x⟧ represents the function defined as the

    _____ integer _____ than or equal to x.

    For example, ⟦-1¼⟧ = _____ , ⟦3⅛⟧ = _____ , and

    ⟦9⟧ = _____ . Complete ordered pairs and graph

    y = ⟦x + 1⟧ on the grid below.

greatest; less

-2; 3

9

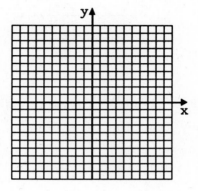

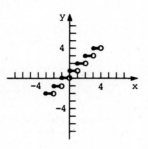

8. Graph y = 2⟦x⟧ + 1 on the grid below.

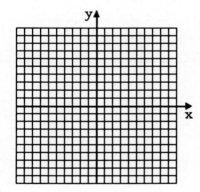

9. A grocery charges a fixed delivery charge of $2,
    plus $1 for delivering groceries a distance of a
    city block or portion of a block.

Complete the graph (distance, cost), started below, for all distances up to five blocks.

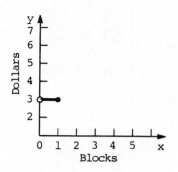

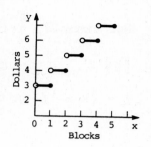

10. A rental firm charges a fixed $7 for a rental, plus $4 per day or any fraction of a day. Graph the function (days, cost).

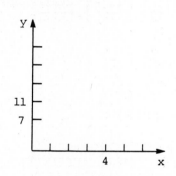

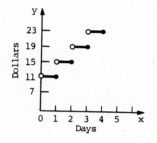

Chapter 8 Test

The answers for these questions are at the back of this Study Guide.

Let $f(x) = 2x - 3$ and $g(x) = -x^2 - 5x + 1$. Find the following.

1. $g(2)$

2. $(f + g)(-3)$

3. $\left(\dfrac{f}{g}\right)(5)$

4. $(f \circ g)(-2)$

5. Graph the quadratic function $f(x) = 4 - x^2$.
    Identify the vertex.

1. _____

2. _____

3. _____

4. _____

5. _____

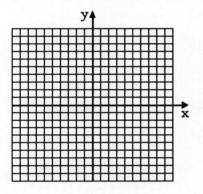

Solve the following.

6. The distance in feet an object moves in
    t seconds is given by

$$d = -4t^2 + 32t + 64.$$

    Find the maximum distance it reaches and
    the time it takes to reach that distance.

6. _____

7. The length and width of a rectangle have a
    sum of 74 inches. What width will lead to
    the maximum area, and what is that area?

7. _____

Use the tests for symmetry to determine all symmetries for each of the following relations.

8. $y = 3x - 4$

9. $y^2 = 25 + x^2$

10. $y^2 = 16 - x$

11. $y = -2x^5 + 2x$

8. _____

9. _____

10. _____

11. _____

12. Give all intervals on which the function
    is increasing, decreasing, or constant.

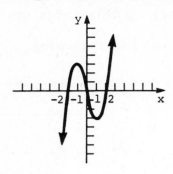

12. _____

Find the center and radius of each circle.

13. $(x - 4)^2 + (y + 3)^2 = 9$ 13. _____

14. $x^2 - x + y^2 = 6$ 14. _____

Graph each relation.

15. $9x^2 + y^2 = 36$

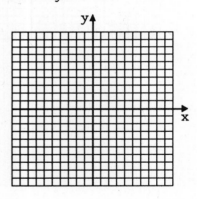

16. $x^2 + y^2 - 4y - 5 = 0$

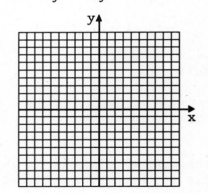

17. $\dfrac{x^2}{9} - \dfrac{y^2}{16} = 1$

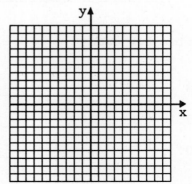

18. $x = \sqrt{y^2 + 1}$

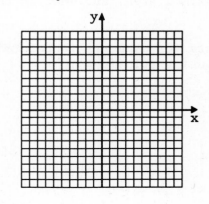

Identify each equation as a *parabola*, *hyperbola*, *ellipse*, or *circle*.

19. $4x^2 + 9y^2 = 36$

20. $2x - y^2 + 4y = 0$

21. $4y^2 - x^2 - 6x = 13$

19. _____

20. _____

21. _____

Graph each function.

22. $f(x) = x^2 + 2x - 3$

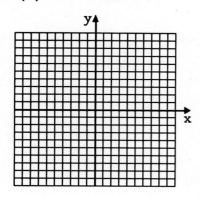

23. $f(x) = |x - 4| + 2$

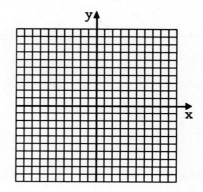

24. $f(x) = [\![x]\!] + 1$

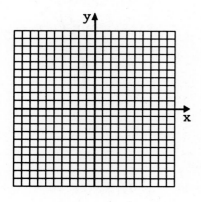

25. $f(x) = \begin{cases} x^2 & \text{if } x < -1 \\ -2x & \text{if } x \geq -1 \end{cases}$

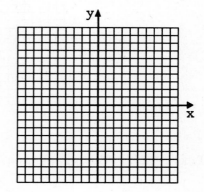

CHAPTER 9   POLYNOMIAL AND RATIONAL FUNCTIONS AND THEIR ZEROS

9.1   Polynomial Functions and Their Graphs

1️⃣   Graph polynomial functions of the form $P(x) = ax^n$.   (Frames 1-5)

2️⃣   Graph polynomial functions of the form $P(x) = ax^n + k$, $P(x) = a(x - h)^n$, and $P(x) = a(x - h)^n + k$.   (Frames 6-12)

3️⃣   Graph polynomial functions that are factored or are easily factorable into linear factors.   (Frames 13-16)

4️⃣   Learn the relationships among x-intercept of a graph, zero of a function, and solution of an equation.   (Frames 17-20)

---

1.   To graph the _____ function $P(x) = x^3$, first complete a table of values, as follows.

x	-2	-1	0	1	2
P(x)	___	___	___	___	___

Use these results to complete ordered pairs and then draw the graph.

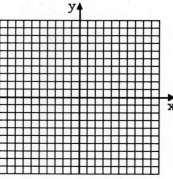

polynomial

-8; -1; 0; 1; 8

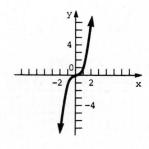

2.   Graph $P(x) = x^4$. The graph includes the points (0, ___), (1, ___), (-1, ___), (2, ___), and (-2, ___).

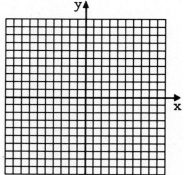

0; 1; 1; 16
16

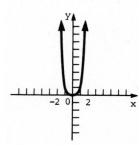

3. For the graph of $P(x) = ax^n$, the absolute value of _____ affects the width of the graph. When $|a|$ _____ 1, the graph is narrower than that of $P(x) = x^n$. When _____, the graph is broader. The graph of $P(x) = -ax^n$ is reflected about the ___-axis as compared to the graph of $P(x) = ax^n$.

a

>

$0 < |a| < 1$

x

4. Graph $P(x) = -\frac{1}{2}x^3$ below. The graph includes the points $(0, \_\_\_)$, $(2, \_\_\_)$, and $(-2, \_\_\_)$.

$0; -4; 4$

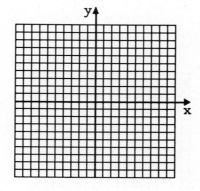

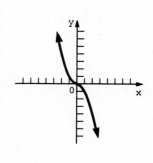

5. Graph $P(x) = 3x^2$ below. The graph includes the points $(0, \_\_\_)$, $(1, \_\_\_)$, $(-1, \_\_\_)$, $(2, \_\_\_)$, and $(-2, \_\_\_)$.

$0; 3; 3; 12$

12

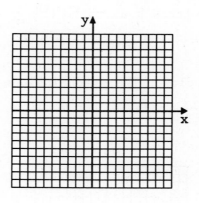

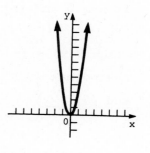

6. As compared to the graph of $P(x) = ax^n$, the graph of $P(x) = ax^n + k$ is shifted k units _____ if $k > 0$ and $|k|$ units _____ if $k < 0$.

up

down

7. Graph $P(x) = x^5 + 2$ below. The graph includes
    the points $(0, \_\_\_)$, $(1, \_\_\_)$, and $(-1, \_\_\_)$.

2; 3; 1

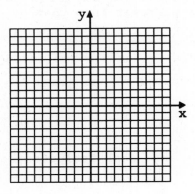

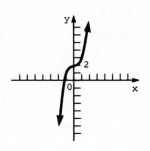

8. Graph $P(x) = x^2 - 4$ below.

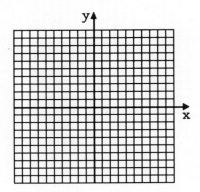

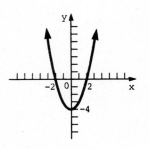

9. The graph of $P(x) = a(x - h)^n$, as compared to that
    of $P(x) = ax^n$, is translated h units to the _____
    if $h > 0$ and $|h|$ units to the _____ if $h < 0$.

right

left

10. Graph $P(x) = (x - 2)^3$ below. The graph includes
     the points $(0, \_\_\_)$, $(1, \_\_\_)$, $(2, \_\_\_)$, $(3, \_\_\_)$,
     and $(4, \_\_\_)$.

-8; -1; 0; 1
8

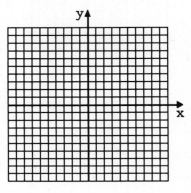

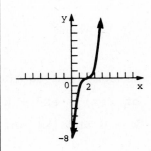

11. The graph of P(x) = _____ shows a com-
    bination of vertical and horizontal shifts.

    $a(x - h)^n + k$

12. Graph P(x) = $2(x + 1)^5 + 3$ below.
    The graph has the same shape as that of
    P(x) = $2x^5$, but is shifted 1 unit _____ and
    3 units _____. The graph includes the points
    (0, ___), (-1, ___), and (-2, ___).

    left
    up
    5; 3; 1

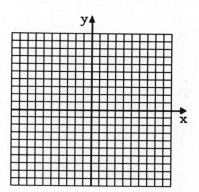

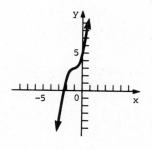

13. To graph a polynomial such as

    $$P(x) = x^2(2x - 3)(x + 2),$$

    first find the zeros (the values of x that make
    P(x) = ____) by setting each _____ equal to
    ____. Here the zeros are ____, ____, and ____.
    These zeros divide the x-axis into four intervals:

    0; factor
    0; 0; 3/2; -2

    $$(-\infty, -2), \text{_____}, \left(0, \frac{3}{2}\right), \text{_____}.$$

    $(-2, 0); \left(\frac{3}{2}, \infty\right)$

    Find the sign of each factor in each interval,
    and then the sign of ____. Decide if the graph
    is above or below the _____.

    P(x)
    x-axis

Interval	Sign of P(x)	Location relative to axis	
$(-\infty, -2)$	_____	above	+
$(-2, 0)$	_____	_____	-; below
$\left(0, \frac{3}{2}\right)$	_____	_____	-; below
$\left(\frac{3}{2}, \infty\right)$	_____	_____	+; above

Locate at least one point in each region and
sketch the graph of P(x).

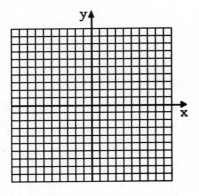

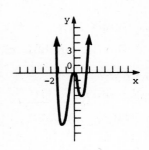

Graph each function in Frames 14–16.

14. P(x) = −3x(x − 1)(x + 3)

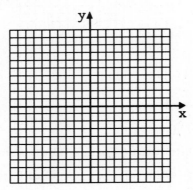

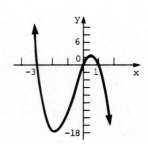

15. P(x) = x²(x + 2)(x − 2)²

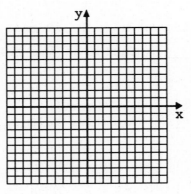

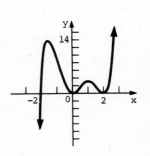

16. $P(x) = x^5 + x^4 - 2x^3$

First factor the polynomial completely.

$$P(x) = x^3(\underline{\hspace{2cm}})$$
$$= x^3(\underline{\hspace{1.5cm}})(\underline{\hspace{1.5cm}})$$

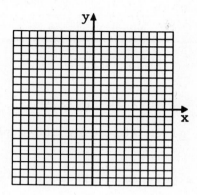

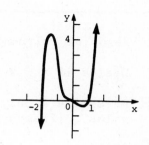

$x^2 + x - 2$

$x + 2; x - 1$

17. If the point $(a, 0)$ is an x-intercept of the graph of the function $y = P(x)$, then ____ is a zero of $P(x)$, and a is a _____ of the equation $P(x) = 0$.

a

solution

In Frames 18–20, determine the x-intercepts of the graphs of each polynomial function. Then determine the zeros of $P(x)$ and the solutions of $P(x) = 0$.

18. $P(x) = (x + 3)(x - 2)(x + 1)$

  x-intercepts of the graph: _____
  _____

  Zeros of $P(x)$: _____

  Solutions of $P(x) = 0$: _____

$(-3, 0); (2, 0);$
$(-1, 0)$

$-3; 2; -1$

$-3; 2; -1$

19. $P(x) = x^4 + 3x^3 - 4x^2$

  x-intercepts of the graph: _____
  _____

  Zeros of $P(x)$: _____

  Solutions of $P(x) = 0$: _____

$(0, 0); (-4, 0);$
$(1, 0)$

$0; -4; 1$

$0; -4; 1$

20. $P(x) = x^5 - 13x^3 + 36x$

    x-intercepts of the graph: _____

    _____

    Zeros of $P(x)$: _____

    Solutions of $P(x) = 0$: _____

(0, 0); (2, 0); (-2, 0)
(3, 0); (-3, 0)
0; 2; -2; 3; -3
0; 2; -2; 3; -3

9.2   Zeros of Polynomials

[1] Decide if x - k is a factor of the polynomial P(x). (Frames 1-9)

[2] Determine a polynomial given its zeros and a particular function value. (Frames 10-16)

[3] Find all the zeros of a polynomial if some of its zeros are given. (Frames 17-21)

[4] Factor a polynomial into linear factors given some of its zeros. (Frames 22-25)

1. According to the _____ theorem, x - k is a factor of P(x) if and only if _____ .

    factor

    $P(k) = 0$

2. Is x + 3 a factor of $x^4 + 6x^3 + 3x^2 - 26x - 24$? Use _____ division and the _____ theorem to decide.

    synthetic; remainder

```
-3) 1   6    3   -26  -24
        -3  -9    __   __
    _____
    1   3   __    __   __
```

    18; 24

    -6; -8; 0

Since the remainder is _____, $P(-3) =$ _____, so x + 3 (*is/is not*) a factor of $x^4 + 6x^3 + 3x^2 - 26x - 24$.

    0; 0

    is

In Frames 3-4 use the factor theorem to find the value of k that makes the second polynomial a factor of the first.

3. $x^3 + 2x + k$; x + 2

    Use synthetic division until you reach a point where you can determine the necessary k

that makes the remainder zero. (Remember that
x − k is a factor of P(x) if P(k) = 0.)

$$-2{\overline{\smash{\big)}\,1\quad 0\quad 2\quad k}}$$
$$\ -2\quad 4\quad \underline{}$$
$$\ \overline{1\quad -2\quad \underline{}\quad \underline{}}$$

Since k − 12 = ____, we have k = ____.

	−12
	6; k − 12
	0; 12

4. $x^3 - 19x + k$; x − 3

$$3{\overline{\smash{\big)}\,\underline{}\quad \underline{}\quad \underline{}\quad \underline{}}}$$
$$\ \overline{\underline{}\quad \underline{}\quad \underline{}}$$
$$\ \overline{\underline{}\quad \underline{}\quad \underline{}\quad \underline{}}$$

k = ____

	1; 0; −19; k
	3; 9; −30
	1; 3; −10; k − 30; 30

5. A polynomial of degree n has at most ____
distinct zeros.

n

6. The number k is said to be a _____ of
_____ n of the polynomial P(x) if
_____ appears in the factored form of P(x)
and there is no power of (x − k) greater than
____ that is also a factor of P(x).

zero
multiplicity
$(x - k)^n$

n

In Frames 7–9, determine the zeros of P(x) and give
their multiplicities.

7. $P(x) = (x - 1)^2(x + 3)^4$

Zero: _____; multiplicity: _____
Zero: _____; multiplicity: _____

1; 2
−3; 4

8. $P(x) = x^3(x + 2)^5$

Zero: _____; multiplicity: _____
Zero: _____; multiplicity: _____

0; 3
−2; 5

9. $P(x) = (x - 2i)^5(x + 2i)^5$

Zero: _____; multiplicity: _____
Zero: _____; multiplicity: _____

2i; 5
−2i; 5

In Frames 10–16, find a polynomial of lowest degree
with real coefficients having the given zeros and
function values.

10. −5 and 4; P(2) = −7

        The factors of the polynomial must be

             (x − ____) and (x − ____),   −5; 4

so that the polynomial has the form P(x) =

a(x + 5)(x − 4) for some nonzero real number a.

We want P(2) = _____. Substitute ____ for x.   −7; 2

     P(2) = a(_____)(_____) = ____   2 + 5; 2 − 4; −7

                  ____a = ____   −14; −7

                    a = ____ $\frac{1}{2}$

     P(x) = _____ $\frac{1}{2}(x + 5)(x − 4)$

     or   _____ $\frac{1}{2}x^2 + \frac{1}{2}x − 10$

11. 1, −2, and 3; P(−1) = 12

                P(x) = _____ $\frac{3}{2}x^3 − 3x^2 − \frac{15}{2}x + 9$

12. 2 − i and 2 + i; P(2) = 2

        We know that x − (____) and x − (____) are   2 − i; 2 + i

factors of the polynomial. Then the polynomial

of lowest degree is

    P(x) = a(x − 2 + i)(x − 2 − i) = a(_____).   x² − 4x + 5

Since P(2) = 2, a = ____ and P(x) = _____.   2; 2x² − 8x + 10

13. 3 + i and 3 − i; P(1) = 5   P(x) = _____   x² − 6x + 10

14. 2 − i, 2 + i, and −2; P(−1) = 2

        We want P(x) = a(x − 2 + i)(x − 2 − i)(x + 2).

First multiplying (x − 2 + i)(x − 2 − i), we get

    P(x) = a(x² − 4x + 5)(x + 2) = a(_____).   x³ − 2x² − 3x + 10

Since P(−1) = 2, P(x) = _____. $\frac{1}{5}x^3 − \frac{2}{5}x^2 − \frac{3}{5}x + 2$

15. 1, −1, and −3; P(2) − 15

$$P(x) = a(x - 1)(x + 1)(\underline{})$$

$$= a(\underline{}).$$

Since P(2) − 15, P(x) = _____.

x + 3
$x^3 + 3x^2 - x - 3$
$x^3 + 3x^2 - x - 3$

16. 1 + i, 1 − i, −2, and $\sqrt{2}$; P(0) = −8$\sqrt{2}$

$$P(x) = a[x - (1 + i)][x - (1 - i)](x + 2)(\underline{})$$

$$= a(x^2 - 2x + 2)(x + 2)(x - \sqrt{2})$$

$$= a(\underline{}).$$

Since P(0) = −8$\sqrt{2}$, P(x) = _____.

$x - \sqrt{2}$
$x^4 - \sqrt{2}x^3 - 2x^2$ $+ (4 + 2\sqrt{2})x - 4\sqrt{2}$
$2x^4 - 2\sqrt{2}x^3 - 4x^2$ $+ (8 + 4\sqrt{2})x - 8\sqrt{2}$

17. If P(x) is a polynomial having only real

_____ and if a + bi is a zero of

P(x), then the conjugate, _____, is also

a zero of P(x).

coefficients
a − bi

In Frames 18–21, one zero is given. Find all others.

18. P(x) − $x^3 + 4x^2 + x - 6$; −3

We know _____ is a factor of P(x) since

____ is a zero of P(x). Use synthetic division

to find another polynomial that is a factor of

P(x).

x + 3
−3

$$\begin{array}{r|rrrr} -3) & 1 & 4 & 1 & -6 \\ & & -3 & \underline{} & \underline{} \\ \hline & 1 & 1 & \underline{} & \underline{} \end{array}$$

−3; 6
−2; 0

The second polynomial is $x^2 + x - 2$ which factors

as (x + 2)(x − 1). The two other zeros are ____

and ____.

−2
1

19. P(x) − $x^3 - x^2 - 4x + 4$; 1   _____

2; −2

20. P(x) − $x^3 - 2x^2 - 14x + 40$; x = 3 + i

We know that x − (_____) is a factor of

3 + i

P(x) since _____ is a zero of P(x). Use
synthetic division to find another polynomial
that is a factor of P(x).

3 + i

$$
\begin{array}{r}
3 + i \overline{)\ 1 \quad -2 \qquad\quad -14 \qquad\quad 40} \\
\ 3 + i \qquad\ \rule{2cm}{0.4pt}\ \ \rule{1cm}{0.4pt} \\
\hline
\ 1 \quad\ 1 + i \qquad\ \rule{2cm}{0.4pt}\ \ \rule{1cm}{0.4pt}
\end{array}
$$

2 + 4i; −40

−12 + 4i; 0

Now, since 3 + i is a zero, we know that _____
is a zero. Use synthetic division with the
quotient $x^2 + (1 + i)x + (-12 + 4i)$ and the zero
_____. The final zero is _____.

3 − i

3 − i; −4

21. $P(x) = x^4 - 6x^3 + 10x^2 + 2x - 15$; 2 + i

Use synthetic division to get

$$
\begin{array}{r}
2 + i \overline{)\ 1 \quad -6 \qquad\ 10 \qquad\quad 2 \qquad -15} \\
\ 2 + i \quad -9 - 2i \ \ \rule{1cm}{0.4pt}\ \ \rule{1cm}{0.4pt} \\
\hline
\ 1 \quad -4 + i \quad 1 - 2i \ \ \rule{1cm}{0.4pt}\ \ \rule{1cm}{0.4pt}.
\end{array}
$$

4 − 3i; 15

6 − 3i; 0

Since 2 + i is a zero of P(x), so is _____. Use
synthetic division with 2 − i on the quotient
above. Doing so gives the polynomial _____
as a result. Now find the zeros of the quadratic polynomial $x^2 - 2x - 3$, which factors as
(_____)(_____). The other zeros are _____
and _____.

2 − i

$x^2 - 2x - 3$

x − 3; x + 1; 3

−1

22. Suppose we know that the polynomial P(x) =
$x^3 - 4x^2 - 2x + 20$ has −2 as a zero. We can
write this polynomial as a product of linear
factors. First use _____ division to
divide $x^3 - 4x^2 - 2x + 20$ by _____.

synthetic

x + 2

$$
\begin{array}{r}
 \overline{)\ 1 \quad -4 \qquad -2 \qquad 20} \\
\ -2 \qquad\ \rule{1cm}{0.4pt}\ \ \rule{1cm}{0.4pt} \\
\hline
\ 1 \quad\ \rule{1cm}{0.4pt} \quad\ 10 \qquad \rule{1cm}{0.4pt}
\end{array}
$$

−2

12; −20

−6; 0

This means that

$$x^3 - 4x^2 - 2x + 20 = (x + 2)(\text{_____}).$$

Use the quadratic formula to solve $x^2 - 6x +$
10 = 0:

$x^2 - 6x + 10$

$$x = \underline{\hspace{1.5cm}} \text{ or } x = \underline{\hspace{1.5cm}}.$$

This means

$$x^3 - 4x^2 - 2x + 20$$

$$= (x + 2)[x - (\underline{\hspace{1cm}})][x - (\underline{\hspace{1cm}})]$$

$$= (x + 2)(\underline{\hspace{2cm}})(x - 3 + i).$$

3 + i; 3 − i
3 + i; 3 − i
x − 3 − i

Write each polynomial of Frames 23–25 as a product of linear factors.

23. $P(x) = x^3 - 3x^2 + 4$, one zero is 2.

$$\underline{\hspace{3cm}}$$

$(x - 2)(x + 1)(x - 2)$

24. $P(x) = 4x^3 + 8x^2 - 13x - 3$, one zero is −3.

Use synthetic division to get

$$4x^3 + 8x^2 - 13x - 3 = (x + 3)(\underline{\hspace{2cm}})$$

$4x^2 - 4x - 1$

Use the quadratic formula to solve $4x^2 - 4x - 1 = 0$:

$$x = \underline{\hspace{1.5cm}} \text{ or } x = \underline{\hspace{1.5cm}},$$

$\dfrac{1 + \sqrt{2}}{2}; \dfrac{1 - \sqrt{2}}{2}$

so

$$4x^3 + 8x^2 - 13x - 3$$

$$= a(x + 3)[x - (\underline{\hspace{1cm}})][x - (\underline{\hspace{1cm}})]$$

$\dfrac{1 + \sqrt{2}}{2}; \dfrac{1 - \sqrt{2}}{2}$

for some constant a. To get $4x^3$ in the product on the right, we must choose a = ___. This gives

4

$$= \underline{\hspace{1cm}}(x + 3)\left[x - \left(\frac{1 + \sqrt{2}}{2}\right)\right]\left[x - \left(\frac{1 - \sqrt{2}}{2}\right)\right]$$

4

$$= (x + 3)(2)\left[x - \left(\frac{1 + \sqrt{2}}{2}\right)\right](2)\left[x - (\underline{\hspace{1cm}})\right]$$

$\dfrac{1 - \sqrt{2}}{2}$

$$= (x + 3)[2x - (\underline{\hspace{1cm}})][2x - (\underline{\hspace{1cm}})]$$

$1 + \sqrt{2}; 1 - \sqrt{2}$

$$= (x + 3)(2x - 1 - \sqrt{2})(\underline{\hspace{2cm}}).$$

$2x - 1 + \sqrt{2}$

25. $P(x) = 18x^3 - 3x^2 - 10x - 2$, one zero is −1/2.

$$P(x) = \underline{\hspace{4cm}}$$

$(2x + 1)(3x - 1 - \sqrt{3}) \cdot$
$(3x - 1 + \sqrt{3})$

9.3 Rational Zeros of Polynomial Functions

1 State and understand the rational zeros theorem. (Frame 1)

2 Find all rational zeros of a polynomial with integer coefficients.
(Frames 2-6)

3 Find all rational zeros of a polynomial with rational number
coefficients. (Frames 7-9)

---

1. Let $P(x) = a_n x^n + a_{n-1} x^{n-1} + \cdots + a_1 x + a_0$, where
$a_n \neq 0$, be a polynomial of degree n with _____   integer
coefficients. By the rational zeros theorem, a
rational number in lowest terms p/q can be a _____   zero
of $P(x)$ only if p is a factor of ____ and q is a a_0
factor of ____. a_n

Find all rational zeros of the polynomials of Frames 2-6.

2. $P(x) = x^3 - 5x^2 + 2x + 8$

If p/q is a rational zero of $P(x)$, then we
know by the theorem that p is a factor of a_0 =
_____ and q is a factor of a_3 = _____. Since   8; 1
$a_0 = 8$, we know that p must be _____. ± 1, ± 2, ± 4, or ± 8
Since $a_3 = 1$, we know that q must be _____ and ± 1
any rational zero p/q of $P(x)$ will come from the
list _____. Use synthetic division ± 1, ± 2, ± 4, or ± 8
to try various values. For example, try 1.

```
1) 1   -5    2    8
        1   -4   -2
   ─────────────────
   1   -4   -2    6      1 (is/is not)          is not
                         a zero of P(x).
```

Now try -1.

```
-1) 1   -5    2    8
        -1    6   -8
    ─────────────────
    1   -6    8    0     -1 (is/is not)          is
                         a zero of P(x).
```

Try 2.

```
2) 1   -5    2    8
        2   -6   -8
   ─────────────────
   1   -3   -4    0      2 (is/is not)          is
                         a zero of P(x).
```

Try -2.

```
-2) 1   -5    2     8
         -2   14   -32
    1   -7   16   -24    -2 (is/is not)         is not
                         a zero of P(x).
```

Test ± 4 and ± 8. The rational zeros are _____ . $-1; 2; 4$

3. $P(x) = x^3 - x^2 - 4x + 4$

     Since $a_0 = 4$, we know p must be _____ . $\pm 1, \pm 2,$ or ± 4

Since $a_3 = 1$, we know that q must be ____ and ± 1

p/q will come from the list _____ . $\pm 1, \pm 2,$ or ± 4

Use synthetic division. Try -1.

```
-1) 1   -1   -4    4
         -1    2    2
    1   -2   -2    6    -1 (is/is not)          is not
                        a zero of P(x).
```

Try 1.

```
1) 1   -1   -4    4
        1    0   -4
   1    0   -4    0    1 (is/is not)            is
                      a zero of P(x).
```

Test ± 2 and ± 4. (Remember if or when you find
three zeros, your task is completed. Why?) The
zeros are _____ . $1, 2, -2$

4. $P(x) = x^4 - 13x^2 + 36$

     p will come from the list $\pm 1, \pm 2, \pm 3, \pm 4, \pm 6,$
$\pm 9, \pm 12, \pm 18,$ or ± 36. To find the possibilities
for p with large numbers, it is sometimes helpful
to write the number as a product of its prime
factors. (In this case $36 = 2 \cdot 2 \cdot 3 \cdot 3$.) Then
form all the possible combinations of prime
factors. q will again be ± 1 so p/q will come
from the same list as the one for p. Begin synthetic division, being careful to use zeros for
the missing powers of x. Try -1.

```
-1) 1    0   -13    0    36
         -1    1   12   -12
    1   -1   -12   12    24    -1 (is/is not)    is not
                              a zero of P(x).
```

Try 1.

```
1) 1    0   -13    0    36
        1    1   -12  -12
   ────────────────────────
   1    1   -12  -12    24
```
1 (*is/is not*)   is not
a zero of $P(x)$.

Try −2.

```
-2) 1    0   -13    0    36
        -2    4   18  -36
   ────────────────────────
   1   -2   -9   18    0
```
−2 (*is/is not*)   is
a zero of $P(x)$.

Continue the trial–and–error process. Now use
the reduced polynomial $x^3 - 2x^2 - 9x + 18$ found
from the synthetic division for −2. You can
stop when you find _____ zeros. The rational   four

zeros are _____ .   −2, 2, 3, −3

5. $P(x) = 2x^3 - 7x^2 - 5x + 4$

    p will come from the list ±1, ±2, or ±4; q
will come from the list ±1, or ±2. Then p/q
will come from the list ±1, ±1/2, ±2, or ±4.
Test the values from this list to find the
rational zeros, if they exist.   _____   −1, 1/2, 4

6. $P(x) = x^3 - 7x^2 + 13x - 6$

    p will come from the list ±1, ±2, ±3, or ±6;
q will be ±1, so we use the list for p.

```
1) 1   -7    13    -6
       1    -6     7
   ─────────────────────
   1   -6    7     1
```
1 (*is/is not*)   is not
a zero of $P(x)$.

```
-1) 1   -7    13    -6
        -1    8   -21
   ─────────────────────
   1   -8    21   -27
```
−1 (*is/is not*)   is not
a zero of $P(x)$.

```
2) 1   -7    13    -6
       2   -10     6
   ─────────────────────
   1   -5    3     0
```
2 (*is/is not*)   is
a zero of $P(x)$.

Find the rest of the zeros by solving

_____ = 0. $x^2 - 5x + 3$

They are

$$x = \underline{\hspace{2cm}} \quad \text{or} \quad x = \underline{\hspace{2cm}}.$$

$$\frac{5 + \sqrt{13}}{2}, \frac{5 - \sqrt{13}}{2}$$

These additional zeros (*are/are not*) rational.

are not

7. The rational zeros theorem may not be applied if a polynomial does not have _____ coeffi-

integer

cients. If fractional coefficients appear, then we must first _____ by a number that will

multiply

clear it of all fractions. Then we use the theorem.

8. Find all rational zeros of

$$P(x) = x^3 - \frac{17}{6}x^2 - \frac{13}{3}x - \frac{4}{3}.$$

We must find the values that make $P(x) = \underline{\hspace{1cm}}$.

0

Multiply both sides by ____ to eliminate all

6

fractions, obtaining

$$6x^3 - 17x^2 - 26x - 8 = 0.$$

The polynomial on the left side will have the same zeros as $P(x)$. From the possible rational zeros, we try $-1/2$.

```
-1/2) 6   -17   -26   -8
            -3    10    8
       6   -20   -16    0
```

Therefore, $-1/2$ (*is/is not*) a zero of $P(x)$.

is

We may now find the remaining two zeros by

factoring _____.

$6x^2 - 20x - 16$

$$6x^2 - 20x - 16 = 2(3x^2 - 10x - 8)$$

$$= 2(\underline{\hspace{1.5cm}})(\underline{\hspace{1.5cm}})$$

$3x + 2$; $x - 4$

Therefore, the three zeros of $P(x)$ are _____,

$-1/2$

_____, and _____.

$-2/3$; 4

9. Find the rational zeros of

$$P(x) = x^3 - \frac{9}{2}x^2 - 3x + \frac{5}{2}.$$

The zeros are _____, _____, and _____.

$1/2$; -1; 5

9.4 Real Zeros of Polynomial Functions

1️⃣ Determine the possible numbers of positive and negative real zeros
of a polynomial function using Descartes' rule of signs. (Frames 1–5)

2️⃣ Locate a real zero of a polynomial function between two numbers using
the intermediate value theorem. (Frames 6–9)

3️⃣ Decide which real numbers are greater than or less than all real zeros
of a polynomial function using the boundedness theorem. (Frames 10–15)

4️⃣ Use approximate real zeros to graph polynomial functions that cannot
be factored with rational coefficients. (Frames 16–19)

---

1. By Descartes' rule of _____, if $P(x)$ is a polynomial with _____ coefficients and terms in
descending powers of x, then

    (a) the number of _____ real zeros of $P(x)$
either equals the number of _____ in
sign occurring in the coefficients of $P(x)$,
or is less than the number of variations
by a positive _____ integer.

    (b) the number of _____ real zeros of $P(x)$
either equals the number of variations in
sign of _____, or else is less than the
number of variations by a positive _____
integer.

signs

real

positive

variations

even

negative

$P(-x)$

even

**Use Descartes' rule of signs to find the number of
positive or negative real zeros for the polynomials
of Frames 2–5.**

2. $P(x) = 6x^3 - 7x^2 + 1$

    There are ____ variations of sign in $P(x)$.   2
Then by Descartes' rule of signs, the number
of positive real zeros is either _____ or   2
$2 - 2 =$ _____. Find $P(-x)$.   0
  $P(-x) = 6(-x)^3 - 7(-x)^2 + 1 =$ _____ $-6x^3 - 7x^2 + 1$
There is ____ variation of sign in $P(-x)$. Then   1
by Descartes' rule of signs, the number of negative real zeros is ____.   1

3. $P(x) = x^3 - 5x^2 - 4x - 1$

    $P(x)$ has _____ variation in sign, so that $P(x)$ | 1

    has ____ positive real zero. | 1

        $P(-x) = (\_\_\_)^3 - 5(\_\_\_)^2 - 4(\_\_\_) - 1$ | $-x$; $-x$; $-x$

        = _____ | $-x^3 - 5x^2 + 4x - 1$

    $P(-x)$ has ____ variations in sign, so that _____ | 2; $P(x)$

    has ____ or ____ negative real zeros. | 2; 0

4. $P(x) = 2x^4 - 3x^3 - 5x^2 + 4x - 1$

      There are ____ or ____ positive real zeros. | 3; 1

        $P(-x) =$ _____ | $2x^4 + 3x^3 - 5x^2 - 4x - 1$

    There is 1 _____ real zero. | negative

5. $P(x) = x^4 - 3x^3 + 11x^2 - 9x - 8$

                  positive zeros: _____ | 3 or 1

                  negative zeros: _____ | 1

6. By the _____ value theorem, if $P(x)$ | intermediate

    defines a polynomial with only real coefficients

    and if $P(a)$ and $P(b)$ have _____ signs, then | opposite

    there is at least one _____ zero between _____ | real; a

    and _____. | b

In Frames 7-9, show that the following polynomials have a real zero between the numbers given.

7. $P(x) = 2x^3 + x^2 - 5x + 2$; 0 and $3/4$

      By the intermediate value theorem, if $P(0)$

    and $P(3/4)$ are _____ in sign, then there | opposite

    is a zero between them.

    First, $P(0) =$ ____ | 2

    and $P\left(\dfrac{3}{4}\right) = 2\left(\_\_\_\right) +$ ____ $-$ ____ $+ 2$ | $\dfrac{27}{64}$; $\dfrac{9}{16}$; $\dfrac{15}{4}$

               = _____. | $-\dfrac{11}{32}$

    Then there (*is/is not*) a zero between 0 and $3/4$. | is

8. $P(x) = 2x^3 + 3x^2 - 5x - 6$; 1 and 2

    $$P(1) = -6 \text{ and } P(2) = \underline{\hspace{1cm}}$$ 12

    Then there (*is/is not*) a zero between 1 and 2.   is

9. $P(x) = 2x^3 + 9x^2 + 13x + 6$; -1.9 and -1.2

    $P(-1.9) \underline{\hspace{0.5cm}} 0 \text{ and } P(-1.2) \underline{\hspace{0.5cm}} 0$ >; <

    There (*is/is not*) a zero between -1.9 and -1.2.   is

10. Suppose that $P(x)$ is a polynomial with real
    coefficients and with a positive leading co-
    efficient. Suppose that $P(x)$ is divided
    synthetically by $x - c$.

    (a) If $c > 0$, and all numbers in the bottom row
         of the synthetic division are nonnegative,
         then $P(x)$ has no zero _____.   greater than c
         c is called a(n) _____ bound.   upper

    (b) If $c < 0$, and all the numbers in the bottom
         row of the synthetic division _____   alternate
         in sign (with 0 considered positive or neg-
         ative as needed), then $P(x)$ has no zero
         _____. c is called a(n)   less than c
         _____ bound.   lower

11. Show that $P(x) = x^3 + 4x^2 + x - 6$ has no zero
    larger than 2.

    To show this, use _____ division.   synthetic

    $$
    \begin{array}{r|rrrr}
    2) & 1 & 4 & 1 & -6 \\
       &   & 2 & \underline{\hspace{0.7cm}} & \underline{\hspace{0.7cm}} \\
    \hline
       & 1 & 6 & \underline{\hspace{0.7cm}} & \underline{\hspace{0.7cm}}
    \end{array}
    $$
    12; 26
    13; 20

    Since 2 is positive and every number in the
    bottom row is _____, there is no zero   positive
    greater than _____.   2

12. Show that $P(x) = x^4 + 2x^3 - 7x^2 - 8x + 12$ has no
    zero greater than 3.

```
3) 1  2  -7  -8  12
      3  15  24  48
   1  5   8  16  60
```
Every number in the bottom row is positive.

13. Show that $P(x) = x^3 + 4x^2 + x - 6$ has no zero less than −5.

```
-5) 1   4   1   -6
       -5   5  -30
    1  -1   6
```

Since −5 is negative and the numbers in the bottom row _____ in sign, there is no zero less than _____.

−36

alternate

−5

14. Show that $P(x) = x^4 + 2x^3 - 7x^2 - 8x + 12$ has no zero less than −6.

```
-6) 1   2  -7   -8   12
       -6  24 -102  660
    1  -4  17 -110  672
```
Signs in the bottom row alternate.

15. Approximate the real zeros of $P(x) = 6x^3 - 7x^2 + 1$ to the nearest tenth.

We found in Frame 2 that $P(x)$ has 2 or 0 positive real zeros and ____ negative real zero. Let us look for the negative zero first. As a start, use synthetic division with −1 as a divisor.

1

```
-1) 6   -7   0   1
       -6  13
    6  -13  13
```

−13
−12

Since the numbers in the bottom row of the synthetic division alternate in sign, we know that _____ is less than any zero of $P(x)$. Then, our negative real zero must be between ____ and ____.

−1
−1; 0

Is the zero between −.5 and 0?   Find P(−.5).

$$P(-.5) = 6(-.125) - 7(.25) + 1 = \underline{\hspace{2cm}}$$ -1.5

$$P(0) = \underline{\hspace{2cm}}$$ 1

Then, we know, since the signs are _____,   opposite

that our negative real zero lies between _____   −.5

and 0. Now test −.3.

$$P(-.3) = 6(-.3)^3 - 7(-.3)^2 + 1 = \underline{\hspace{1.5cm}}$$ $.208$

Since P(−.3) ___ 0 and P(−.5) ___ 0, the negative   >; <

zero lies between _____ and _____. Try −.4.   −.3; −.5

$$P(-.4) = \underline{\hspace{2cm}}$$ $-.504$

P(_____) is closer to 0 than P(_____), so to the   −.3; −.4

nearest tenth our zero is _____. Continue in this   −.3

way to find the two positive real zeros, _____.   1 and .5

16. Earlier in this chapter, we graphed polynomial

functions that could be factored. However, we

cannot factor

$$P(x) = 6x^3 - 23x^2 + 12x + 20,$$

so we must graph the function using theorems of

this section.

First note that we have _____ variations of   2

sign in P(x) and since

$$P(-x) = \underline{\hspace{4cm}},$$ $-6x^3 - 23x^2 - 12x + 20$

we have _____ variation in sign for P(−x). We   1

have two or no positive real zeros and one neg-

ative real zero. Let's evaluate some points,

using the shortened version of synthetic division.

x				P(x)	Ordered pair	
	6	−23	12	20		
−1	6	−29	41	−21	_____	(−1, −21)
0	6	−23	12	20	_____	(0, 20)
1	6	−17	−5	15	_____	(1, 15)
2	6	−11	−10	0	_____	(2, 0)
3	6	−5	−3	11	_____	(3, 11)
4	6	1	16	84	_____	(4, 84)

Now we analyze our chart. The first thing we
notice is that ____ is a zero of P(x). Notice   2
that the row of the synthetic division of –1 has
alternating signs, which means that ____ is less   –1
than any zero of P(x). Also, P(–1) and P(0) vary
in sign so our negative zero is between ____ and   –1
____.   0

The last row of the synthetic division of 4
has all nonnegative numbers so ____ is larger   4
than any zero of P(x). Let's do one further
test between 2 and 3. (We choose this area because 0 can be negative or positive as needed
and P(2) – 0.) Testing 5/2 (which is 2.5), we
get

$$
\begin{array}{r}
5/2\overline{)6 \quad\; -23 \quad\;\; 12 \quad\quad 20} \\
15 \quad -20 \quad\;\;\underline{} \\
\overline{6 \quad\; -8 \quad\;\; -8 \quad\;\;\underline{}}.
\end{array}
$$

          –20
          0

This is the third zero. Sketch the graph and
check your answer. Notice that for x < –1 the
values will become increasingly more negative
and for x > 3 the values will become increasingly
more ____.   positive

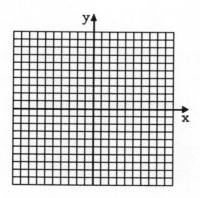

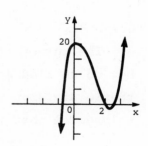

17. Graph P(x) – $x^3 + x^2 - 2x$.

P(x) has ____ variation in sign, so we expect   1
to find one real positive zero.

P(–x) = _____ $-x^3 + x^2 + 2x$

which also has one variation in sign, so we expect

to find _____ real negative zero. We may have a
polynomial of third degree with only two real
zeros, but we must work carefully. Let's find a
few points.

x				P(x)	Ordered pair
	1	1	−2	0	
−2	1	−1	0	0	_____
−1	1	0	−2	2	_____
0	1	1	−2	0	_____
1	1	2	0	0	_____
2	1	3	4	8	_____

one

(−2, 0)
(−1, 2)
(0, 0)
(1, 0)
(2, 8)

Sketch the graph, testing the area between
x = 0 and x = 1.

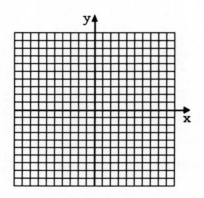

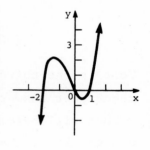

18. Graph P(x) = x³ + 1

The number −1 is a zero of this function, but
how do we find any other zeros? We can use synthetic division with _____ to find the quadratic
polynomial that is the quotient.

−1

$$\begin{array}{r}
-1)\ \overline{\begin{array}{rrrr} 1 & 0 & 0 & 1 \end{array}} \\
\begin{array}{rrrr} & -1 & 1 & -1 \end{array} \\
\overline{\begin{array}{rrrr} 1 & -1 & 1 & 0 \end{array}}
\end{array}$$

Then P(x) = (x + 1)(_____). (We could have
used our knowledge of factoring the sum of two
_____.) Using the quadratic formula on
_____ = 0, we find complex solutions.
Then _____ is the only real zero of P(x). Plot
a few points and sketch the graph.

$x^2 - x + 1$

cubes
$x^2 - x + 1$

−1

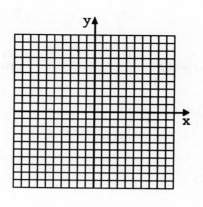

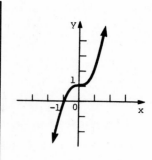

19. Graph $P(x) = 2x^3 + x^2 - 5x + 2$.

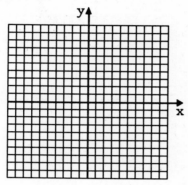

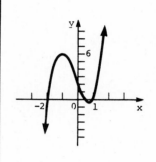

9.5 Rational Functions

[1] Define *rational function*. (Frame 1)

[2] Graph rational functions using reflection and translation. (Frames 2-4)

[3] Find asymptotes of a rational function. (Frames 5-10)

[4] Graph rational functions where the degree of the numerator is less than the degree of the denominator. (Frames 11-13)

[5] Graph rational functions where the degrees of the numerator and denominator are equal. (Frames 14 and 15)

[6] Graph rational functions where the degree of the numerator is greater than the degree of the denominator. (Frame 16)

[7] Graph a rational function that is not in lowest terms. (Frames 17 and 18)

1. A function of the form

$$f(x) = \frac{p(x)}{q(x)},$$

where $p(x)$ and $q(x)$ are _____ functions   polynomial
is called a _____ function.   rational

In Frames 2—4, use reflections and translations to graph each rational function.

2. $f(x) = \dfrac{3}{x} + 1$

Compared to the graph of $f(x) = 1/x$, the graph of this function will be translated 1 unit _____,   up

and each point will be _____ times as far away   3

from the x-axis. The graph includes the points

$(1, \underline{\quad})$, $(3, \underline{\quad})$, $(-1, \underline{\quad})$ and $(-3, \underline{\quad})$.   4; 2; -2; 0

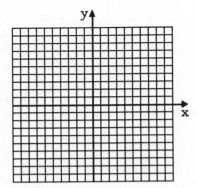
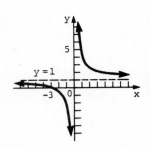

3. $f(x) = \dfrac{1}{x - 2}$

Compared to the graph of $f(x) = 1/x$, the graph of this function will be translated ____ units _____.   2; right

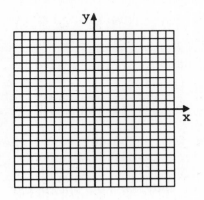
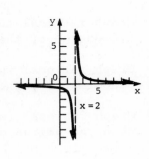

4. $f(x) = -\dfrac{4}{x} - 1$

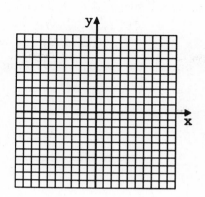

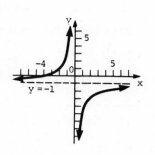

5. Vertical asymptotes are found by setting the

_____ of a rational function equal to

_____ and then solving. _____, or in

some cases _____, asymptotes are found by

considering what happens to f(x) as _____ gets

larger and larger, written _____.

denominator

zero; Horizontal

oblique

$|x|$

$|x| \to \infty$

**Identify all vertical and horizontal asymptotes in
Frames 6–8.**

6. $f(x) = \dfrac{6x - 5}{2x + 7}$

To find the vertical asymptote, set _____ = 0,

and solve for _____. To find the horizontal

asymptote, divide each term in the numerator and

denominator by ____, the largest power of x in

the expression. As $|x|$ gets larger, the value

of f(x) approaches _____.

                          Vertical: _____

                       Horizontal: _____

2x + 7

x

x

3

x = -7/2

y = 3

7. $f(x) = \dfrac{9}{x - 8}$

                          Vertical: _____

                       Horizontal: _____

x = 8

y = 0

8. $f(x) = \dfrac{3x - 5}{9x + 11}$

                          Vertical: _____

                       Horizontal: _____

x = -11/9

y = 1/3

9. Find all asymptotes of $f(x) = \dfrac{2x^2 - 1}{x + 3}$.

The vertical asymptote is _____, found by $x = -3$
setting the denominator equal to zero. Dividing
each term by _____, the largest power of x in x^2
the expression, indicates that there is no
_____ asymptote. (Note that the degree   horizontal
of the numerator is _____ than the degree   greater
of the denominator.)

  If we divide the numerator by the denominator,
$f(x) =$ _____. Then, as $|x|$ gets larger $2x - 6 + \dfrac{17}{x + 3}$
and larger, the graph of f(x) gets closer to the
oblique asymptote, the line _____. $y = 2x - 6$

10. The graph of a rational function (*may/may not*)   may not
intersect a vertical asymptote; it (*may/may not*)   may
intersect a nonvertical asymptote.

11. Graph $f(x) = \dfrac{3}{2x + 1}$.

The vertical asymptote is _____. $x = -\dfrac{1}{2}$
The horizontal asymptote is _____. $y = 0$
f(0) = _____, so the y-intercept is _____.   3; (0, 3)
The x-intercept is found by solving f(x) = 0,
which gives 3 ≠ 0, so there (*is/is not*) an   is not
x-intercept. The graph (*does/does not*) inter-   does not
sect its horizontal asymptote, y = 0. Plot a
point using an x-value in each interval determined
by the x-intercept(s) and vertical asymptotes.
The intervals for this function are _____ and $(-\infty, -1/2)$
_____. $(-1/2, \infty)$

Complete the sketch.

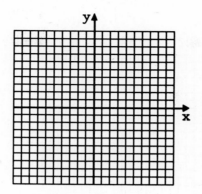

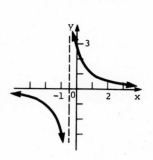

12. Graph $f(x) = \dfrac{2}{(x + 1)(x - 1)}$.

Vertical asymptote(s): _____

Horizontal asymptote: _____

y-intercept: _____

x-intercept: _____

Plot appropriate points and complete the sketch.

$x = -1, \ x = 1$

$y = 0$

$(0, -2)$

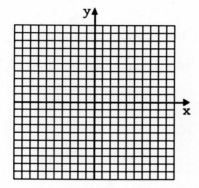

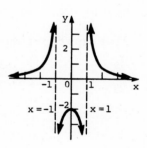

13. Graph $f(x) = \dfrac{x - 1}{(3x + 1)(x - 2)}$

Vertical asymptote(s): _____

Horizontal asymptote: _____

y-intercept: _____

x-intercept: _____

The graph crosses its horizontal asymptote at

_____. The x-intercept and vertical asymptotes

divide the x-axis into the following intervals:

_____.

$x = -\dfrac{1}{3}; \ x = 2$

$y = 0$

$\left(0, \dfrac{1}{2}\right)$

$(1, 0)$

$(1, 0)$

$\left(-\infty, -\dfrac{1}{3}\right); \ \left(-\dfrac{1}{3}, 1\right);$
$(1, 2); \ (2, \infty)$

Plot points using an x-value in each interval
and complete the sketch.

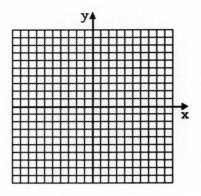

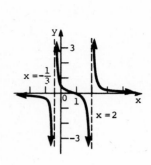

14. Graph $f(x) = \dfrac{3x + 4}{x + 2}$.

Vertical asymptote: _____ $x = -2$

Horizontal asymptote: _____ $y = 3$

y-intercept: _____ $(0, 2)$

x-intercept: _____ $\left(-\dfrac{4}{3}, 0\right)$

The graph (*does*/*does not*) intersect its horizontal   does not

asymptote. We know this because _____. $f(x) = 3$ has no
solution

Plot appropriate points and complete the sketch.

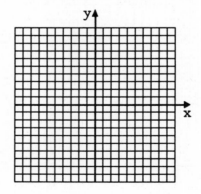

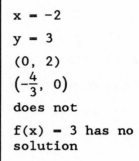

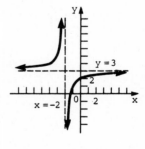

15. Graph $f(x) = \dfrac{x(x + 3)}{(x - 1)^2}$.

Vertical asymptote: _____ $x = 1$

Horizontal asymptote: _____ $y = 1$

y-intercept: _____ $(0, 0)$

x-intercepts: _____ $(0, 0); \ (-3, 0)$

Solve _____ to determine where the graph crosses its horizontal asymptote; it crosses at _____. The graph includes the points $\left(-4,\; \underline{\hspace{1cm}}\right)$, $\left(-2,\; \underline{\hspace{1cm}}\right)$, $\left(\frac{1}{2},\; \underline{\hspace{1cm}}\right)$, and $\left(2,\; \underline{\hspace{1cm}}\right)$. Complete the sketch.

$f(x) = 1$

$\left(\frac{1}{5},\; 1\right)$

$\frac{4}{25}$; $-\frac{2}{9}$; 3; 10

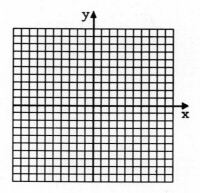

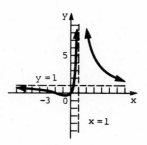

16. Graph $f(x) = \dfrac{x^2 + 2}{x - 1}$.

    Vertical asymptote: _____
    Oblique asymptote: _____
    y-intercept: _____
    Since the numerator, _____ has _____ real zeros, there (*is*/*is not*) an x-intercept. Complete the sketch.

    $x - 1$
    $y = x + 1$
    $(0, -2)$
    $x^2 + 2$; no
    is not

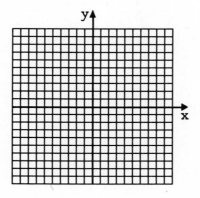

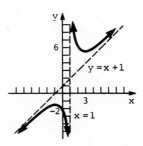

17. Graph $f(x) = \dfrac{x^2 - 9}{x + 3}$.

    First factor the numerator and write f(x) in lowest terms.

    $f(x) = \underline{\hspace{1.5cm}}$, $x \neq \underline{\hspace{1cm}}$.

    A "_____" will appear in the graph at $x = \underline{\hspace{0.8cm}}$.

    $x - 3$; -3
    hole; -3

Sketch the graph.

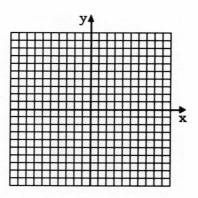

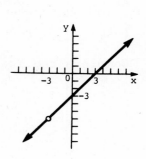

18.   Graph $f(x) = \dfrac{25 - x^2}{5 - x}$.

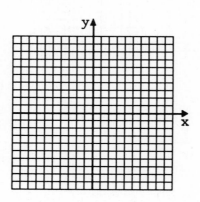

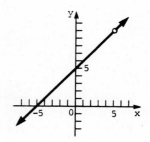

Chapter 9 Test

The answers for these questions are at the back of this Study Guide.

Graph.

1. $f(x) = (2 + x)^2$

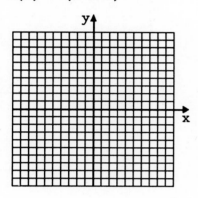

2. $f(x) = x(x - 1)(x + 3)$

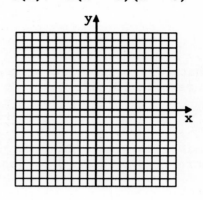

3. Find a lowest degree polynomial with 2, 3, and −4 as zeros.

3. _____

4. Is 4 a zero of
    $$P(x) = x^4 - 3x^3 + 5x^2 - x + 12?$$

4. _____

5. Is $x + 3$ a factor of
    $$P(x) = 2x^3 + 5x^2 - 4x - 3?$$

5. _____

6. Find a polynomial of degree 3 with $\sqrt{2}$, $-\sqrt{2}$, and 4 as zeros, and $P(1) = -2$.

6. _____

7. Find all zeros of
    $$P(x) = x^4 - 3x^3 + 2x^2 + 2x - 4$$
    given that $1 + i$ is a zero.

7. _____

8. Factor $P(x) = x^3 + 2x^2 - x - 2$, given that −2 is a zero.

8. _____

List all rational numbers that can possibly be zeros of P(x) and
find all rational zeros of the polynomials in Problems 9 and 10.

9. $P(x) = 2x^3 + 3x^2 - 3x - 2$

9. _____

_____

10. $P(x) = x^4 - 5x^3 + 11x^2 - 13x + 10$

10. _____

_____

11. Show that $P(x) = 2x^3 - 3x^2 - 3x + 2$ has
a real zero in $[0, 1]$.

11. _____

12. Show that $P(x) = x^3 + 2x^2 - 5x - 6$ has no
real zero greater than 3 or less than -4.

12. _____

13. Find the number of positive and negative real
zeros of $P(x) = 4x^4 - 8x^3 + 17x^2 - 2x - 14$.

13. _____

14. Show that $P(x) = x^3 - 3x^2 - x + 6$ has a real
zero in $[-2, -1]$ and approximate it to the
nearest tenth.

14. _____

Graph each of the following.

15. $P(x) = 8x^3 - 6x^2 - 11x + 3$

16. $f(x) = \dfrac{1}{2 - x}$

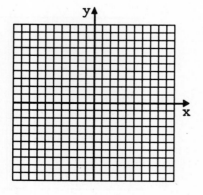

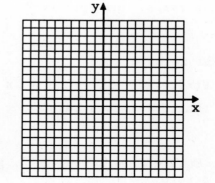

17. $f(x) = \dfrac{2x - 1}{x + 1}$

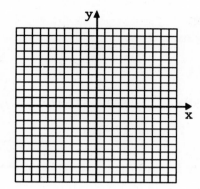

18. $f(x) = \dfrac{x^2 - 4}{x - 1}$

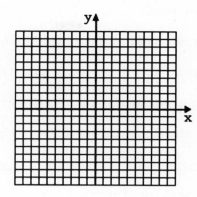

19. $f(x) = \dfrac{3x}{x^2 - 1}$

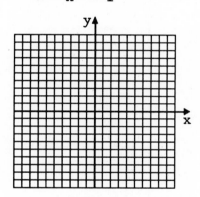

20. $f(x) = \dfrac{x^2 - 1}{x - 1}$

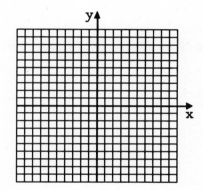

CHAPTER 10 EXPONENTIAL AND LOGARITHMIC FUNCTIONS

10.1 Inverse Functions

1   Decide whether a function is one-to-one. (Frames 1-6)

2   Show that two functions are inverses. (Frames 7-11)

3   Find the equation of the inverse of a function. (Frames 12-16)

4   Graph the inverse of f from the graph of f. (Frames 17-21)

---

1. In a function, for each value of ____ there is exactly one value of ____. If any two different values of x always produce two different values of ____, the function is _____.

        x

        y

        y; one-to-one

Identify the functions of Frames 2-4 that are one-to-one.

2. $y = -6x + 5$ _____

        one-to-one

3. $y = x^2 - 6$ _____

        not one-to-one

4. $y = \sqrt[3]{x + 2}$ _____

        one-to-one

5. A function is one-to-one if a _____ line cuts the graph of the function in no more than _____ point.

        horizontal

        one

6. Which of these functions are one-to-one?

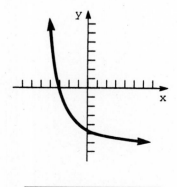   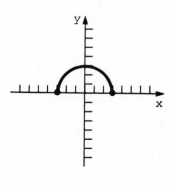

_____   _____

        one-to-one;
        not one-to-one

7. Any function that is always increasing or
_____ on its domain is _____ and
(has/does not have) an inverse.

decreasing; one-to-one
has

8. Recall the definition of the composition of
two functions: if f and g are functions, then

$(f \circ g)(x) =$ _____ and $(g \circ f)(x) =$ _____ .

$f[g(x)]$; $g[f(x)]$

9. Functions f and g are _____ if

$(f \circ g)(x) =$ ____ and $(g \circ f)(x) =$ ____ .

The function g can be written _____, and f can
be written _____.

inverses

x; x

f^{-1}

g^{-1}

10. Are $f(x) = 5x - 9$ and $g(x) = \dfrac{x + 9}{5}$ inverses of
each other?

To find out, find $(f \circ g)(x)$ and _____.

$(f \circ g)(x) = f[g(x)] = 5($____$) - 9 =$ ____

$(g \circ f)(x) = g[f(x)] = \dfrac{(\underline{\quad}) + 9}{5} =$ ____

Since both compositions equal x, the functions
(are/are not) inverses.

$(g \circ f)(x)$

$\dfrac{x + 9}{5}$; x

5x - 9; x

are

11. Are $f(x) = \sqrt[3]{x + 6}$ and $g(x) = x^3 - 6$ inverses
of each other?

_____

yes

12. The function $f(x) = 5x - 17$ (is/is not)
one-to-one, and thus has an _____ function.
To find the equation of the inverse, exchange
____ and ____. Write $y = 5x - 17$ as

_____.

Solve for y: y = _____.

Thus, $f^{-1}(x) =$ _____.

is

inverse

x; y

x = 5y - 17

$\dfrac{x + 17}{5}$

$\dfrac{x + 17}{5}$

Find the inverses of each one-to-one function in Frames 13-16.

13. $f(x) = 3x + 10$ $f^{-1}(x) = $ _____

$\dfrac{x - 10}{3}$

14. $f(x) = x^3 - 1$ $f^{-1}(x) = $ _____

$\sqrt[3]{x + 1}$

15. $y = x^2 - 4$ _____

not one-to-one

16. $y = \dfrac{2x - 5}{3x + 1}$ $f^{-1}(x) = $ _____

$\dfrac{5 + x}{2 - 3x}$

17. To obtain the graph of an inverse function, use the fact that the points (a, b) and (b, a) are "mirror images" with respect to the line _____. $y = x$
     The figure shows four points; find their mirror images with respect to the line $y = x$.

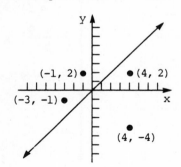

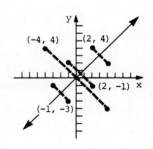

In each of Frames 18-20, the graph of a one-to-one function f is given. Sketch the graph of f^{-1}.

18.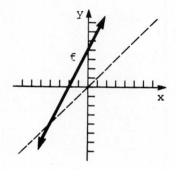

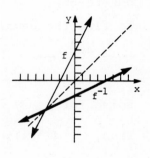

19.

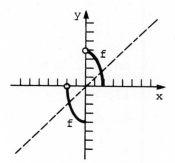

20.

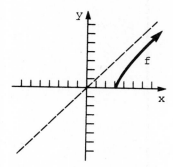

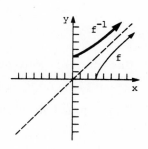

21. For inverse functions f and f⁻¹, the domain of f
is equal to the _____ of f⁻¹, and the range
of f is equal to the _____ of f⁻¹.

range

domain

10.2 Exponential Functions

[1] Learn the additional properties of exponential functions.
(Frames 1 and 2)

[2] Solve exponential equations. (Frames 3–9)

[3] Graph exponential functions. (Frames 10–13)

[4] Solve compound interest problems. (Frames 14–18)

[5] Use exponential functions in growth and decay applications.
(Frames 19–22)

1. So far we have defined a^m only for _____
exponents. In this section we will extend this
definition to include all _____ number exponents.

rational

real

2. If $a > 0$, $a \neq 1$, and if x, y, and z are any real numbers, we define a^x so that

    (a) a^x is a _____ _____ number.   unique; real

    (b)   If $x = y$, then _____ = _____. If a^x; a^y
          $a^x = a^y$, then ____ = ____.   x; y

    (c)   If $a > 1$ and if ____ $< y <$ ____, then   x; z
          _____ $< a^y <$ ____. a^x; a^z

    (d)   If $0 < a < 1$ and if $x < y < z$, then
          a^x ___ a^y ___ a^z.   >; >

3. Use these assumptions to solve the equation
$2^x = 1/16$. First write $1/16$ as $1/16 = 2^{---}$.   −4
The equation thus becomes

$$2^{---} = 2^{---},$$ x; −4

from which $x =$ _____.   −4

4. To solve $9^x = 243$, write 9 as _____, and 243 as 3^2
3^{---}. The equation thus becomes   5

$$(3^{---})^x = 3^{---},$$ 2; 5

  or
$$3^{---} = 3^{---},$$ 2x; 5

from which $2x =$ ____,   5
  or $x =$ _____.   5/2

Solve each of the equations in Frames 5–8.

5. $4^x = 64$ $x =$ _____   3

6. $16^x = 32$

$$(2^{---})^x = 2^{---}$$ 4; 5
$$2^{---} = 2^{---}$$ 4x; 5
$$x =$$ _____   5/4

7. $\left(\frac{1}{4}\right)^x = 32$ $x =$ _____   −5/2

8. $\left(\frac{1}{9}\right)^x = 27$ x = _____   −3/2

9. Using the assumptions in Frame 2 about real
    number exponents, we can define an _____   exponential
    function.

         y = _____, where a > 0, a ≠ 1. a^x

10. To graph an exponential function such as
                $y = 3^x$,
     first complete a table of values for the function,
     and then use these values to sketch the graph.
     Complete the chart below, and then sketch the
     graph by drawing a smooth _____ through the   curve
     points.

x	3^x
−3	____
___	1/9
−1	____
0	____
1	____
___	9

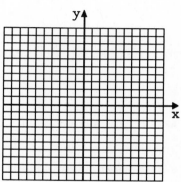

1/27
−2
1/3
1
3
2

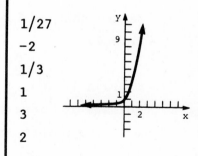

This graph is typical of $y = a^x$ for _____.   a > 1

11. To graph $y = \left(\frac{1}{3}\right)^x$ complete a table of values
     again, as below. Sketch the graph.

x	$\left(\frac{1}{3}\right)^x$
___	9
___	3
0	____
1	____
2	____

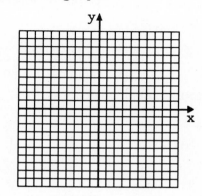

−2
−1
1
1/3
1/9

This graph is typical of the graph of $y = a^x$
for ____ < a < ____.   0; 1

12.   Graph $y = 4^{-x}$.

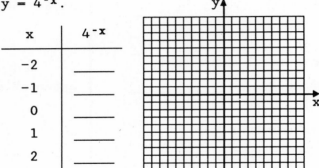

x	4^{-x}
−2	_____
−1	_____
0	_____
1	_____
2	_____

16
4
1
1/4
1/16

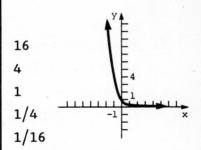

13.   In general, the graph of $f(x) = a^x$ contains
      the point _____. If $a > 1$, f is a(n)
      (*increasing/decreasing*) function and if $0 < a < 1$,
      f is a(n) (*increasing/decreasing*) function.
      The ____-axis is a _____ asymptote.
      The domain is _____ and the range is _____.

(0, 1)

increasing

decreasing

x; horizontal

$(-\infty, \infty)$; $(0, \infty)$

14.   We are able to find compound interest with the
      formula
      $$A = P\left(1 + \frac{i}{m}\right)^{\underline{\quad}},$$

      where A is the _____ amount on deposit if ____
      dollars are invested, i is the _____ rate,
      m is the number of times per year that interest
      is _____, and n is the number of _____.

nm

final; P

interest

compounded; years

**Find the compound interest in Frames 15–17. (Use a
calculator.)**

15. $15,000 is deposited at 14% interest for 6 years,
      with interest compounded annually.
      Here P = _____, i = _____, m = ____ (since
      interest is compounded _____ time per year), and
      n = ____. The total amount on deposit is
      $$A = 15,000(\underline{\qquad})^{\underline{\quad}}$$
      $$= 15,000(1.14)^{\underline{\quad}}$$
      Now use a calculator.
      $$A = 15,000(\underline{\qquad}) = \underline{\qquad}$$
      The interest earned is
      $$\underline{\qquad} - \$15,000 = \underline{\qquad}.$$

15,000; .14; 1

one

6

1 + .14; 6

6

2.19497; 32,924.55

$32,924.55; $17,924.55

16. $8500 is deposited at 12% compounded quarterly for 3 years.

    Here P = _____; i = _____, m = _____, and n = _____.   8500; .12; 4; 3

    A = _____ $12,118.97

    interest = _____ $3618.97

17. $11,600 is deposited at 18% compounded monthly for 2 1/2 years.

    A = _____ $18,131.73

    interest = _____ $6531.73

18. How much should be deposited now at 8% interest compounded semiannually to yield $10,000 in 4 years?

    Here, we solve for _____.   P

    $$A = P\left(1 + \frac{i}{m}\right)^{nm}$$

    _____ = P(_____)‾   10,000; $1 + \frac{.08}{2}$; 8

    _____ = P   7306.90

    The amount that should be deposited now

    (or the _____ _____) is   present value

    _____. $7306.90

19. Situations about growth or decay often involve the number _____. To nine decimal places,   e

    e = _____.   2.718281828

20. Suppose the population of a city is given by

    $$P(t) = 5000e^{.04t},$$

    where t is time in years. Find the population for the following values of t.

    (a) t = 0

    P(0) = 5000e——   .04(0)

    = 5000e——   0

    = 5000(___)   1

    = _____   5000

    The population is _____ when t = 0.   5000

(b)   t = 5

$$P(5) = 5000e^{\underline{\qquad}}$$ .04(5)

$$= 5000e^{.2}$$

Find a value of $e^{.2}$ from Table 2 or from a

calculator. Either way, $e^{.2}$ = _____,   1.22140

and

$$P(5) = 5000(\underline{\qquad}) = \underline{\qquad}.$$ 1.22140; 6107

The population in 5 years will be about

_____ people.   6100

21. Suppose the sales of an unadvertised product at

time t in years is

$$A(t) = 1000(2)^{-.2t}.$$

Find the sales when t = 0.   _____   1000

22. In Frame 21, find the sales of the product when

(a)   t = 5   _____   500

(b)   t = 10   _____   250

(c)   t = 15.   _____   125

10.3 Logarithmic Functions

1   Define a logarithm. (Frame 1)

2   Convert between exponential and logarithmic statements. (Frames 2–13)

3   Solve logarithmic equations of the form $\log_a b = k$ for a, b, or k. (Frames 14–29)

4   Graph logarithmic functions. (Frames 30–36)

5   Use logarithmic functions in applications. (Frames 37 and 38)

---

1. The exponential statement $x = a^y$ can be converted

to the _____ statement   logarithmic

_____ and $y = \log_a x$

read "y is the _____ of x to the base ___."   logarithm; a

2. $5^3 = 125$ can be written in logarithmic form as

$$\log\underline{\quad}\;\underline{\quad\quad} = \underline{\quad}.$$

5; 125; 3

Write the expressions in each of Frames 3–8 in logarithmic form.

3. $4^2 = 16$ $\qquad$ $\log\underline{\quad}\;\underline{\quad\quad} = \underline{\quad\quad}$ $\qquad$ 4; 16; 2

4. $6^3 = 216$ $\qquad$ $\log\underline{\quad}\;\underline{\quad\quad} = \underline{\quad\quad}$ $\qquad$ 6; 216; 3

5. $3^{-2} = \frac{1}{9}$ $\qquad$ $\log\underline{\quad}\;\underline{\quad\quad} = \underline{\quad\quad}$ $\qquad$ 3; $\frac{1}{9}$; -2

6. $2^{-3} = \frac{1}{8}$ $\qquad$ $\log\underline{\quad}\;\underline{\quad\quad} = \underline{\quad\quad}$ $\qquad$ 2; $\frac{1}{8}$; -3

7. $\left(\frac{1}{3}\right)^{-2} = 9$ $\qquad$ $\log\underline{\quad}\;\underline{\quad\quad} = \underline{\quad\quad}$ $\qquad$ 1/3; 9; -2

8. $\left(\frac{3}{4}\right)^{-2} = \frac{16}{9}$ $\qquad$ $\log\underline{\quad}\;\underline{\quad\quad} = \underline{\quad\quad}$ $\qquad$ 3/4; $\frac{16}{9}$; -2

9. We can convert a logarithmic expression such as

$$\log_{12} 144 = 2$$

into exponential form by writing $12^{\underline{\quad}} = 144$.

2

Convert the expressions of Frames 10–13 into exponential form.

10. $\log_8 64 = 2$ $\qquad$ $\underline{\quad\quad} = \underline{\quad\quad}$ $\qquad$ 8^2; 64

11. $\log_3 \frac{1}{3} = -1$ $\qquad$ $\underline{\quad\quad} = \underline{\quad\quad}$ $\qquad$ 3^{-1}; $\frac{1}{3}$

12. $\log_6 \frac{1}{36} = -2$ $\qquad$ $\underline{\quad\quad} = \underline{\quad\quad}$ $\qquad$ 6^{-2}; $\frac{1}{36}$

13. $\log_{10} .01 = -2$ $\qquad$ $\underline{\quad\quad} = \underline{\quad\quad}$ $\qquad$ 10^{-2}; .01

14. To solve the logarithmic equation

$$\log_3 x = 4,$$

write the statement in exponential form as

$\underline{\quad\quad} = \underline{\quad\quad}$

3^4; x

or

$x = \underline{\quad\quad}.$ $\qquad$ Solution set: $\underline{\quad\quad}$

81; $\{81\}$

15. To solve the equation

$$x = \log_5 125,$$

write _____ = 125

and then   5—— = 5——

from which   x = ____.   Solution set: _____

5^x

x; 3

3; {3}

16. To solve the equation $\log_x 9 = 2$, write the
statement in exponential form as

____ - ____

or   x = ____.   Solution set: _____

Recall that the use of a _____ must be a

_____ number. Although _____ is also a

solution of $x^2 = 9$, it is not a valid answer to

this problem.

x^2; 9

3; {3}

logarithm

positive; −3

Solve each of the equations in Frames 17–29.

17. $\log_x 64 = 3$ _____ - _____   x = _____ x^3; 64; 4

18. $\log_3 x = -2$ _____ - _____   x = _____ 3^{-2}; x; $\frac{1}{9}$

19. $\log_{10} 100 = x$ _____ - _____   x = _____ 10^x; 100; 2

20. $\log_5 x = -3$ _____ - _____   x = _____ 5^{-3}; x; $\frac{1}{125}$

21. $\log_x 16 = -4$ _____ - _____   x = _____ x^{-4}; 16; $\frac{1}{2}$

22. $\log_2 64 = x$ _____ - _____   x = _____ 2^x; 64; 6

23. $\log_x 27 = -3$ _____ - _____   x = _____ x^{-3}; 27; $\frac{1}{3}$

24. $\log_9 x = \frac{3}{2}$ _____ - _____   x = _____ $9^{3/2}$; x; 27

25. $\log_x 3 = 1$ _____ - _____   x = _____ x^1 or x; 3; 3

26. $\log_{32} 2 = x$ _____ - _____   x = _____ 32^x; 2; $\frac{1}{5}$

27. $\log_x 2 = \frac{2}{3}$ _____ - _____   x = _____ $x^{2/3}$; 2; $2^{3/2}$

28. $\log_m 5 = -1$ _____ - _____   m = _____ m^{-1}; 5; $\frac{1}{5}$

29. $\log_3 \sqrt{27} = x$ _____ - _____   x = _____ 3^x; $\sqrt{27}$; $\frac{3}{2}$

30. If a > 0, a ≠ 1, and x > 0, then

$$f(x) = \log_a x$$

is the _____ function with base ___.

logarithmic; a

31. Since exponential functions and logarithmic
functions are _____ of each other, the

inverses

domain of a logarithmic function is the set of
all _____ real numbers, and the range

positive

of a logarithmic function is the set of all
_____ numbers.

real

32. Logarithms can be found for _____

positive

numbers only.

33. To graph $f(x) = \log_2 x$, complete this table.

x	1/4	1/2	1	2	4
f(x)	−2	___	___	___	___

−1; 0; 1; 2

Graph $f(x) = \log_2 x$.

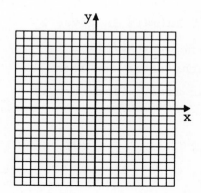

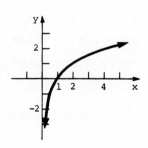

34. Graph $f(x) = \log_{1/2} x$.

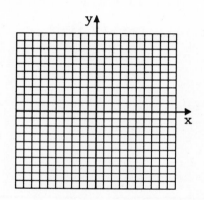

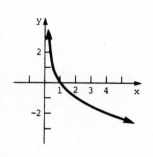

35. Graph $f(x) = \log_3 (x + 1)$.

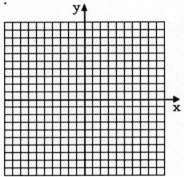

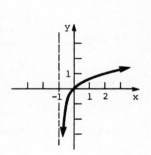

36. Graph $f(x) = \log_2 |x|$.

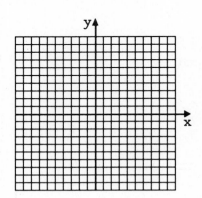

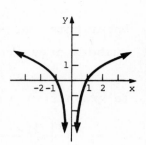

37. Suppose a population of animals is given by

$$P(t) = 150 \cdot \log_{10} (2t + 30),$$

where t is time in months after the animals were introduced into an area. Find each of the following.

(a) $P(35) = 150 \cdot \log_{10} (2 \cdot \underline{\qquad} + 30)$ 35

$= 150 \cdot \log_{10} (\underline{\qquad})$ 100

$= 150 \cdot (\underline{\qquad})$ 2

$= \underline{\qquad}$ 300

There were $\underline{\qquad}$ animals after 35 months.   300

(b) $P(50)$, given that $\log_{10} 130 \approx 2.1139$

$P(50) \approx \underline{\qquad}$ 317

38. Suppose $P(t) = 500 \cdot \log_{10} (t + 1)$. Find

(a) $P(0) = \underline{\qquad}$ 0

(b) $P(9) = \underline{\qquad}$ 500

(c) $P(99) = \underline{\qquad}$.   1000

10.4 Properties of Logarithms

$\boxed{1}$ Use the property for the logarithm of a product. (Frames 1–6)

$\boxed{2}$ Use the property for the logarithm of a quotient. (Frames 7–11)

$\boxed{3}$ Use the property for the logarithm of a power. (Frames 12–17)

$\boxed{4}$ Use the properties of logarithms to write logarithmic expressions in alternative forms. (Frames 18–31)

---

1. For all positive numbers x, y, and b, where $b \neq 1$,

    $\log_b xy = $ _____ + _____ .

    $\log_b x; \; \log_b y$

2. With this result,

    $\log_2 5y = \log_2$ ____ $+ \log_2$ ____ .

    $5; \; y$

3. Also, $\log_7 3k = \log_7$ ____ $+ \log_7$ ____ .

    $3; \; k$

Use the property for the logarithm of a product to complete the expressions in Frames 4–6.

4. $\log_k 7rm = \log_k$ ____ $+ \log_k$ ____ $+ \log_k$ ____

    $7; \; r; \; m$

5. $\log_2 8k = $ _____ + _____

    $\log_2 8; \; \log_2 k$

    Since $\log_2 8 = $ ____ , we have

    3

    $\log_2 8k = $ ____ $+ \log_2 k$.

    3

6. $\log_4 16m = \log_4$ ____ $+ \log_4 m$

    16

    $= $ ____ $+ \log_4 m$

    2

7. If x, y, and b are positive numbers, $b \neq 1$, then

    $\log_b \frac{x}{y} = $ _____ $-$ _____ .

    $\log_b x; \; \log_b y$

Use the property for the logarithm of a quotient to complete the expressions of Frames 8–11.

8. $\log_7 \frac{3}{k} = \log_7$ ____ $- \log_7$ ____

    $3; \; k$

9. $\log_{12} \dfrac{mn}{r} = \log_{12}$ ____ $+ \log_{12}$ ____ $- \log_{12}$ ____

                 m; n; r

10. $\log_2 \dfrac{r}{8} = \log_2$ ____ $- \log_2$ ____

            $=$ _____ $-$ ____

                 r; 8

                 $\log_2 r$; 3

11. $\log_3 \dfrac{9r}{2} = \log_3$ ____ $+ \log_3$ ____ $- \log_3$ ____

            $=$ ____ $+ \log_3$ ____ $- \log_3$ ____

                 9; r; 2

                 2; r; 2

12.   If x and b are positive, $b \neq 1$, and if r is any real number, then

                   $\log_b x^r =$ _____ .

                 $r(\log_b x)$

Use the property for the logarithm of a power to complete the expressions of Frames 13–17.

13. $\log_3 x^5 =$ ____ $(\log_3$ ____ $)$

                 5; x

14. $\log_6 5^x =$ ____ $(\log_6$ ____ $)$

                 x; 5

15. $\log_9 2^{3/5} =$ ____ $(\log_9$ ____ $)$

                 $\dfrac{3}{5}$; 2

16. $\log_2 12^{1/3} =$ ____ $(\log_2$ ____ $)$

                 $\dfrac{1}{3}$; 12

17. $\log_5 \sqrt[3]{125} = \log_5 125$——— $=$ ____ $(\log_5$ ____ $)$

            $=$ ____ $($ ____ $) =$ ____

                 1/3; $\dfrac{1}{3}$; 125

                 $\dfrac{1}{3}$; 3; 1

Use the three properties of logarithms to express each of the following as a sum, difference, or product of logarithms, or as a single number if possible.

18. $\log_5 \dfrac{3}{4} =$ _____ $-$ _____

                 $\log_5$ 3; $\log_5$ 4

19. $\log_7 \dfrac{m}{kr} =$ _____ $- ($ _____ $+$ _____ $)$

                 $\log_7$ m; $\log_7$ k; $\log_7$ r

20. $\log_3 9^{3/2} =$ ____ $(\log_3$ ____ $) =$ ____ $($ ____ $) =$ ____

                 $\dfrac{3}{2}$; 9; $\dfrac{3}{2}$; 2; 3

21. $\log_4 \sqrt{ab} = \log_4 (ab)$———

    $= \underline{\quad} (\log_4 \underline{\quad} + \log_4 \underline{\quad})$

                 1/2

                 $\frac{1}{2}$; a; b

22. $\log_6 \sqrt[3]{mk} = \log_6 (mk)$———

    $= \underline{\quad} (\log_6 \underline{\quad} + \log_6 \underline{\quad})$

                 1/3

                 $\frac{1}{3}$; m; k

Write each of the expressions of Frames 23—26 as a single logarithm.

23. $\log_5 m + \log_5 k + \frac{1}{2} \log_5 s = \log_5 \underline{\quad\quad}$

                 $mk\sqrt{s}$

24. $\log_a 3 - \log_a 2 + \log_a 5 = \log_a \underline{\quad\quad}$

                 $\frac{15}{2}$

25. $2 \log_k m - \log_k \sqrt{n} + \frac{1}{2} \log_k p = \log_k \underline{\quad\quad\quad}$

                 $\dfrac{m^2\sqrt{p}}{\sqrt{n}}$

26. $\log_k (x + 1) - \log_k 3x = \log_k \underline{\quad\quad\quad}$

                 $\dfrac{x + 1}{3x}$

Given that $\log_4 3 = .7925$ and $\log_4 7 = 1.4037$ evaluate the expressions in Frames 27—29.

27. $\log_4 21 = \log_4 (\underline{\quad} \cdot \underline{\quad})$

     $= \log_4 \underline{\quad} + \log_4 \underline{\quad}$

      $= \underline{\quad\quad} + \underline{\quad\quad}$

       $= \underline{\quad\quad}$

                 3; 7

                 3; 7

                 .7925; 1.4037

                 2.1962

28. $\log_4 \dfrac{3}{7} = \underline{\quad\quad}$

                 $-.6112$

29. $\log_4 49 = \underline{\quad\quad}$

                 2.8074

Decide whether the statements in Frames 30 and 31 are *true* or *false*.

30. $\log_5 12 - \log_5 3 = \log_5 9$

                 false

31. $\dfrac{\log_4 16}{\log_4 64} = \dfrac{2}{3}$

                 true

10.5   Evaluating Logarithms; Natural Logarithms

1   Evaluate common logarithms by using a calculator.   (Frames 1–5)

2   Find the antilogarithm of a common logarithm.   (Frames 6–10)

3   Use common logarithms in an application.   (Frames 11–16)

4   Evaluate natural logarithms using a calculator.   (Frames 17–21)

5   Find the antilogarithm of a natural logarithm.   (Frames 22–25)

6   Use natural logarithms in an application.   (Frames 26 and 27)

7   Use the change-of-base rule.   (Frames 28–33)

---

1. Logarithms to base 10 are called _____ log-
    arithms, and $\log_{10} x$ is abbreviated _____,
    where the base is understood to be _____.

    common
    log x
    10

2. To evaluate common logarithms using a calculator,
    enter the number and then touch the _____ key.

    log

Evaluate the common logarithms in Frames 3–5.

3. log 207.3 = _____

    2.3166

4. log 22,000 = _____

    4.3424

5. log .3127 = _____

    −.5049

6. The number whose common logarithm is k is called
    the _____ of k, abbreviated _____.

    antilogarithm; antilog

7. The common antilogarithm is found with most cal-
    culators by entering the number, then touching
    the _____ key, or it may be found by touching a
    combination of the _____ key and the _____ key.

    10^x
    INV; log

Find the common antilogarithm of each number in Frames 8–10. Round to four decimal places.

8. 3.1123   The common antilogarithm is _____.   1295.0901

9. .4307   The common antilogarithm is _____.   2.6959

10. −1.7533   The common antilogarithm is _____.   .0176

Given that pH $= -\log \left[H_3O^+ \right]$, find the pH in Frames 11–13 and find the hydronium ion concentration in Frames 14–16.

11. $\left[H_3O^+ \right] = 5.2 \times 10^{-3}$ pH = _____   2.3

12. $\left[H_3O^+ \right] = 1.7 \times 10^{-5}$ pH = _____   4.8

13. $\left[H_3O^+ \right] = 2.8 \times 10^{-6}$ pH = _____   5.6

14. pH = 6.2 $\left[H_3O^+ \right]$ = _____ 6.3×10^{-7}

15. pH = 8.3 $\left[H_3O^+ \right]$ = _____ 5.0×10^{-9}

16. pH = 2.1 $\left[H_3O^+ \right]$ = _____ 7.9×10^{-3}

17. Logarithms to base e are called _____ log-   natural

     arithms. The natural logarithm of x is written

     _____. To seven decimal places e = _____.   ln x; 2.7182818

18. To evaluate natural logarithms using a calculator,

     enter the number, then touch the _____ key.   ln

Evaluate the natural logarithms in Frames 19–21.
Round to four decimal places.

19. ln .6334   −.4567

20. ln 314.5   5.7510

21. ln 8.63   2.1552

22. To find natural antilogarithms using a calcu-
    lator, enter the number, then touch the _____
    key or touch a combination of the _____ key
    and the _____ key.

    e^x

    INV

    ln

**Find the natural antilogarithm of each number in
Frames 23–25. Give answers to three significant
digits.**

23. 4.0027   The natural antilogarithm is _____.   54.7

24. .0713   The natural antilogarithm is _____.   1.07

25. -2.3764   The natural antilogarithm is _____.   .0929

26. The number of years, $N(r)$, since two independ-
    ently evolving languages split off from a common
    ancestral language is approximated by
    $$N(r) = -5000 \ln r,$$
    where r is the percent of words from the ances-
    tral language common to both languages now. Find
    N if r = 30%.

    $N(\underline{\hspace{0.5cm}}) = -5000 \ln \underline{\hspace{0.5cm}}$ .3; .3

    $= -5000(\underline{\hspace{1cm}})$ -1.2040

    $= \underline{\hspace{1.5cm}}$ 6019.9

    Approximately _____ years have passed since the
    two languages separated.   6000

27. Refer to Frame 26. Find r if two languages split
    about 1500 years ago.

    $\underline{\hspace{1.5cm}} = -5000 \ln r$ 1500

    $\underline{\hspace{1.5cm}} = \ln r$ -.03

    $\underline{\hspace{1.5cm}} = r$ .74

    Approximately _____% of the words from the an-
    cestral language are common to both languages
    now.   75

28. The change-of-_____ rule states that if $a > 0$, $a \neq 1$, $b > 0$, $b \neq 1$, and $x > 0$, then

$$\log_a x = \underline{\hspace{2cm}}.$$

base
$\dfrac{\log_b x}{\log_b a}$

29. Any _____ number other than ____ can be used for b, but the two practical bases are _____ and _____.

positive; 1
e
10

Use the change-of-base rule to find each logarithm in Frames 30-32.

30. $\log_3 5 = \dfrac{\overline{\hspace{1.5cm}}}{\underline{\hspace{1cm}}} = \underline{\hspace{1.5cm}}$

$\dfrac{\log 5}{\log 3}$; 1.4650

31. $\log_8 17 = \underline{\hspace{1.5cm}}$

1.3625

32. $\log_{12} 2 = \underline{\hspace{1.5cm}}$

.2789

10.6 Exponential and Logarithmic Equations

1̄ Solve exponential equations. (Frames 1-4)

2̄ Solve logarithmic equations. (Frames 5-12)

3̄ Solve continuous compounding problems. (Frames 13-17)

4̄ Solve exponential growth and decay problems. (Frames 18 and 19)

1. To solve equations involving logarithms and exponents, we need two properties. Each has two parts.

   (i)   For all $b > 0$ and $b \neq 1$:

      (a) $x = y$ if $b^{\underline{\hspace{0.8cm}}} = b^{\underline{\hspace{0.8cm}}}$.

      (b) $b^x = b^y$ if ____ = ____.

   (ii)   For all $b > 0$, $b \neq 1$, and all _____ real numbers x and y:

      (a) $x = y$ if $\log_b$ ____ $= \log_b$ ____.

      (b) $\log_b x = \log_b y$ if ____ = ____.

x; y
x; y
positive
x; y
x; y

2. Solve $5^x = 15$.

Since 5 and 15 are not both powers of the same number, we cannot use Property (i/ii). But we can use Property (i/ii). We take base 10 logarithms of both sides. We have

$$\log \underline{\quad} = \log \underline{\quad}$$

or

$$\underline{\quad} \log \underline{\quad} = \log \underline{\quad}.$$

From this

$$x = \frac{\log \underline{\quad}}{\log \underline{\quad}}.$$

After dividing,

$$x = \underline{\qquad}.$$

Solution set: $\underline{\qquad}$

i
ii
5^x; 15
x; 5; 15
15
5
1.6825
{1.6825}

3. Solve $3^{x+1} = 20$.

$$(\underline{\qquad}) \log \underline{\quad} = \log \underline{\quad}$$

or

$$x + 1 = \underline{\qquad}$$

$$x = \frac{\log \underline{\quad}}{\log \underline{\quad}} - 1$$

$$= \underline{\qquad} - 1$$

$$= \underline{\qquad} - 1$$

$$= \underline{\qquad}$$

Solution set: $\underline{\qquad}$

x + 1; 3; 20
$\dfrac{\log 20}{\log 3}$
20
3
$\dfrac{1.3010}{.4771}$
2.7269
1.7269
{1.7269}

4. Solve the equation $2^{-y+1} = 15$.

Solution set: $\underline{\qquad}$

{-2.9073}

5. To solve $\log (x + 3) = \log (3x + 5)$, we use part (b) of Property (ii) to write

$$\underline{\qquad} = \underline{\qquad}$$

or

$$x = \underline{\quad}.$$

Solution set: $\underline{\quad}$

x + 3; 3x + 5
-1
{-1}

6. To solve $\log (3x - 1) - \log x = \log 2$, use the properties of logarithms to write

$$\log \underline{\qquad} = \log 2.$$

$\dfrac{3x - 1}{x}$

This leads to the equation

_____ = 2

$\dfrac{3x-1}{x}$

from which x = ____.

1

Solution set: _____

$\{1\}$

7. To solve $\log_3 x + \log_3 (x-2) = 1$, first write
$1 = \log_3$ ____. Since

3

$\log_3 x + \log_3 (x-2) = \log_3$ _____,

$x(x-2)$

the original equation can be written as

$\log_3$ _____ $= \log_3$ ____,

$x(x-2)$; 3

or

_____ = ____.

$x(x-2)$; 3

Expanding, we get

_____ = 0.

$x^2 - 2x - 3$

The solutions of this equation are x = ____ or

3

x = ____. Now check these proposed answers in

−1

the original equation. If we let x = 3,

$\log_3$ ___ $+ \log_3$ ___ $= 1$,

3; 1

which is (*true/false*). If we try x = −1, we

true

see that log x, or log (−1) does not exist.

Hence the solution set here is _____.

$\{3\}$

8. To solve $\log (x+2) + \log (x-1) = 1$, first
note that $1 = \log$ _____. Thus

10

$\log$ (_____)(_____) $= \log$ _____

$x+2$; $x-1$; 10

or

(_____)(_____) = _____.

$x+2$; $x-1$; 10

Simplifying,

_____ = 0,

$x^2 + x - 12$

from which x = ____ or x = ____. The only

−4; 3

valid solution for the given equation is

x = ____. So the solution set is _____.

3; $\{3\}$

9. To solve $\log x + \log (x+2) = 1$, since
$1 = \log 10$, write the quadratic equation

_____ = 0.

$x^2 + 2x - 10$

This equation can be solved by the quadratic
formula.

$$x = \frac{-(\quad) \pm \sqrt{(\quad)^2 - 4(\quad)(\quad)}}{2(\quad)}$$

$x = $ _____ or $x = $ _____

The only valid solution is _____.

Solution set: _____

2; 2; 1; −10

1

$-1 + \sqrt{11}$; $-1 - \sqrt{11}$

$-1 + \sqrt{11}$

$\{-1 + \sqrt{11}\}$

Solve the equations in Frames 10–12.

10. $\log x + \log 2x = 2$

$\log$ _____ $= \log$ _____

$x^2 = $ _____

The only valid answer here is $x = $ _____.

Solution set: _____

$2x^2$; 100

50

$5\sqrt{2}$

$\{5\sqrt{2}\}$

11. $\log (x + 2) + \log (x + 5) = 1$

Solution set: _____

$\{0\}$

12. $\log (x - 1) = \log (x + 2)$

Solution set: _____

$\emptyset$

13. At least in theory, interest on a bank deposit could be compounded every instant. The formula for this _____ compounding is

$A = $ _____

where P _____ are deposited at a rate of

_____ r, compounded _____ for

_____ years. A is the _____ _____ on deposit.

continuous

Pe^{rt}

dollars

interest; continuously

t; final; amount

Find the compound interest in Frames 14–16.

14. $12,000 at 10% compounded continuously for 5 years

$A = P$_____

$= ($ _____ $)e^{.10(\text{---})}$

$= 12,000e^{\text{---}}$

$= 12,000($ _____ $)$

$= $ _____

e^{rt}

12,000; 5

.5

1.64872

19,784.64

Now find the interest.

interest = _____ − $12,000 $19,784.64

= _____ $7784.64

15. $3150 at 12% compounded continuously for 10
years.

interest = _____ $7308.37

16. $19,200 at 14.6% compounded continuously for
4 1/2 years

interest = _____ $17,836.74

17. How long would it take for the money in an
account that is compounded continuously at 6%
interest to triple?

Here, we want to solve for ____. Since we want   t

to triple the principal, let A = ____P.   3

_____ = Pe——   3P; .06t

_____ = _____   3; e$^{.06t}$

Take natural logarithms on both sides.

ln 3 = ln _____   e$^{.06t}$

ln 3 = _____   .06t

$\dfrac{\underline{\qquad}}{\underline{\qquad}}$ = t $\dfrac{\ln 3}{.06}$

_____ = t   18.3102

It will take about ____ years for the amount to   18
triple.

18. Suppose the population of a city is given by

P(t) = 25,000e$^{.04t}$,

where t is time in years. How long will it take
for the population of the city to triple?

At time t = 0, the population of the city is

P(____) = 25,000e$^{.04(\underline{\quad})}$ 0; 0

= 25,000(_____)   e^0 or 1

= _____.   25,000

When the population triples, it will equal

____ × 25,000 = _____.   3; 75,000

To find the time for the population to triple,
we must solve for t in the equation

$$\underline{\hspace{2cm}} = 25{,}000e^{.04t}.$$ 75,000

First, divide both sides by _____ to get   25,000

$$\underline{\hspace{1.5cm}} = \underline{\hspace{2cm}}.$$ 3; $e^{.04t}$

Now take natural logarithms of each side.

$$\ln 3 = \ln \underline{\hspace{1.5cm}}$$ $e^{.04t}$

$$\ln 3 = \underline{\hspace{1.5cm}}$$ .04t

$$\frac{\underline{\hspace{1.5cm}}}{\underline{\hspace{1cm}}} = t$$ $\dfrac{\ln 3}{.04}$

$$\underline{\hspace{1.5cm}} = t$$ 27.5

The population will triple in about _____ years.   27.5

19. The amount, y, in grams, of a radioactive substance present at time t in seconds is given by

$$y = 80e^{-.05t}.$$

Find the half-life of the substance.

The half-life is the time it takes until
only _____ the substance is left. There are   half

_____ grams present initially (let t = ____ to   80; 0

see this), so the half-life is found by solving

$$\underline{\hspace{1.5cm}} = 80e^{-.05t}$$ 40

or $$\frac{1}{2} = \underline{\hspace{2cm}}.$$ $e^{-.05t}$

Take natural logarithms on each side to get

$$t = \underline{\hspace{1.5cm}}.$$ 13.9

The half-life is about _____ seconds.   14

Chapter 10 Test

The answers for these questions are at the back of this Study Guide.

1. True or false: $y = 3x + 2$ defines
   a one—to—one function.

   1. _____

2. Find the inverse $f^{-1}(x)$ of the
   one—to—one function

   $$f(x) = \sqrt[5]{2 - 4x}.$$

   2. _____

3. Graph the inverse of f on the same
   axes as f is graphed below.

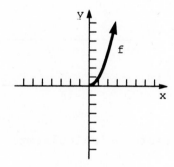

Graph each of the following.

4. $f(x) = 4^x$

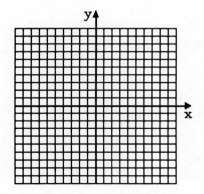

5. $f(x) = \left(\frac{1}{2}\right)^x$

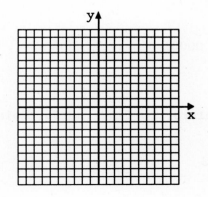

6. $f(x) = \log_{1/3} x$

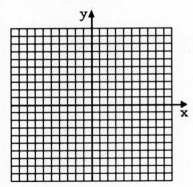

7. $f(x) = \log_2 x$

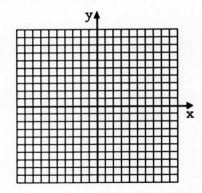

Given that log M = 3.1405 and log N = .8921, evaluate each of the following.

8. $\log \dfrac{N}{M}$

8. _____

9. log MN

9. _____

10. $\log N^3$

10. _____

Find each of the following without the use of a table or calculator.

11. $\log_3 81$

11. _____

12. $\log_4 \dfrac{1}{16}$

12. _____

13. $\log_4 8$

13. _____

14. $\log_{10} .00001$

14. _____

Use a calculator to find the following to four decimal places.

15. ln 27.3

15. _____

16. log .87

16. _____

17. $\log_4 17$ 17. _____

18. $\log_7 42$ 18. _____

Solve each of the following equations.

19. $\log_4 x = -\dfrac{3}{2}$ 19. _____

20. $\log_{1/25} x = -\dfrac{1}{2}$ 20. _____

21. $5^x = 14$ 21. _____

22. $\log 3x = \log 4 + \log (x - 2)$ 22. _____

23. $\log_5 (3x + 1) + \log_5 (x - 3) = 3$ 23. _____

Solve the following problems.

24. Suppose that a colony of bacteria is
     growing according to

$$Q(t) = 500e^{.04t},$$

     where $Q(t)$ is the number of bacteria
     present at time t (measured in hours).

     (a) How many bacteria will be present
          at time t = 0?   24. (a) _____

     (b) How many bacteria will be present
          at time t = 5?   (b) _____

     (c) When will the culture contain 1000
          bacteria?   (c) _____

25. Suppose that $2000 is deposited at 9%
     annual interest and no withdrawals are
     made. How much will be in the account
     after 3 years if interest is compounded

     (a) monthly   (b) continuously?   25. (a) _____

                                                        (b) _____

CHAPTER 11   SYSTEMS OF EQUATIONS AND INEQUALITIES

11.1   Linear Systems of Equations in Two Variables

1   Solve linear systems by the elimination method. (Frames 1–4)

2   Solve linear systems by the substitution method. (Frames 5–7)

3   Identify linear systems that are inconsistent or that have dependent equations. (Frames 8–13)

4   Solve linear systems that have fractions. (Frames 14 and 15)

---

1. A _____ of equations is made up of _____ or more equations. To solve a system, find all values that make all the equations true at the same _____. One method of solution, the elimination method, is shown in the next few frames.

system; two
time

2. To solve the system

$$x - 2y = 7$$
$$2x + 4y = 6,$$

eliminate the variable _____ by multiplying both sides of the first equation by _____ and _____ the two equations.

$$\overline{}$$
$$\underline{2x + 4y = 6}$$
$$\overline{}$$

Then y = _____. Substituting _____ for y in the first equation gives

$$x - 2(-1) = 7$$
$$x = \underline{}.$$

Check the answer by substituting the values for x and y in *both* equations.

$$(\underline{}) - 2(\underline{}) = \underline{}$$
$$2(\underline{}) + 4(\underline{}) = \underline{}$$

The solution set is _____.

x
-2; adding
-2x + 4y = -14
8y = -8
-1; -1
5
5; -1; 7
5; -1; 6
{(5, -1)}

Solve the systems of Frames 3 and 4.

3. $4x + 2y = 20$
   $3x + y = 14$

   It is easier this time to eliminate the variable y. Multiply the second equation by _____ and add.

   $$4x + 2y = 20$$
   $$\underline{\hspace{3cm}}$$
   $$-2x \quad\quad = -8$$
   $$x = \underline{\hspace{1cm}}$$

   Substitute ____ for x in either of the original equations to find that y = ____. The solution set for the system is _____.

   -2

   $-6x - 2y = -28$

   4

   4
   2
   $\{(4, 2)\}$

4. $-3x + 2y = 23$
   $2x + 3y = 2$

   (Hint: Multiply each equation by a different number in order to eliminate one of the variables.)

   Solution set: _____

   $\{(-5, 4)\}$

5. As an alternative to the elimination method, systems can also be solved by _____. For example, to solve the system

   $$3x + 4y = 17$$
   $$y = -5x$$

   by the substitution method, use the first equation and replace y with _____. This gives

   $$3x + 4y = \underline{\hspace{1.5cm}}$$
   $$3x + 4(\underline{\hspace{1cm}}) = 17$$
   $$\underline{\hspace{1.5cm}} = 17$$
   $$x = \underline{\hspace{1.5cm}}$$

   To find y, use the equation y = _____ and replace x with _____.

   $$y = -5x$$
   $$y = -5(\underline{\hspace{1cm}})$$
   $$y = \underline{\hspace{1cm}}$$

   The solution set is _____.

   substitution

   -5x

   17
   -5x
   -17x
   -1

   -5x
   -1

   -1
   5
   $\{(-1, 5)\}$

Solve the systems of Frames 6 and 7 by the substitution
method.

6. 8x - 3y = -1
      x = 4 - y

   Replace x with _____ in the first equation.   4 - y

$$8(\underline{\hspace{1cm}}) - 3y = -1$$ 4 - y

$$32\underline{\hspace{1cm}} - 3y = -1$$ - 8y

$$32 - \underline{\hspace{1cm}} = -1$$ 11y

$$-11y = \underline{\hspace{1cm}}$$ -33

$$y = \underline{\hspace{1cm}}$$ 3

   Use x = _____ to find that x = _____. The   4 - y; 1

   solution set is _____. $\{(1, 3)\}$

7. 2x - 5y = -43
   x + 2y = 10

   First, solve the second equation for x.

$$x = \underline{\hspace{2cm}}$$ 10 - 2y

   Now replace x with _____ in the first equation   10 - 2y

   and complete the solution.

                Solution set: _____ $\{(-4, 7)\}$

8. If a system of equations has no solution at all,

   the system is called _____.   inconsistent

9. If both equations of the system represent the

   same straight line, the equations are called

   _____.   dependent

10. To solve the system

            2x + 3y = 4
            4x + 6y = 5,

   multiply the first equation by _____, obtaining   -2

   the system

           _____   -4x - 6y = -8
           4x + 6y = 5.

   Then add the equations.

           _____   0 = -3

The result, 0 = -3, is a (*true/false*) statement,   false

so the system has _____ solution, and thus is   no

_____.   inconsistent

11. Find the solution set for the system

$$-x + 2y = 3$$
$$3x - 6y = 4.$$

The solution set is _____, so the system is   ∅

_____.   inconsistent

12. To eliminate x in the system

$$4x - y = 8$$
$$8x - 2y = 16,$$

multiply the first equation by _____, and add,   -2

giving

$$\underline{}$$ -8x + 2y = -16
$$8x - 2y = 16$$

_____.   0 = 0

The sum, 0 = 0, is (*true/false*), so the   true

equations of the system represent the same

straight lines. These equations are _____.   dependent

The solution set can be written by expressing x

in terms of y. Choose either equation and solve

for _____.   The solution set is written as   x

$$\{(\underline{}, y)\}.$$ $\dfrac{8 + y}{4}$

13. Solve the system

$$-2x - 4y = -8$$
$$x + 2y = 4.$$

The equations are _____, and the solution   dependent

set is an _____ set of ordered pairs. By   infinite

expressing x in terms of y, it can be written

$$\{(\underline{}, \underline{})\}.$$ 4 - 2y; y

14. To solve the system

$$\frac{x}{4} + \frac{y}{3} = 2$$
$$\frac{x}{3} + \frac{y}{4} = \frac{25}{12},$$

write the equations without fractions by multiplying each equation by _____, the LCD of all   12
the fractions. This gives

$$\underline{\hspace{4cm}}$$
$$\underline{\hspace{4cm}}.$$ $3x + 4y = 24$
                                                   $4x + 3y = 25$

Solve this system. The solution set is _____. $\{(4,\ 3)\}$

15. Solve the system

$$\frac{1}{3}x + \frac{1}{2}y = 1$$
$$\frac{1}{6}x + \frac{3}{2}y = -2$$

The solution set is _____. $\{(6,\ -2)\}$

11.2   Linear Systems of Equations in Three Variables

[1]   Solve linear systems with three equations and three variables
      by the elimination method. (Frames 1–5)

[2]   Solve linear systems with three equations and three variables
      where some of the equations have missing terms. (Frames 6–8)

[3]   Solve linear systems with three equations and three variables that
      are inconsistent or that include dependent equations. (Frames 9–11)

---

1. To begin to solve the linear system

$$2x + 3y - 4z = -5$$
$$3x + 2y + 3z = 13$$
$$x + 3y + 3z = 6,$$

select any _____ of the equations and eliminate   two
a _____ from them. For example, select the   variable
first two equations and eliminate x. To do this,
multiply the first equation by _____ and the   3
second equation by _____. This gives -2

$$\underline{\hspace{4cm}}$$
$$\underline{\hspace{4cm}}$$ $6x + 9y - 12z = -15$
                                                       $-6x - 4y - 6z = -26$
$$\underline{\hspace{4cm}} \quad (*)$$ $5y - 18z = -41$

Now eliminate ____ from another pair of equations.   x
To use the second and third, multiply the third
equation by _____. This gives -3

$$3x + 2y + 3z = 13$$

_____

_____.

Now form a system using the last equation together with equation (*) from above.

_____

_____

Solve this system to obtain y = ____ and z = ____.
Substituting these values into any of the equations above gives x = ____. The solution set is

_____.

$-3x - 9y - 9z = -18$
$-7y - 6z = -5$
$5y - 18z = -41$
$-7y - 6z = -5$
$-1; \; 2$
3
$\{(3, -1, 2)\}$

2. To solve the system

$$2x - 3y + 4z = -1$$
$$3x + 2y - 5z = -2$$
$$-x + 4y + 3z = 28,$$

eliminate x from the first two equations by
multiplying the first equation by ____ and the
second equation by _____, and adding. This
results in the equation

_____.

Now eliminate ____ from the second and third
equations, by multiplying the third equation by
____ and adding. This gives

_____.

Now form the system

_____
_____.

Solving this system gives y = ____ and z = ____.
Use these numbers to find that x = ____.

Solution set: _____

3
-2
$-13y + 22z = 1$
x
3
$14y + 4z = 82$
$-13y + 22z = 1$
$14y + 4z = 82$
$5; \; 3$
1
$\{(1, 5, 3)\}$

Solve the systems in Frames 3–5.

3. $2x + 2y + 3z = 6$
    $3x - 2y + 4z = 12$
    $3x + y + 2z = 1$

Solution set: _____

$\{(-2, -1, 4)\}$

4. $2x - y + z = -3$
    $x + 2y - z = 2$
    $3x - 5y + 4z = -9$

           Solution set: _____ $\{(-1, 2, 1)\}$

5. $2x - 3y + 3z = 4$
    $3x - 2y + 4z = 11$
    $4x - 2y + 3z = 5$

           Solution set: _____ $\{(-1, 3, 5)\}$

6. The system

$$x - y = -3$$
$$y + 2z = 8$$
$$2x + z = 1$$

has equations with missing terms. These systems
can be solved more quickly than the ones above.
To solve this system, first eliminate y by adding
the first two equations.

      _____ $x + 2z = 5$

This gives the system

$$x + 2z = 5$$
_____. $2x + z = 1$

Solve this system to get x = _____, z = _____. $-1; 3$
By substitution into the first equation of the
given system, y = _____. 2

           Solution set: _____ $\{(-1, 2, 3)\}$

Solve the systems in Frames 7 and 8.

7. $x + 4y = 16$
    $y - z = 3$
    $3x + 2z = 12$

           Solution set: _____ $\{(4, 3, 0)\}$

8. $2x + y = -1$
    $y + 3z = -2$
    $-x + 5z = 3$

           Solution set: _____ $\{(2, -5, 1)\}$

9. Recall from Section 11.1 that if, at any point
   in the solution of a system, a statement results
   that is false, such as 0 = 4, then the system is
   _____. If a statement results that is   inconsistent
   true, such as 0 = 0, then the equations are
   _____.   dependent

**Write *inconsistent system* or *dependent equations* for
the systems of Frames 10 and 11.**

10. 3x + 2y + 4z = 8
     2x - y + 4z = 4
     6x + 4y + 8z = 10   _____   inconsistent system

11. 4x - y + z = 6
     3x + 2y - 2z = 4
     8x - 2y + 2z = 12   _____   dependent equations

11.3 Applications of Linear Systems of Equations

[1] Solve geometry problems using two variables. (Frames 1 and 2)

[2] Solve money problems using two variables. (Frames 3–5)

[3] Solve mixture problems using two variables. (Frames 6 and 7)

[4] Solve distance–rate–time problems using two variables. (Frame 8)

[5] Solve problems with three unknowns using a system of three
     equations. (Frames 9–11)

---

1. **A triangle has two equal sides. The third side
   has a length which is 6 meters less than twice
   the length of one of the equal sides. The perimeter of the triangle is 30 meters. Find the
   length of each side.**

   The formula for the perimeter of a triangle is
   $P =$ _____, where a, b, and c represent the   a + b + c
   lengths of the sides of the triangle.

   Let x = the length of each equal side and y =
   the length of the third side. Since the perimeter is _____ meters,   30

_____ = x + ____ + ____   30; x; y

or   30 = _____.   2x + y

The third side is 6 meters less than twice the
length of one of the equal sides, so that

y = _____.   2x – 6

Find the values of x and y. x = ____, y = ____   9; 12

So, the length of the two equal sides is ____   9

meters and the length of the third side is ____   12

meters.

2. **The length of a rectangle is twice the width.
    The perimeter is 30 meters. Find the length and
    width of the rectangle.**

length = _____   10 meters

width = _____   5 meters

3. **George has a total of 18 dimes and nickels, with
    a total value of $1.65. How many of each kind of
    coin does he have?**

If x = the number of dimes, and y = the number of
nickels, then "there is a total of 18 coins" can
be written

_____.   x + y = 18

George has ____ dimes, and each is worth _____,   x; .10

making the total value of his dimes _____. In   .10x

the same way, the total value of his nickels is

_____. We know that the total value of all his   .05y

money is _____. Finally,   1.65

_____.   .10x + .05y = 1.65

Solve this system of equations.

x = _____   15

y = _____.   3

George has _____ dimes and _____ nickels.   15; 3

4. **Tricia has 152 bills, some one–dollar bills and
    some five–dollar bills. The total value of her
    money is $592. How many of each type of bill
    does she have?**

Let x = the number of ones and y = the number of fives. The system is

_____

_____ .

| | x + y = 152 |
| | x + 5y = 592 |

Solve this system.

x = _____

y = _____ .

| | 42 |
| | 110 |

Tricia has _____ ones and _____ fives.

| | 42; 110 |

5. Tickets to the class play cost $1.25 for children and $2.25 for adults. A total of $205.75 was collected from 119 people. How many were adults and how many were children?

_____ adults

_____ children

| | 57 |
| | 62 |

6. How many ounces each of 30% acid and 90% acid must be mixed together to obtain 30 ounces of 70% acid?

Let x = the amount of 30% acid and y = the amount of 90% acid. Complete the following chart.

Kind of acid	Total ounces	Ounces of acid
30%		
90%		
70%		

| x; .30x |
| y; .90y |
| 30; .70(30) |

Use this chart to write the two equations

_____

_____ .

| x + y = 30 |
| .30x + .90y = .70(30) |

Solving the system gives x = _____ and y = _____ .

So, _____ of 30% acid must be mixed with

_____ of 90% acid.

| 10; 20 |
| 10 ounces |
| 20 ounces |

7. How many liters of 20% acid and 50% acid must be added together to get 120 liters of 30% acid?

_____ liters of 20%

_____ liters of 50%

| 80 |
| 40 |

8. A plane goes 1000 miles in the same time that a
    car goes 300 miles. If the speed of the plane is
    20 miles per hour more than three times the speed
    of the car, find the speed of each.

	Distance	Rate	Time
Plane			
Car			

Since the times are equal, _____ = _____.

Multiply both sides by xy to get _____.

Since the speed of the plane is 20 miles per hour
more than the speed of the car, ____ = _____.

Solve the system

_____

_____.

                    Speed of the plane: _____

                    Speed of the car: _____

1000; x; $\dfrac{1000}{x}$

300; y; $\dfrac{300}{y}$

$\dfrac{1000}{x}$; $\dfrac{300}{y}$

$1000y = 300x$

x; $y + 20$

$1000y = 300x$

$x = y + 20$

200 miles per hour

60 miles per hour

9. Find three numbers whose sum is 30, if the first
    is two less than the second, and the second is
    two less than the third.

If x = the first number, y = the second, and
z = the third, the system of equations is

_____

_____

_____

Upon solving this system, x = ____, y = ____,
and z = ____. The solution set is _____.

The numbers are ____, ____, and ____.

$x + y + z = 30$

$x = y - 2$

$y = z - 2$

8; 10

12; {(8, 10, 12)}

8; 10; 12

10. The sum of three numbers is 30. The middle num-
     ber is 3 less than the largest number and 1 more
     than twice the smallest number. Find the numbers.

Let x = smallest number,

    y = _____ number,

and z = _____ number.

middle

largest

Now write a system of equations.

"The sum of three numbers is _____" becomes

$$\underline{\hspace{3cm}} = 30.$$

"The middle number is ____ less than the largest"
is
$$y = \underline{\hspace{2cm}}.$$

"The middle number is ____ more than _____ the
smallest" is
$$y = \underline{\hspace{2cm}}.$$

Solve this system to get _____.

The numbers are ____, ____, and ____.

30
$x + y + z$
3
$z - 3$
1; twice
$1 + 2x$
$\{(5, 11, 14)\}$
5; 11; 14

11. A pension fund invests $110,000. Part of the
money is put in bonds paying 10%. Ten thousand
dollars more than this amount goes into stocks
paying 15%. Finally, ten thousand dollars more
than the total of the other two investments goes
into a real estate deal paying 12%. Find the
amount invested at each rate.

Let x = amount invested at 10%,

y = amount invested at _____,

and z = amount invested at _____.

$$x + y + z = \underline{\hspace{2cm}}$$
$$y = x + \underline{\hspace{2cm}}$$
$$z = \underline{\hspace{1cm}} + \underline{\hspace{1cm}} + \underline{\hspace{1.5cm}}$$

Solve the system.

Amount at 10%: _____

Amount at 15%: _____

Amount at 12%: _____

15%
12%
110,000
10,000
x; y; 10,000
$20,000
$30,000
$60,000

11.4   Nonlinear Systems of Equations

1   Solve a nonlinear system by substitution. (Frames 1-5)

2   Use the elimination method to solve a system with two second-degree equations. (Frames 6-8)

3   Solve a system that requires a combination of methods. (Frames 9-11)

---

1. A system of equations in which at least one equation cannot be put into the form $Ax + By = C$ is called a _____ system of equations.

nonlinear

2. The substitution method is useful in solving a nonlinear system that includes a _____ equation. To solve the system

linear

$$x^2 + y^2 = 25$$
$$2x - y = 2,$$

we can first solve $2x - y = 2$ for y, obtaining $y =$ _____. Substitute this for y in the first equation.

$2x - 2$

$$x^2 + (\text{_____})^2 = 25$$

$2x - 2$

or

$$x^2 + (\text{_____}) = 25,$$

$4x^2 - 8x + 4$

which becomes

$$\text{_____} = 0.$$

$5x^2 - 8x - 21$

The last equation can be factored.

$$(x - 3)(\text{_____}) = 0$$

$5x + 7$

From this,

$$x = \text{\_\_\_\_} \quad \text{or} \quad x = \text{_____}.$$

$3; -\dfrac{7}{5}$

To find y, use the equation $y =$ _____.

$2x - 2$

If $x = 3$, then $y = 2(\text{\_\_\_\_}) - 2 = \text{\_\_\_\_}$.

$3; 4$

If $x = -\dfrac{7}{5}$, then $y = 2(\text{\_\_\_\_}) - 2 = \text{_____}$.

$-\dfrac{7}{5}; -\dfrac{24}{5}$

The solution set of the system is

$$\{(\text{\_\_\_\_}, \text{\_\_\_\_}), (\text{\_\_\_\_}, \text{\_\_\_\_})\}.$$

$3; 4; -\dfrac{7}{5}; -\dfrac{24}{5}$

Graph each equation of this system on the following axes to show that there are two points of

_____.

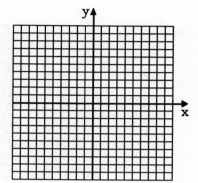

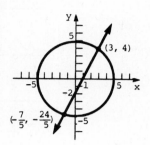

3. Solve the following system.

$$y = x^2 - 4x + 5$$
$$x - 2y = -1$$

We can solve the second equation for _____,

obtaining x = _____. Substitute this into the
first equation.

$$y = (\underline{\hspace{1cm}})^2 - 4(\underline{\hspace{1cm}}) + 5$$
$$y = \underline{\hspace{1.5cm}} + \underline{\hspace{1.5cm}} + 5$$
$$y = \underline{\hspace{2.5cm}}$$

or

$$0 = \underline{\hspace{2.5cm}}$$

which gives $(\underline{\hspace{0.8cm}})(\underline{\hspace{0.8cm}}) = 0$.

Hence y = _____ or y = _____.

Find the corresponding values of x and complete
the solution set.

$$\{(\underline{\hspace{0.8cm}}, 2), (\underline{\hspace{0.8cm}}, \tfrac{5}{4})\}$$

x

2y − 1

2y − 1; 2y − 1

$4y^2 - 4y + 1$; $-8y + 4$

$4y^2 - 12y + 10$

$4y^2 - 13y + 10$

y − 2; 4y − 5

2; $\dfrac{5}{4}$

3; $\dfrac{3}{2}$

Solve the systems of Frames 4 and 5 by substitution.

4. $x^2 - y^2 = 8$
    $x + 3y = 6$

From the second equation,

$$x = \underline{\hspace{2cm}}.$$

Now substitute.

$$(6 - 3y)^2 - y^2 = 8$$

$$(\underline{\hspace{3cm}}) - y^2 = 8$$

$$\underline{\hspace{3cm}} = 0$$

or

$$\underline{\hspace{3cm}} = 0$$

or

$$(\underline{\hspace{1.5cm}})(\underline{\hspace{1.5cm}}) = 0$$

$$y = \underline{\hspace{1cm}} \quad \text{or} \quad y = \underline{\hspace{1cm}}$$

Solution set: $\left\{ \left(\underline{\hspace{1cm}}, \frac{7}{2}\right), \left(\underline{\hspace{1cm}}, 1\right) \right\}$

$6 - 3y$

$36 - 36y + 9y^2$

$8y^2 - 36y + 28$

$2y^2 - 9y + 7$

$2y - 7;\ y - 1$

$\frac{7}{2};\ 1$

$-\frac{9}{2};\ 3$

5. $y^2 - x^2 = 16$
   $2x^2 + y^2 = 43$

Solution set:

$\{(3, \underline{\hspace{1cm}}), (3, \underline{\hspace{1cm}}), (-3, \underline{\hspace{1cm}}), (-3, \underline{\hspace{1cm}})\}$ $5;\ -5;\ 5;\ -5$

6. If both equations are nonlinear, the \underline{\hspace{2cm}}
   method often can be used. For example, to solve
   the system

$$x^2 + y^2 = 25$$
$$x^2 - y^2 = 7,$$

   add, obtaining

$$x^2 + y^2 = 25$$
$$\underline{x^2 - y^2 = 7}$$
$$\underline{\hspace{3cm}}$$

   from which $x = \underline{\hspace{1cm}}$ or $x = \underline{\hspace{1cm}}$.

   If $x = 4$, then $(\underline{\hspace{1cm}})^2 + y^2 = 25$, and

$$y^2 = \underline{\hspace{1cm}}.$$

   Then, $y = \underline{\hspace{1cm}}$ or $y = \underline{\hspace{1cm}}$.

   In the same way, if $x = -4$, then $y = \underline{\hspace{1cm}}$ or

   $y = \underline{\hspace{1cm}}$.

   Write the solution set of the system.

$\{(4, \underline{\hspace{1cm}}), (4, \underline{\hspace{1cm}}), (-4, \underline{\hspace{1cm}}), (-4, \underline{\hspace{1cm}})\}$

   Graph both equations of this system.

   $x^2 - y^2 = 7$ has x-intercepts at \underline{\hspace{2cm}} and

   \underline{\hspace{2cm}}. It has \underline{\hspace{1cm}} y-intercepts.

elimination

$2x^2 = 32$

$4;\ -4$

4

9

$3;\ -3$

3

-3

$3;\ -3;\ 3;\ -3$

$(\sqrt{7}, 0)$

$(-\sqrt{7}, 0);$ no

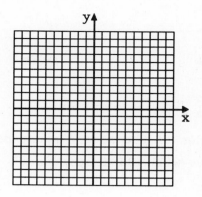

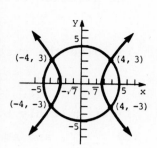

7. Solve the following system.

$$x^2 + 2y^2 = 9$$
$$6x^2 - y^2 = 2$$

Multiply the second equation by _____, and add.

$$x^2 + 2y^2 = 9$$

2

_____

$$12x^2 - 2y^2 = 4$$

_____

$$13x^2 = 13$$

From the last equation, x = _____ or x = _____.

1; −1

Find the corresponding values of y to complete

the solution set.

$\{(1, \_\_\_), (1, \_\_\_), (-1, \_\_\_), (-1, \_\_\_)\}$

2; −2; 2; −2

8. $x^2 + y^2 = 9$
    $2x^2 - y^2 = 3$

Use the elimination method.

$$x^2 + y^2 = 9$$
$$2x^2 - y^2 = 3$$

_____

$3x^2 = 12$

$$x^2 = \_\_\_\_$$

4

$$x = \_\_\_\_ \quad \text{or} \quad x = \_\_\_\_$$

2; −2

To find y, substitute the x-values into either

equation.

$$2^2 + y^2 = 9$$

$$y^2 = \_\_\_\_$$

5

$$y = \sqrt{5} \quad \text{or} \quad y = -\sqrt{5}$$

and

$$(-2)^2 + y^2 = 9$$

$$y = \_\_\_\_ \quad \text{or} \quad y = \_\_\_\_$$

$\sqrt{5}$; $-\sqrt{5}$

The solution set is _____.

$\{(2, \sqrt{5}), (2, -\sqrt{5}),$
$\{(2, \sqrt{5}), (-2, -\sqrt{5})\}$

9. To solve the system

$$x^2 - xy + y^2 = 3$$
$$x^2 + y^2 = 5,$$

first multiply both sides of the second equation

by _____, and add. This will eliminate the x^2 -1

and _____ terms. y^2

$$x^2 - xy + y^2 = 3$$
$$\underline{-x^2 \qquad \underline{\quad} = \underline{\quad}}$$ $-y^2$; -5

Adding gives   _____, $-xy = -2$

or $xy = $ ____. 2

From $xy = 2$, we can solve for x or y. Let us

solve for y to get

$$y = \underline{\quad}.$$ $\dfrac{2}{x}$

Substitute this result into the given equation

$x^2 + y^2 = 5$.

$$x^2 + (\underline{\quad})^2 = 5$$ $\dfrac{2}{x}$

$$x^2 + \underline{\quad} = 5$$ $\dfrac{4}{x^2}$

Multiply both sides by _____, getting x^2

_____, $x^4 + 4 = 5x^2$

or   _____ $= 0$. $x^4 - 5x^2 + 4 = 0$

Factor.

$(\underline{\qquad})(\underline{\qquad}) = 0$ $x^2 - 4$; $x^2 - 1$

This gives two equations,

$$x^2 - 4 = 0 \quad\text{or}\quad \underline{\quad} = 0.$$ $x^2 - 1$

$(\underline{\quad})(\underline{\quad}) = 0$ or $(\underline{\quad})(\underline{\quad}) = 0$ $x + 2$; $x - 2$; $x + 1$; $x - 1$

$x = $ ____ or $x = $ ____ or $x = $ ____ or $x = $ ____ -2; 2; -1; 1

Since $y = 2/x$, the solution set of the given

system is

$\{(2, \underline{\quad}), (-2, \underline{\quad}), (1, \underline{\quad}), (-1, \underline{\quad})\}$. 1; -1; 2; -2

10. $x^2 + 2xy + y^2 = 20$
     $x^2 - xy + y^2 = 5$

Eliminate the x^2 and y^2 variables by

addition.

$$x^2 + 2xy + y^2 = 20$$
$$\underline{-x^2 + xy - y^2 = -5}$$

$3xy = 15$

Solve for y.

$$y = \underline{} \text{ with } x \neq 0$$

$\dfrac{5}{x}$

Substitute in the first equation.

$$x^2 + 2x(\underline{}) + (\underline{})^2 = 20$$

$\dfrac{5}{x}; \dfrac{5}{x}$

$$x^2 + 10 + \underline{} = 20$$

$\dfrac{25}{x^2}$

$$x^4 + 10x^2 + 25 = 20x^2$$

$$\underline{} = 0$$

$x^4 - 10x^2 + 25$

$$(x^2 - 5)(x^2 - 5) = 0$$

$$x = \underline{} \quad \text{or} \quad x = \underline{}$$

$\sqrt{5}; -\sqrt{5}$

Substitute this value for x into $y = 5/x$ to find y.

Solution set: $\underline{}$

$\{(\sqrt{5}, \sqrt{5}),$
$(-\sqrt{5}, -\sqrt{5})\}$

11. $x^2 + y^2 = 9$
     $y - |x| = -3$

The substitution method is required here. The equation $y - |x| = -3$ can be rewritten as $y + 3 = |x|$. Then, by the definition of absolute value,

$$x = y + 3 \quad \text{or} \quad x = -(\underline{}) = \underline{}.$$

$y + 3; -y - 3$

Substitute into $x^2 + y^2 = 9$.

$$(y + 3)^2 + y^2 = 9 \quad \text{or} \quad (\underline{})^2 + y^2 = 9$$

$-y - 3$

$$y^2 + 6y + 9 + y^2 = 9 \quad \text{or} \quad \underline{} + y^2 = 9$$

$y^2 + 6y + 9$

Since these two equations are the same, either equation becomes

$$\underline{} = 0$$

$2y^2 + 6y$

$$\underline{} = 0$$

$2y(y + 3)$

$$y = 0 \quad \text{or} \quad y = \underline{}$$

-3

Thus the solution set is $\underline{}$.

$\{(3, 0), (-3, 0),$
$(0, -3)\}$

11.5   Second-Degree Inequalities, Systems, and Linear Programming

1   Graph second-degree inequalites. (Frames 1-3)

2   Graph the solution set of a system of inequalities. (Frames 4-10)

3   Solve linear programming problems by graphing. (Frames 11 and 12)

---

1. Second-degree inequalities are graphed in a manner much like that of linear inequalities. For
example, in order to graph the solution set of
$y < x^2 - 3$, the boundary, with equation $y = x^2 - 3$,
is a _____ with vertex _____, opening
_____, graphed as a _____ curve. Choose
a point on the exterior of the parabola, say
(3, 0). This ordered pair (*does/does not*) satisfy $y < x^2 - 3$. Therefore, shade (*outside/inside*)
the parabola.

parabola; (0, -3)

upward; dashed

does

outside

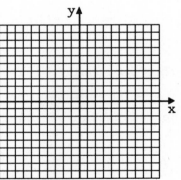

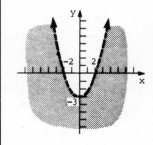

2. $x^2 - y^2 \geq 9$

The boundary is the (*solid/dashed*) _____
with x-intercepts (___, 0) and (___, 0). Choose
a test point, off the graph, to determine which
region to shade.

solid; hyperbola

-3; 3

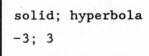

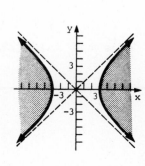

3. $(x - 1)^2 + (y + 2)^2 < 16$

The boundary is the (*solid/dashed*) _____ with
_____ at (1, −2). Choose a test point off the
graph to determine which region to shade.

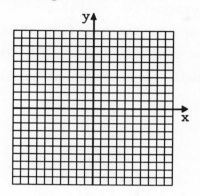

dashed; circle

center

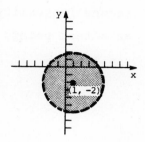

4. A system containing at least one inequality is
called a system of _____. To graph such
a system, find the _____ of the graphs of
each of the inequalities.

inequalities

intersection

**Graph the solution sets of the systems of inequalities
in Frames 2—6.**

5. $y - \frac{1}{2}x > 1$

$y + \frac{1}{2}x \leq 1$

First, note that this is a system of linear
inequalities. We graph the boundaries, being
careful to use dashed lines when appropriate.

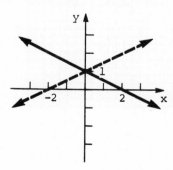

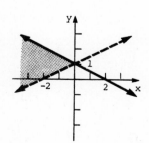

Use test points to shade the proper area.

6. $x^2 + y^2 \leq 9$
   $y \geq 2^x$

First sketch the graphs, starting with the

_____ with center at _____ and radius _____   circle; (0, 0); 3

as the boundary of one inequality of the system

and an _____ graph as the boundary of the   exponential

second inequality of the system. Now use (0, 0)

as a test point, and shade the appropriate region.

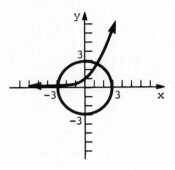

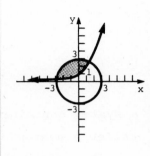

7. $x^2 - y^2 \leq 1$
   $y < 1$

The two equations to be graphed are a

_____ with x-intercepts (1, 0) and   hyperbola

(-1, 0) and the _____ line y = 1.   horizontal

Graph the system on the grid below.

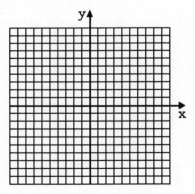

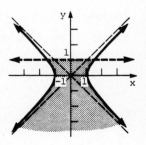

8. $x < 3$
    $y < 3$
    $y \leq |x| - 2$

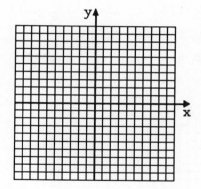

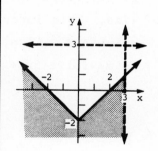

9. $\dfrac{x^2}{4} + \dfrac{y^2}{16} \leq 1$
    $y \leq |x - 1|$
    $y < x^2$

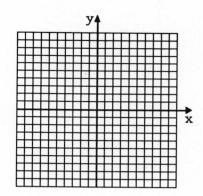

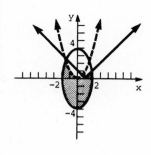

10. $x^2 + y^2 \geq 16$ and
     $x^2 + y^2 \leq 25$

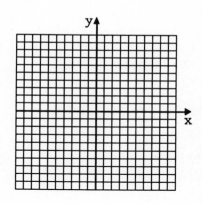

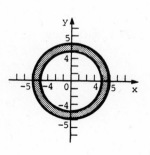

11. The method of linear programming requires that we
    find the maximum or _____ values of a linear   minimum
    expression, subject to restrictions, or _____ .   constraints
    For example, let us find the maximum value of the
    expression 5x + 2y subject to

$$2x + 3y \leq 6$$
$$4x + \ y \leq 6$$
$$x \geq 0$$
$$y \geq 0.$$

First, graph the intersection of these inequalities. This intersection is the region of _____
solutions. Also, locate all _____ points.
To maximize 5x + 2y, try all _____ points.

feasible

corner, or vertex

corner, or vertex

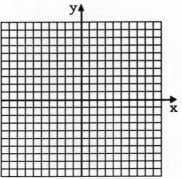

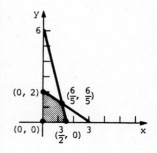

Vertex point	Value of 5x + 2y	
$(0, \underline{\quad})$	$5 \cdot 0 + 2 \cdot 0 = 0$	0
$(\underline{\quad}, 2)$	$5 \cdot 0 + 2 \cdot 2 = 4$	0
$\left(\frac{6}{5}, \frac{6}{5}\right)$	$5 \cdot \frac{6}{5} + 2 \cdot \frac{6}{5} = \underline{\quad} = 8.4$	$\frac{42}{5}$
$\left(\frac{3}{2}, \underline{\quad}\right)$	$5 \cdot \frac{3}{2} + 2 \cdot 0 = \underline{\quad} = \underline{\quad}$	$0; \frac{15}{2}; 7.5$

The maximum is _____ at the vertex point _____.

$8.4; \left(\frac{6}{5}, \frac{6}{5}\right)$

12. Find the minimum value of 2x + y subject to

$$3x - y \geq 12$$
$$x + y \leq 15$$
$$x \geq 2$$
$$y \geq 5.$$

The minimum value is _____ at the vertex point

_____.

$\frac{49}{3}$

$\left(\frac{17}{3}, 5\right)$

Chapter 11 Test

The answers for these questions are at the back of this Study Guide.

Solve each system.

1. $3x - y = -7$
   $-2x + 3y = 14$

   1. _____

2. $2x - 3y = -11$
   $3x + 2y = 16$

   2. _____

3. $5x + 2y = 1$
   $10x = y - 13$

   3. _____

4. $6x + y = 4$
   $y = 2 - 6x$

   4. _____

5. $\dfrac{x}{4} + \dfrac{y}{6} = -1$
   $\dfrac{3x}{2} + \dfrac{5y}{6} = -7$

   5. _____

6. $\dfrac{x}{3} + \dfrac{2y}{5} = 4$
   $-\dfrac{x}{6} + \dfrac{y}{5} = 0$

   6. _____

7. $\dfrac{1}{x} + \dfrac{2}{y} = -1$
   $-\dfrac{3}{x} + \dfrac{4}{y} = -7$

   7. _____

8. $\dfrac{2}{x} + \dfrac{4}{y} = 6$
   $\dfrac{3}{x} + \dfrac{9}{y} = 12$

   8. _____

9. $3x + y + 2z = -1$
      $2x + 2y + z = 2$
      $-2x - y + 2z = 0$

9. _____

10. $x - 2y + 4z = -10$
      $-3x + 6y - 12z = 20$
      $2x + 5y + z = 12$

10. _____

11. $2x - 3z = -4$
      $2y + z = 8$
      $x + y = 4$

11. _____

Translate each problem into a system of equations and then solve.

12.   A chemist needs to mix a 70% solution with
      12 liters of a 40% solution to get a 50%
      solution. How many liters of the 70% so-
      lution are needed?

12. _____

13.   Two cars start from points 760 miles apart
      and travel toward each other. They meet
      after 8 hours. Find the average speed of
      the slower car if one travels 15 miles per
      hour faster than the other.

13. _____

14.   The perimeter of a triangle is 30 meters.
      The sum of the smallest and largest sides
      is 21 meters, while the difference between
      the middle and smallest side is 3 meters.
      Find the lengths of the three sides.

14. _____

Solve each nonlinear system.

15. $xy = 1$
      $y = x$

15. _____

16. $x^2 + y^2 = 9$
      $y + x = 3$

16. _____

17. $x^2 + y^2 = 25$
      $2x^2 - y^2 = 2$

17. _____

Graph each of the following.

18. $y \leq x^2 - 2x + 3$

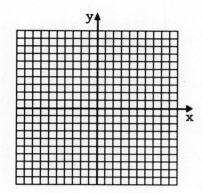

19. $x^2 + y^2 \leq 9$

$\dfrac{x^2}{25} + \dfrac{y^2}{4} \geq 1$

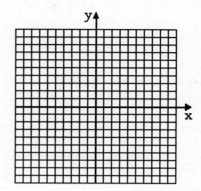

20. Find $x \geq 0$ and $y \geq 0$ such that

$$2x + 3y \leq 6$$
$$4x + y \leq 6$$

and $5x + 2y$ is maximized.

20. _____

CHAPTER 12 MATRICES AND DETERMINANTS

12.1 Matrices and Determinants

1   Understand the terminology of matrices. (Frames 1–4)

2   Evaluate 2 × 2 determinants. (Frames 5–9)

3   Use expansion by minors about the first column to evaluate determinants. (Frames 10–12)

4   Use expansion by minors about any row or column to evaluate determinants. (Frames 13–17)

5   Evaluate 4 × 4 or larger determinants. (Frames 18 and 19)

---

1. A _____ (plural: _____) is a rectan-   matrix; matrices
    gular array of numbers enclosed by brackets or
    _____ in which the _____ of   parentheses; position
    each number is meaningful. Matrices are clas-
    sified by _____, that is, by the number of   size
    _____ and _____ in the matrix. A matrix   rows; columns
    with m rows and n columns is an _____ matrix.   m × n
    The numbers in a matrix are its _____.   elements

Find the size of each matrix in Frames 2–4.

2. $\begin{bmatrix} 9 & 7 & -10 & 14 \\ 5 & -3 & 4 & 8 \end{bmatrix}$ Size: _____   2 × 4

3. $\begin{bmatrix} 3 & -2 \\ 2 & -3 \end{bmatrix}$ Size: _____   2 × 2

    Since the number of rows equals the number of
    columns, this matrix is a _____ matrix.   square

4. $\begin{bmatrix} 1 \\ 9 \end{bmatrix}$ Size: _____   2 × 1

    This matrix has only one _____, so it is   column
    called a _____ matrix.   column

5. Every square matrix is associated with a real
   number called the _____ of the matrix.   determinant
   While _____ are enclosed by square brackets,   matrices
   _____ are denoted by vertical bars.   determinants

6. The determinant of a 2 × 2 matrix

   $$\begin{bmatrix} a & b \\ c & d \end{bmatrix}$$

   is defined as

   $$\begin{vmatrix} a & b \\ c & d \end{vmatrix} = \underline{\hspace{1cm}} .$$ $ad - bc$

7. Using this definition,

   $$\begin{vmatrix} -1 & 2 \\ 3 & -4 \end{vmatrix} = (\underline{\hspace{0.4cm}})(\underline{\hspace{0.4cm}}) - (\underline{\hspace{0.4cm}})(\underline{\hspace{0.4cm}}) = \underline{\hspace{0.4cm}} .$$ -1; -4; 3; 2; -2

8. Also,

   $$\begin{vmatrix} 2 & -1 \\ -1 & 5 \end{vmatrix} = (\underline{\hspace{0.4cm}})(\underline{\hspace{0.4cm}}) - (\underline{\hspace{0.4cm}})(\underline{\hspace{0.4cm}}) = \underline{\hspace{0.4cm}} .$$ 2; 5; -1; -1; 9

9. $\begin{vmatrix} -3 & 0 \\ 2 & 4 \end{vmatrix} = (\underline{\hspace{0.4cm}}) - (\underline{\hspace{0.4cm}}) = \underline{\hspace{0.4cm}}$ -12; 0; -12

10. We can also evaluate 3 × 3 and larger determi-
    nants by _____ by minors about the first   expanding
    column.

11. Find the determinant

    $$\begin{vmatrix} -3 & 0 & 4 \\ 2 & -1 & 3 \\ -2 & 1 & 5 \end{vmatrix}$$

    by expanding by minors about the first column.
    We must find the minor of each entry in the
    _____ column. The minor of −3 is the deter-   first
    minant of the 2 × 2 matrix remaining when the
    _____ and _____ containing −3 are eliminated.   row; column

    The minor of −3 is $\begin{vmatrix} \underline{\hspace{0.6cm}} & \underline{\hspace{0.6cm}} \\ \underline{\hspace{0.6cm}} & \underline{\hspace{0.6cm}} \end{vmatrix}$. -1; 3
                                                                        1; 5

    In the same way we can find the minors of 2 and
    −2.

The minor of 2 is $\begin{vmatrix} \underline{} & \underline{} \\ \underline{} & \underline{} \end{vmatrix}$.

0; 4
1; 5

The minor of -2 is $\begin{vmatrix} \underline{} & \underline{} \\ \underline{} & \underline{} \end{vmatrix}$.

0; 4
-1; 3

The desired determinant is given by

$(\underline{})\begin{vmatrix} -1 & 3 \\ 1 & 5 \end{vmatrix} - (\underline{})\begin{vmatrix} 0 & 4 \\ 1 & 5 \end{vmatrix} - (\underline{})\begin{vmatrix} 0 & 4 \\ -1 & 3 \end{vmatrix}$.

-3; -; 2; +; -2

We now have

$= (\underline{})(\underline{}) - (\underline{})(\underline{}) + (\underline{})(\underline{})$

-3; -8; 2; -4; -2; 4

$= \underline{}.$

24

12. Find the determinant

$$\begin{vmatrix} 2 & -3 & 4 \\ -1 & 5 & 7 \\ 0 & 4 & 3 \end{vmatrix}$$

by expansion about the first column.   _____

-51

13.   Find the determinant

$$\begin{vmatrix} 2 & 0 & 3 \\ -4 & 3 & 1 \\ 3 & -2 & 1 \end{vmatrix}$$

by expanding about the second row.

$-(-4)\begin{vmatrix} 0 & 3 \\ -2 & 1 \end{vmatrix} + (\underline{})\begin{vmatrix} \underline{} & \underline{} \\ \underline{} & \underline{} \end{vmatrix} - 1\begin{vmatrix} 2 & 0 \\ 3 & -2 \end{vmatrix}$

3; 2; 3
    3; 1

where we used a + or - sign according to the
following sign array.

$$\begin{matrix} + & - & + \\ - & + & - \\ + & - & + \end{matrix}$$

We now have

$= (\underline{})(\underline{}) + (\underline{})(\underline{}) - (\underline{})(\underline{})$

4; 6; 3; -7; 1; -4

$= \underline{}.$

7

**Evaluate each of the determinants of Frames 14–17 by
expanding about any row or column.**

14. $\begin{vmatrix} -2 & 1 & 0 \\ 3 & -3 & 2 \\ -1 & 2 & -3 \end{vmatrix}$ _____

-3

15. $\begin{vmatrix} 3 & -2 & 2 \\ 4 & -3 & 0 \\ -1 & 2 & -3 \end{vmatrix}$ _____ 13

16. $\begin{vmatrix} 2 & -1 & 4 \\ 0 & 3 & 2 \\ -1 & 4 & 3 \end{vmatrix}$ _____ 16

17. $\begin{vmatrix} 3 & -2 & 1 \\ 6 & -4 & 2 \\ 0 & 1 & 3 \end{vmatrix}$ _____ 0

18. The same methods can be used to evaluate larger determinants. For example, let us evaluate

$$\begin{vmatrix} 2 & -1 & 4 & 1 \\ 0 & -2 & 3 & 7 \\ -4 & 1 & 0 & 2 \\ -2 & -1 & 3 & 5 \end{vmatrix}$$

by expansion about the first column. This gives

$$2\begin{vmatrix} \underline{} & \underline{} & \underline{} \\ \underline{} & \underline{} & \underline{} \\ \underline{} & \underline{} & \underline{} \end{vmatrix} - 0\begin{vmatrix} -1 & 4 & 1 \\ 1 & 0 & 2 \\ -1 & 3 & 5 \end{vmatrix}$$

-2; 3; 7
1; 0; 2
-1; 3; 5

$$+ (-4)\begin{vmatrix} -1 & 4 & 1 \\ -2 & 3 & 7 \\ -1 & 3 & 5 \end{vmatrix} - (-2)\begin{vmatrix} \underline{} & \underline{} & \underline{} \\ \underline{} & \underline{} & \underline{} \\ \underline{} & \underline{} & \underline{} \end{vmatrix}.$$

-1; 4; 1
-2; 3; 7
1; 0; 2

Evaluating the 3 × 3 determinants gives

$$2(\underline{}) - 0(\underline{}) + (-4)(\underline{}) - (-2)(\underline{})$$

12; -19; 15; 35

$$= \underline{}.$$

34

19. Find

$$\begin{vmatrix} 3 & -1 & 4 & 2 \\ 0 & -3 & 7 & 1 \\ 2 & 0 & -5 & 3 \\ 1 & -4 & 2 & 1 \end{vmatrix}.$$

_____ -183

12.2 Properties of Determinants

[1]   Use the properties of determinants to simplify the process of
       evaluating determinants. (Frames 1–12)

---

1. We shall summarize the properties of _____.   determinants
    For convenience, we let $\delta(A)$ represent the deter-
    minant of A.

    (1) If every element in a _____ or column of A   row
         is ____, then $\delta(A)$ = ____.   0; 0

    (2) If corresponding rows and _____ of a   columns
         determinant are _____, the determi-   interchanged
         nant is _____.   unchanged

    (3) If any two rows (or columns) of a matrix are
         _____, the sign of the determinant   interchanged
         is _____.   changed

    (4) If every element of a row (or column) of a
         matrix is multiplied by ____, the determi-   k
         nant of the new matrix is ____ times the   k
         determinant of the original matrix.

    (5) If two rows (or columns) of a matrix are
         _____, the determinant equals ____.   identical; 0

    (6) If a multiple of a row (or column) of a
         matrix is added to the corresponding ele-
         ments of another row (or column), the
         determinant is _____.   unchanged

**Tell why each of the determinants of Frames 2–4 has a
value of 0.**

2. $\begin{vmatrix} 1 & 0 & 1 & 1 \\ 0 & 0 & 3 & 2 \\ -1 & 0 & 6 & 5 \\ 2 & 0 & 3 & 3 \end{vmatrix}$

    Use Property ____.   The determinant is 0 because   1
    the elements of one (*row/column*) are all ____.   column; 0

3. $\begin{vmatrix} -2 & 1 & 3 \\ 6 & 4 & 5 \\ 2 & -1 & -3 \end{vmatrix}$

Use Property 6 and add the third row to the first row, leaving the third row as is. We get

$\begin{vmatrix} \underline{} & \underline{} & \underline{} \\ 6 & 4 & 5 \\ 2 & -1 & -3 \end{vmatrix}.$

0; 0; 0

The determinant is 0 because the elements of one _____ are all 0.

row

4. $\begin{vmatrix} 1 & -4 & 1 \\ 3 & 1 & 3 \\ 2 & 0 & 2 \end{vmatrix}$

Use Property ____. The determinant is 0 because two _____ are identical.

5

columns

Use the appropriate properties from this section to tell why each of the statements of Frames 5–9 is true. Do not evaluate the determinants.

5. $\begin{vmatrix} 3 & -1 \\ 5 & 7 \end{vmatrix} = \begin{vmatrix} 3 & 5 \\ -1 & 7 \end{vmatrix}$

This is a direct result of Property ____ since the rows and columns have been _____.

2

interchanged

6. $\begin{vmatrix} 1 & 4 \\ 1 & 4 \end{vmatrix} = 0$

This determinant is 0 by Property ____.

5

7. $\begin{vmatrix} 6 & -3 & 1 \\ 4 & 0 & 2 \\ 7 & 1 & 2 \end{vmatrix} = -\begin{vmatrix} 1 & -3 & 6 \\ 2 & 0 & 4 \\ 2 & 1 & 7 \end{vmatrix}$

This is a direct result of Property ____ since the first and third columns have been _____.

3

interchanged

8. $\begin{vmatrix} 2 & -6 & 4 \\ 1 & 0 & -1 \\ 7 & 2 & 3 \end{vmatrix} = 2\begin{vmatrix} 1 & -3 & 2 \\ 1 & 0 & -1 \\ 7 & 2 & 3 \end{vmatrix}$

This is a direct result of Property ____ since   4

the first row of the left determinant is ____   2

times the first row of the right determinant.

9. $\begin{vmatrix} 4 & -1 \\ 3 & 1 \end{vmatrix} = \begin{vmatrix} 19 & 4 \\ 3 & 1 \end{vmatrix}$

This is a direct result of Property ____. Each   6

element in the _____ row was multiplied by   second

____, with the result added to the _____ ____.   5; first; row

**Use Property 6 to find the value of each of the deter-
minants of Frames 10–12.**

10. $\begin{vmatrix} 7 & 28 \\ 3 & 12 \end{vmatrix}$

Using Property ____ multiply the first column by   6

_____ and add the corresponding elements to the   -4

second column, leaving the first column intact.

The determinant is ____.   0

11. $\begin{vmatrix} -1 & 0 & 2 & 3 \\ 6 & 1 & 0 & 2 \\ -5 & 0 & 10 & 15 \\ 5 & 3 & 6 & 1 \end{vmatrix} = $ _____   0

12. $\begin{vmatrix} 5 & 3 & -1 & 0 \\ 2 & 1 & 6 & -1 \\ -3 & 2 & 5 & 0 \\ 2 & 1 & 4 & 1 \end{vmatrix}$

Our goal is to change some row or column to

one in which every element but one is ____. The   0

fourth column is the likely one to choose. Adding

the second and fourth rows and leaving the fourth

row intact, we get

$$\begin{vmatrix} 5 & 3 & -1 & 0 \\ \underline{} & \underline{} & \underline{} & \underline{} \\ -3 & 2 & 5 & 0 \\ 2 & 1 & 4 & 1 \end{vmatrix}.$$

	4; 2; 10; 0

Expand by minors about the _____ column.

fourth

$$= 1 \begin{vmatrix} 5 & 3 & -1 \\ 4 & 2 & 10 \\ -3 & 2 & 5 \end{vmatrix}.$$

Multiply the last row by ____ and add it to the _____ row.

−2

second

$$\begin{vmatrix} 5 & 3 & -1 \\ \underline{} & \underline{} & \underline{} \\ -3 & 2 & 5 \end{vmatrix}$$

10; −2; 0

Multiply the first row by ____ and add it to the _____ row.

5

third

$$\begin{vmatrix} 5 & 3 & -1 \\ 10 & -2 & 0 \\ \underline{} & \underline{} & \underline{} \end{vmatrix}$$

22; 17; 0

$$= -1 \begin{vmatrix} 10 & -2 \\ 22 & 17 \end{vmatrix} = -1(170 + 44) = \underline{}$$

−214

12.3 Solution of Linear Systems of Equations by Determinants—Cramer's Rule

[1] Understand the derivation of Cramer's rule. (Frames 1 and 2)

[2] Apply Cramer's rule to a linear system with two equations and two variables. (Frames 3–7)

[3] Apply Cramer's rule to a linear system with three equations and three variables. (Frames 8–12)

---

1. If the _____ method is used to solve the general system of two equations with two variables,

elimination

$$a_1x + b_1y = c_1$$
$$a_2x + b_2y = c_2,$$

the result will be a _____ that can be used for any system of two equations with two unknowns.

formula

The result is sumarized as _____ _____ for 2 × 2 systems.

Cramer's; rule

2. According to Cramer's rule, if

$$D = \underline{\hspace{2cm}} \quad (D \underline{\hspace{1cm}} 0)$$

$$D_x = \underline{\hspace{2cm}}$$

$$D_y = \underline{\hspace{2cm}},$$

then $x = \dfrac{\overline{}}{\underline{}}$ and $y = \dfrac{\overline{}}{\underline{}}$.

$\begin{vmatrix} a_1 & b_1 \\ a_2 & b_2 \end{vmatrix}; \;\neq$

$\begin{vmatrix} c_1 & b_1 \\ c_2 & b_2 \end{vmatrix}$

$\begin{vmatrix} a_1 & c_1 \\ a_2 & c_2 \end{vmatrix}$

$D_x; \; D_y$

$D; \; D$

3. We can solve a system of equations such as

$$2x + 3y = 1$$
$$3x + 5y = 3$$

using Cramer's rule. To use Cramer's rule, write the determinants,

$$D = \begin{vmatrix} \underline{} & \underline{} \\ \underline{} & \underline{} \end{vmatrix}, \; D_x = \begin{vmatrix} \underline{} & \underline{} \\ \underline{} & \underline{} \end{vmatrix}.$$

2; 3; 1; 3
3; 5; 3; 5

To find D_x, replace the first column of the determinant for D with the _____ from the system of equations. Find D_y.

constants

$$D_y = \begin{vmatrix} \underline{} & \underline{} \\ \underline{} & \underline{} \end{vmatrix}$$

2; 1
3; 3

The definition of determinant gives

$$D = \underline{\hspace{1cm}}, \; D_x = \underline{\hspace{1cm}}, \; \text{and } D_y = \underline{\hspace{1cm}}.$$

1; −4; 3

According to Cramer's rule,

$$x = \dfrac{\overline{}}{D} \text{ and } y = \dfrac{\overline{}}{D}.$$

Here

$$x = \dfrac{\overline{}}{\underline{}} \text{ and } y = \dfrac{\overline{}}{\underline{}}$$

or

$$x = \underline{\hspace{1.5cm}} \text{ and } y = \underline{\hspace{1.5cm}}.$$

$D_x; \; D_y$

−4; 3
1; 1

−4; 3

4. To solve the system

$$2x + y = -2$$
$$4x - 3y = 11,$$

start with

$$D = \begin{vmatrix} \underline{\quad} & \underline{\quad} \\ \underline{\quad} & \underline{\quad} \end{vmatrix}, \quad D_x = \begin{vmatrix} \underline{\quad} & \underline{\quad} \\ \underline{\quad} & \underline{\quad} \end{vmatrix},$$

$$D_y = \begin{vmatrix} \underline{\quad} & \underline{\quad} \\ \underline{\quad} & \underline{\quad} \end{vmatrix}.$$

| 2; 1; −2; 1 |
| 4; −3; 11; −3 |

| 2; −2 |
| 4; 11 |

From this,

$$D = \underline{\quad}, \quad D_x = \underline{\quad}, \quad \text{and } D_y = \underline{\quad},$$

−10; −5; 30

so

$$x = \frac{\underline{\quad\quad}}{\underline{\quad}} \quad \text{and} \quad y = \frac{\underline{\quad\quad}}{\underline{\quad}}$$

−5; 30
−10; −10

or

$$x = \underline{\quad} \quad \text{and} \quad y = \underline{\quad}.$$

$\frac{1}{2}$; −3

Use Cramer's rule to solve the systems in Frames 5–7.

5. $4x - 3y = 2$
    $-4x + 9y = 0$

$$D = \underline{\quad}, \quad D_x = \underline{\quad}, \quad D_y = \underline{\quad}$$

$$x = \underline{\quad}, \quad y = \underline{\quad}$$

Solution set: _____

24; 18; 8
$\frac{3}{4}$; $\frac{1}{3}$
$\left\{\left(\frac{3}{4}, \frac{1}{3}\right)\right\}$

6. $2x + y = 5$
    $-3x + 2y = 1$

$$x = \underline{\quad}, \quad y = \underline{\quad}$$

Solution set: _____

$\frac{9}{7}$; $\frac{17}{7}$
$\left\{\left(\frac{9}{7}, \frac{17}{7}\right)\right\}$

7. $4x - 3y = 7$
    $8x - 6y = 12$

Here $D = \underline{\quad}$, so Cramer's rule does not apply.

0

8. We can also use Cramer's rule to solve a system of three equations in three unknowns. For example, to solve the system

$$x + y + z = 3$$
$$2x - 3y - 4z = -3$$
$$-3x + 2y + 2z = -4,$$

    start with the determinants

$$D = \begin{vmatrix} 1 & \rule{1cm}{0.4pt} & 1 \\ 2 & \rule{1cm}{0.4pt} & -4 \\ -3 & \rule{1cm}{0.4pt} & 2 \end{vmatrix}, \quad D_x = \begin{vmatrix} \rule{1cm}{0.4pt} & 1 & 1 \\ \rule{1cm}{0.4pt} & -3 & -4 \\ \rule{1cm}{0.4pt} & 2 & 2 \end{vmatrix},$$

1; 3
−3; −3
2; −4

$$D_y = \begin{vmatrix} 1 & \rule{1cm}{0.4pt} & 1 \\ 2 & \rule{1cm}{0.4pt} & -4 \\ -3 & \rule{1cm}{0.4pt} & 2 \end{vmatrix}, \quad D_z = \begin{vmatrix} 1 & 1 & \rule{1cm}{0.4pt} \\ 2 & -3 & \rule{1cm}{0.4pt} \\ -3 & 2 & \rule{1cm}{0.4pt} \end{vmatrix}.$$

3; 3
−3; −3
−4; −4

Here $D =$ ____, $D_x =$ ____, $D_y =$ ____, and

$D_z =$ ____.

5; 10; −15

20

By Cramer's rule, we have

$$x = \frac{\rule{1cm}{0.4pt}}{\rule{1cm}{0.4pt}}, \quad y = \frac{\rule{1cm}{0.4pt}}{\rule{1cm}{0.4pt}}, \quad \text{and} \quad z = \frac{\rule{1cm}{0.4pt}}{\rule{1cm}{0.4pt}}.$$

10; −15; 20
5; 5; 5

Hence $x =$ _____, $y =$ _____, and $z =$ _____.

2; −3; 4

Solution set: _____

$\{(2, -3, 4)\}$

Solve the systems of Frames 9–12 by Cramer's rule.

9. $3x + 2y - z = -3$
$2x + y + 4z = 9$
$x + 3y + 2z = -8$

$D =$ _____, $D_x =$ _____, $D_y =$ _____, $D_z =$ _____

−35; −105; 175; −70

Solution set: _____

$\{(3, -5, 2)\}$

10. $2x + y - 4z = -7$
$3x + 2y - z = 8$
$-x - 3y + 2z = 11$

$D =$ _____, $D_x =$ _____, $D_y =$ _____, $D_z =$ _____

25; 150; −75; 100

Solution set: _____

$\{(6, -3, 4)\}$

11. $2x + y + 4z = 9$
$4x + 2y + 8z = 12$
$3x + y - 2z = 5$

$D =$ _____, which means that Cramer's rule

(*does/does not*) apply.

0

does not

12. $3x - 2y + 8z = -7$
$-4x + 2y - 12z = 2$
$2x + y - 3z = -5$

Solution set: _____

$\{(-3, 7, 2)\}$

12.4 Solution of Linear Systems of Equations by Matrices

1. Write the augmented matrix for a system of equations. (Frames 1 and 2)

2. Use row transformations to solve a system with two equations. (Frames 3–10)

3. Use row transformations to solve a system with three equations. (Frames 11–20)

4. Use row transformation to solve inconsistent systems or systems with dependent equations. (Frames 21–24)

---

1. Systems of equations can also be solved by matrix methods. To begin, write the _____ matrix of the _____.

    augmented

    system

2. Write the augmented matrix of the following system.

$$2x + y = 3$$
$$x - 3y = 5$$

_____

$$\begin{bmatrix} 2 & 1 & 3 \\ 1 & -3 & 5 \end{bmatrix}$$

3. To use the augmented matrix, we need the three basic row _____.

    transformations

    (a) Any two ____ of a matrix may be _____.

    rows; interchanged

    (b) The elements in any row may be _____ by any nonzero _____ _____.

    multiplied

    real; number

    (c) Any row may be changed by _____ to its elements a _____ of the corresponding elements of another _____.

    adding

    multiple

    row

4. Let us solve the system of Frame 2 using the Gauss–Jordan method. The augmented matrix of this system is

_____.

$$\begin{bmatrix} 2 & 1 & 3 \\ 1 & -3 & 5 \end{bmatrix}$$

5. To begin, we need to get ____ in the first row,
    the _____ column. To do so here, multiply
    the elements of the first row by _____. This
    This gives the following matrix.

$$\left[\begin{array}{cc|c} \underline{\hspace{1cm}} & \underline{\hspace{1cm}} & \underline{\hspace{1cm}} \\ 1 & -3 & 5 \end{array}\right]$$

1

first

1/2

1; 1/2; 3/2

6. Now we need ____ in the _____ row, the first
    column. To get 0, add to the elements in the
    _____ row the result of multiplying each
    element in the first row by _____.

$$\left[\begin{array}{cc|c} 1 & 1/2 & 3/2 \\ \underline{\hspace{1cm}} & \underline{\hspace{1cm}} & \underline{\hspace{1cm}} \end{array}\right]$$

0; second

second

−1

0; −7/2; 7/2

7. Next, we need ____ in the _____ row , the
    second column. Multiply the elements of the
    _____ by _____.

$$\left[\begin{array}{cc|c} 1 & 1/2 & 3/2 \\ \underline{\hspace{1cm}} & \underline{\hspace{1cm}} & \underline{\hspace{1cm}} \end{array}\right]$$

1; second

second; −2/7

0; 1; −1

8. Finally, we need 0 in the _____ row, the
    _____ column. Multiply the elements of
    the second row by _____ and add to the ele-
    ments of the first row.

$$\left[\begin{array}{cc|c} \underline{\hspace{1cm}} & \underline{\hspace{1cm}} & \underline{\hspace{1cm}} \\ 0 & 1 & -1 \end{array}\right]$$

first;
second
−1/2

1; 0; 2

9. This matrix corresponds to the system
$$x = \underline{\hspace{1cm}}$$
$$y = \underline{\hspace{1cm}}.$$
    The solution set is _____.

2
−1
{(2, −1)}

10. Use matrix methods to solve the following system.

$$3x + 2y = -5$$
$$2x - y = -8$$

Solution set: _____

$\{(-3,\ 2)\}$

11. To use matrix methods to solve the system

$$2x - y + z = 8$$
$$2x + 2y - z = -1$$
$$x - y + 2z = 9,$$

first write the augmented matrix for the system.

_____

$$\begin{bmatrix} 2 & -1 & 1 & 8 \\ 2 & 2 & -1 & -1 \\ 1 & -1 & 2 & 9 \end{bmatrix}$$

12. Start by getting ____ in the first row, the first column.

$$\begin{bmatrix} \underline{} & \underline{} & \underline{} & \underline{} \\ 2 & 2 & -1 & -1 \\ 1 & -1 & 2 & 9 \end{bmatrix}$$

1

1; -1/2; 1/2; 4

13. Now get 0 in the second row, the first column.

$$\begin{bmatrix} 1 & -1/2 & 1/2 & 4 \\ \underline{} & \underline{} & \underline{} & \underline{} \\ 1 & -1 & 2 & 9 \end{bmatrix}$$

0; 3; -2; -9

14. Get 0 in the third row, the _____ column.

$$\begin{bmatrix} 1 & -1/2 & 1/2 & 4 \\ 0 & 3 & -2 & -9 \\ \underline{} & \underline{} & \underline{} & \underline{} \end{bmatrix}$$

first

0; -1/2; 3/2; 5

15. Get 1 in the _____ row, the _____ column.

$$\begin{bmatrix} 1 & -1/2 & 1/2 & 4 \\ \underline{} & \underline{} & \underline{} & \underline{} \\ 0 & -1/2 & 3/2 & 5 \end{bmatrix}$$

second; second

0; 1; -2/3; -3

16. We need ____ in the _____ row, the _____ column.

$$\begin{bmatrix} 1 & -1/2 & 1/2 & 4 \\ 0 & 1 & -2/3 & -3 \\ \underline{\quad} & \underline{\quad} & \underline{\quad} & \underline{\quad} \end{bmatrix}$$

0; third; second

0; 0; 7/6; 7/2

17. Now get ____ in the _____ row, the _____ column.

$$\begin{bmatrix} 1 & -1/2 & 1/2 & 4 \\ 0 & 1 & -2/3 & -3 \\ \underline{\quad} & \underline{\quad} & \underline{\quad} & \underline{\quad} \end{bmatrix}$$

1; third; third

0; 0; 1; 3

18. Get 0 in the second row, the _____ column.

$$\begin{bmatrix} 1 & -1/2 & 1/2 & 4 \\ \underline{\quad} & \underline{\quad} & \underline{\quad} & \underline{\quad} \\ 0 & 0 & 1 & 3 \end{bmatrix}$$

third

0; 1; 0; -1

19. Get ____ in the first row, the second column.

$$\begin{bmatrix} \underline{\quad} & \underline{\quad} & \underline{\quad} & \underline{\quad} \\ 0 & 1 & 0 & -1 \\ 0 & 0 & 1 & 3 \end{bmatrix}$$

0

1; 0; 1/2; 7/2

20. Finally, get 0 in the _____ row, the _____ column.

$$\begin{bmatrix} \underline{\quad} & \underline{\quad} & \underline{\quad} & \underline{\quad} \\ 0 & 1 & 0 & -1 \\ 0 & 0 & 1 & 3 \end{bmatrix}$$

first; third

1; 0; 0; 2

21. This matrix corresponds to the system

$$x = \underline{\quad}$$
$$y = \underline{\quad}$$
$$z = \underline{\quad}.$$

2

-1

3

The solution set is _____.

$\{(2, -1, 3)\}$

22. _____ systems or systems with   Inconsistent

        _____ equations can be recognized when   dependent

        solving these systems with row transformations.

23. Use the Gauss-_____ method to solve the   Jordan

        system

$$9x - 15y = 4$$
$$-3x + 5y = 6.$$

Write the augmented matrix.

$$\begin{bmatrix} \underline{} & \underline{} & | & \underline{} \\ \underline{} & \underline{} & | & \underline{} \end{bmatrix}$$

     9; -15; 4

     -3; 5; 6

Multiply the _____ in the second row by   elements

3 and add the result to the _____   corresponding

elements of the _____ row.   first

$$\begin{bmatrix} \underline{} & \underline{} & | & \underline{} \\ -3 & 5 & | & 6 \end{bmatrix}$$

     0; 0; 22

The first row corresponds to the equation

_____, which has no solution. The   0x + 0y = 22

system is _____.   Inconsistent

24. Use the Gauss-Jordan method to solve the system

$$x + 2y - 8z = 5$$
$$4x + 2y - 2z = -4.$$

Write the augmented matrix.

$$\begin{bmatrix} \underline{} & \underline{} & \underline{} & | & \underline{} \\ \underline{} & \underline{} & \underline{} & | & \underline{} \end{bmatrix}$$

     1; 2; -8; 5

     4; 2; -2; -4

Complete the following matrices.

$$\begin{bmatrix} 1 & 2 & \underline{} & | & \underline{} \\ 0 & -6 & \underline{} & | & \underline{} \end{bmatrix}$$

     -8; 5

     30; -24

$$\begin{bmatrix} \underline{} & \underline{} & -8 & | & \underline{} \\ 0 & 1 & -5 & | & \underline{} \end{bmatrix}$$

     1; 2; 5

     4

$$\begin{bmatrix} 1 & 0 & \underline{} & \Big| & \underline{} \\ \underline{} & \underline{} & -5 & \Big| & 4 \end{bmatrix}$$

2; −3

0; 1

It is not possible to go further with the

_____ method.

Gauss–Jordan

The equations that correspond to the final matrix

are

_____ = ____

x + 2z; −3

and   _____ = ____ .

y − 5z; 4

Solve these equations for x and y.

x = _____

−2z − 3

y = _____

5z + 4

The solution can be written with z arbitrary as

_____ . The system has an

$\{(-2z-3,\ 5z+4,\ z)\}$

infinite number of solutions.

Chapter 12 Test

The answers for these questions are at the back of this Study Guide.

Evaluate the following determinants.

1. $\begin{vmatrix} -3 & 4 \\ 6 & 8 \end{vmatrix}$

1. _____

2. $\begin{vmatrix} 2 & 9 \\ 4 & 0 \end{vmatrix}$

2. _____

3. $\begin{vmatrix} 5 & -2 \\ 1 & 7 \end{vmatrix}$

3. _____

4. $\begin{vmatrix} 2 & -3 & -2 \\ -1 & -4 & -3 \\ -1 & 0 & 2 \end{vmatrix}$

4. _____

5. $\begin{vmatrix} 3 & 3 & -1 \\ 2 & 6 & 0 \\ -6 & -6 & 2 \end{vmatrix}$

5. _____

6. $\begin{vmatrix} -2 & 4 & 1 \\ 3 & 0 & 2 \\ -1 & 0 & 3 \end{vmatrix}$

6. _____

Decide if each statement in Frames 7-10 is true or false. Use a property of determinants to justify each true statement.

7. $\begin{vmatrix} 4 & 6 \\ 3 & 5 \end{vmatrix} = \begin{vmatrix} 4 & 3 \\ 6 & 5 \end{vmatrix}$

7. _____

8. $3\begin{vmatrix} 6 & 0 & 2 \\ 4 & 1 & 3 \\ 2 & 8 & 6 \end{vmatrix} = \begin{vmatrix} 18 & 0 & 6 \\ 12 & 3 & 9 \\ 6 & 24 & 18 \end{vmatrix}$

8. _____

9. $\begin{vmatrix} 8 & 2 & -5 \\ -3 & 1 & 4 \\ 2 & 0 & 5 \end{vmatrix} = \begin{vmatrix} 8 & -5 & 2 \\ -3 & 4 & 1 \\ 2 & 5 & 0 \end{vmatrix}$

9. _____

10. $\begin{vmatrix} -4 & 2 & 3 \\ 0 & 1 & 6 \\ -4 & 2 & 3 \end{vmatrix} = \begin{vmatrix} 2 & 8 & -3 \\ 1 & 6 & 5 \\ 1 & 6 & 5 \end{vmatrix}$

10. _____

Find the value of each determinant by using the property which says that the determinant of a matrix is unchanged if a multiple of a row (or column) of the matrix is added to the corresponding elements of another row (or column).

11. $\begin{vmatrix} 8 & 7 \\ 4 & -1 \end{vmatrix}$

11. _____

12. $\begin{vmatrix} -2 & 4 & 1 \\ 2 & 1 & 5 \\ 4 & 0 & 2 \end{vmatrix}$

12. _____

Solve by Cramer's rule.

13. $3x - 4y = 1$
    $2x + 3y = 12$

13. _____

14. $7x - y = -10$
    $-x + 3y = 10$

14. _____

15. $-x + y = -1$
    $y - z = 6$
    $x + z = -1$

15. _____

16. $x + y - z = 6$
    $2x - y + z = -9$
    $x - 2y + 3z = 1$

16. _____

Solve by the Gauss–Jordan method.

17. $4x + y = 2$
    $x - 2y = 14$

17. _____

18. $x + y = -3$
    $2x - 5y = -6$

18. _____

19. $2x - y + 3z = 0$
    $x + 2y - z = 5$
    $2y + z = 1$

19. _____

20. $x - 2y + z = 5$
    $2x + y - z = 2$
    $-2x + 4y - 2z = 2$

20. _____

CHAPTER 13   FURTHER TOPICS IN ALGEBRA

13.1   Introduction to Sequences; Arithmetic Sequences

1.   Define *sequence*. (Frame 1)

2.   Find terms of a sequence given the general term or a recursion formula. (Frames 2–17)

3.   Find the common difference and terms for an arithmetic sequence. (Frames 18–24)

4.   Find specific terms of an arithmetic sequence. (Frames 25–32)

---

1. If the domain of a function is the set of _____ _____, then the _____ elements can be ordered, as f(1), f(2), f(3), and so on. This ordered list of numbers is called a _____. It is customary to use the letter ____ to repre-sent the variable and _____ to represent the functional value instead of f(n).

positive
integers; range

sequence
n
a_n

2. The elements of the sequence are called the _____ of the sequence.

terms

3. For example, $a_n = 5n + 3$ represents a _____. Here $a_1 = $ ____, $a_2 = $ ____, $a_3 = $ ____, $a_4 = $ ____, and so on. The range of such a function when written in the order _____, _____, _____, _____, ..., is called a _____. The range of this sequence is the infinite sequence

   _____, _____, _____, _____, ...

sequence
8; 13; 18; 23

a_1; a_2; a_3; a_4
sequence

8; 13; 18; 23

For the sequences of Frames 4–12, find the first four terms, a_1, a_2, a_3, and a_4.

4. $a_n = 7n - 5$

   $a_1 = 7(\_\_\_\_) - 5 = \_\_\_$ 1; 2
   $a_2 = 7(\_\_\_\_) - 5 = \_\_\_$ 2; 9
   $a_3 = 7(\_\_\_\_) - 5 = \_\_\_$ 3; 16
   $a_4 = 7(\_\_\_\_) - 5 = \_\_\_$ 4; 23

5. $a_n = -3n + 4$

   $a_1 = -3(\underline{\hphantom{xx}}) + 4 = \underline{\hphantom{xx}}$ 1; 1

   $a_2 = -3(\underline{\hphantom{xx}}) + 4 = \underline{\hphantom{xx}}$ 2; -2

   $a_3 = -3(\underline{\hphantom{xx}}) + 4 = \underline{\hphantom{xx}}$ 3; -5

   $a_4 = -3(\underline{\hphantom{xx}}) + 4 = \underline{\hphantom{xx}}$ 4; -8

6. $a_n = \dfrac{n}{2n + 1}$

   $a_1 = \dfrac{\underline{\hphantom{xx}}}{2(\underline{\hphantom{xx}}) + 1} = \underline{\hphantom{xx}}$ $\dfrac{1}{1}$; 1/3

   $a_2 = \underline{\hphantom{xx}}$, $a_3 = \underline{\hphantom{xx}}$, $a_4 = \underline{\hphantom{xx}}$ 2/5; 3/7; 4/9

7. $a_n = n^2 + 3n$

   $a_1 = (\underline{\hphantom{xx}})^2 + 3(\underline{\hphantom{xx}}) = \underline{\hphantom{xx}}$ 1; 1; 4

   $a_2 = \underline{\hphantom{xx}}$, $a_3 = \underline{\hphantom{xx}}$, $a_4 = \underline{\hphantom{xx}}$ 10; 18; 28

8. $a_n = \dfrac{n^2 - 1}{n^2 + 1}$

   $a_1 = \underline{\hphantom{xx}}$, $a_2 = \underline{\hphantom{xx}}$, $a_3 = \underline{\hphantom{xx}}$, $a_4 = \underline{\hphantom{xx}}$ 0; 3/5; 4/5; 15/17

9. $a_n = 4^n$

   $a_1 = \underline{\hphantom{xx}}$, $a_2 = \underline{\hphantom{xx}}$, $a_3 = \underline{\hphantom{xx}}$, $a_4 = \underline{\hphantom{xx}}$ 4; 16; 64; 256

10. $a_n = 2^{-n}$

   $a_1 = \underline{\hphantom{xx}}$, $a_2 = \underline{\hphantom{xx}}$, $a_3 = \underline{\hphantom{xx}}$, $a_4 = \underline{\hphantom{xx}}$ 1/2; 1/4; 1/8; 1/16

11. $a_n = (n - 1)(n + 1)$

   $a_1 = \underline{\hphantom{xx}}$, $a_2 = \underline{\hphantom{xx}}$, $a_3 = \underline{\hphantom{xx}}$, $a_4 = \underline{\hphantom{xx}}$ 0; 3; 8; 15

12. $a_n = (-1)^n(2n - 1)$

   $a_1 = \underline{\hphantom{xx}}$, $a_2 = \underline{\hphantom{xx}}$, $a_3 = \underline{\hphantom{xx}}$, $a_4 = \underline{\hphantom{xx}}$ -1; 3; -5; 7

13. A _____ formula defines the _____ term   recursion; nth
    of a sequence in terms of the previous term.

For the sequences of Frames 14–17, find the first
four terms, a_1, a_2, a_3, and a_4.

14. $a_1 = 4$; $a_n = 3a_{n-1} - 2$

   $a_1 = 4$

   $a_2 = 3a_1 - 2 = 3(\underline{}) - 2 = \underline{}$ 4; 10

   $a_3 = 3a_2 - 2 = 3(\underline{}) - 2 = \underline{}$ 10; 28

   $a_4 = 3a_3 - 2 = 3(\underline{}) - 2 = \underline{}$ 28; 82

15. $a_1 = 3$; $a_n = a_{n-1} + n^2$

   $a_1 = 3$

   $a_2 = (\underline{}) + (\underline{})^2 = \underline{}$ 3; 2; 7

   $a_3 = (\underline{}) + (\underline{})^2 = \underline{}$ 7; 3; 16

   $a_4 = (\underline{}) + (\underline{})^2 = \underline{}$ 16; 4; 32

16. $a_1 = \dfrac{1}{2}$; $a_n = \dfrac{1}{a_{n-1} + 1}$

   $a_1 = \dfrac{1}{2}$

   $a_2 = \dfrac{1}{\underline{} + 1} = \dfrac{1}{\underline{}} = \underline{}$ 1/2; 3/2; 2/3

   $a_3 = \dfrac{1}{\underline{} + 1} = \dfrac{1}{\underline{}} = \underline{}$ 2/3; 5/3; 3/5

   $a_4 = \dfrac{1}{\underline{} + 1} = \dfrac{1}{\underline{}} = \underline{}$ 3/5; 8/5; 5/8

17. $a_1 = 5$; $a_n = 2n - a_{n-1}$

   $a_1 = 5$

   $a_2 = 2(\underline{}) - \underline{} = \underline{}$ 2; 5; −1

   $a_3 = 2(\underline{}) - \underline{} = \underline{}$ 3; −1; 7

   $a_4 = 2(\underline{}) - \underline{} = \underline{}$ 4; 7; 1

18.   A sequence in which each term after the first
   is obtained by adding a fixed number to the preceding term is called an _____ sequence   arithmetic
   (or arithmetic _____). The fixed num-   progression
   ber is called the _____ _____.   common; difference

19. 3, 5, 7, 9, 11, 13, 15 is an _____   arithmetic

sequence having first term a_1 = ____ and common   3

difference d = ____. (To find the common dif-   2

ference when given an arithmetic sequence such

as this, select any term except the _____ and   first

_____ the preceding term.)   subtract

**Find the first term and common difference for any of
the sequences of Frames 20–24 that are arithmetic.**

20. 9, 17, 25, 33, 41, 49

a_1 = ____, d = 33 – ____ = ____   9; 25; 8

21. 2, 4, 8, 16, 32

Since 16 – ____ = ____, while 32 – ____ = ____,   8; 8; 16; 16

each term cannot be obtained by adding the same

number to the preceding term. Hence this se-

quence (*is*/*is not*) an arithmetic sequence.   is not

22. 12, 9, 6, 3, 0, –3

a_1 = ____, d = ____   12; –3

23. 3, $3\frac{1}{2}$, 4, $4\frac{1}{2}$, 5

a_1 = ____, d = ____   3; 1/2

24. 1/2, 1/3, 1/4, 1/5, 1/6

a_1 = ____, d = ____   Not arithmetic

25. The nth term, or general term, of an arithmetic

sequence is given by the formula

a_n = _____. $a_1 + (n - 1)d$

26. The tenth term of the arithmetic sequence having
     first term 8 and common difference 3 is given by

$$a_{\underline{}} = \underline{} + (\underline{} - 1)(\underline{})$$ 10; 8; 10; 3

$$= \underline{} + \underline{}$$ 8; 27

$$= \underline{}.$$ 35

27. To find a_{15} for the arithmetic sequence 9, 11,
     13, 15, 17, ..., first find that $a_1 = \underline{}$ and   9

     $d = \underline{}$. Then   2

$$a_{15} = \underline{} + (\underline{} - 1)(\underline{})$$ 9; 15; 2

$$= \underline{}.$$ 37

In Frames 28-32, find the indicated term for the
given arithmetic sequence.

28. -8, -5, -2, 1, ...; find a_{18}.

$$a_1 = \underline{}, \quad d = \underline{}$$ -8; 3

$$a_{18} = \underline{} + (\underline{})(\underline{}) = \underline{}$$ -8; 17; 3; 43

29. $a_1 = -5$, $d = -3$; find a_{21}.

$$a_{21} = \underline{} + (\underline{})(\underline{}) = \underline{}$$ -5; 20; -3; -65

30. $a_1 = 10$, $d = 1/2$; find a_{15}.

$$a_{15} = \underline{} + (\underline{})(\underline{}) = \underline{}$$ 10; 14; 1/2; 17

31. $a_8 = 6$, $a_9 = 10$; find a_{31}.
     First find $\underline{}$ and $\underline{}$. Here a_1; d

$$d = a_9 - \underline{} = \underline{} - \underline{} = \underline{}.$$ a_8; 10; 6; 4

     Use the formula for the nth term to find a_1. We
     know $a_8 = \underline{}$, $d = \underline{}$, and, for a_8, $n = \underline{}$.   6; 4; 8
     Hence

$$\underline{} = a_1 + (\underline{})(\underline{}),$$ 6; 7; 4

$$a_1 = \underline{}.$$ -22

     Now find a_{31}.

$$a_{31} = \underline{} + (\underline{})(\underline{}) = \underline{}$$ -22; 30; 4; 98

32. $a_{11} = 12$, $a_{12} = 9$; find a_{26}.

  $d =$ ____ , $a_1 =$ ____ , $a_{26} =$ ____ -3; 42; -33

13.2 Geometric Sequences

1 Define *geometric sequence*. (Frames 1 and 2)

2 Find the common ratio for a geometric sequence. (Frames 3–7)

3 Find specific terms of a geometric sequence. (Frames 8–19)

---

1. A sequence in which each term after the first is
    some constant multiple of the preceding term is
    called a _____ sequence. The constant   geometric
    multiplier is called the _____ ratio.   common

2. The sequence 1, 2, 4, 8, 16 is a _____   geometric
    sequence with first term ____ and common ratio   1
    ____.   2

**Find the first term and common ratio of any of the
sequences of Frames 3–7 that are geometric.**

3. 10, 5, 5/2, 5/4
    $a_1 =$ ____ To find the constant ratio, r, take   10
    any term except the first and _____ it by   divide
    the preceding term. If we choose the term 5
    here, we can find r by writing
        $r = ($____$) \div ($____$) =$ ____ .   5; 10; 1/2

4. 8, −8, 8, −8, 8
        $a_1 =$ ____ , $r =$ ____   8; −1

5. 6, 8, 12, 18
        $a_1 =$ ____ , $r =$ ____   Not geometric

6. 1, −2, 4, −8, 16
        $a_1 =$ ____ , $r =$ ____   1; −2

7. 1/2, 1/4, 1/8, 1/16

   $a_1 =$ ____ , $r =$ ____

   1/2; 1/2

8. The nth term of the geometric sequence having first term a_1 and common ratio r is given by

   $a_n =$ ____ $\cdot r^{\text{——}}$.

   a_1; $n - 1$

9. We can use this formula to find the nth term of the sequence of Frame 7. This sequence has $a_1 =$ ____ and $r =$ ____, so that the nth term is

   1/2; 1/2

   $a_n =$ ____ $\cdot ($____$)^{n-1}$.

   1/2; 1/2

   Simplify to get

   $a_n =$ _____.

   $(1/2)^n$

10. Find the nth term of the geometric sequence with first term 2/3 and common ratio 9. The nth term is

    $a_n =$ ____ $\cdot ($____$)^{n-1}$.

    2/3; 9

11. The formula for the nth term can be used to find any specific term. For example, the fifth term of the geometric sequence having first term 3 and common ratio 2 is given by

    $a_5 =$ ____$($____$)^{\text{——}}$

    3; 2; 5 - 1 (or 4)

    $=$ ____.

    48

Find the indicated term for the geometric sequences of Frames 12–19.

12. 3, 6, 12, 24; find a_8.

    Here $a_1 =$ ____ and $r =$ ____. Thus a_8 is given by

    3; 2

    $a_8 =$ ____$($____$)^{\text{——}}$

    3; 2; 7

    $=$ ____.

    384

13. $1/5$, $1/15$, $1/45$, ...; find a_6.

    $a_1 = $ _____ , $r = $ _____

            $a_6 = $ ___ (___) ⎯⎯

               $= $ ⎯⎯⎯⎯

> $1/5$; $1/3$
>
> $1/5$; $1/3$; 5
>
> $1/1215$

14. $a_1 = -2$, $r = -2$; find a_5.

            $a_5 = $ ___ (___) ⎯⎯

               $= $ ___

> -2; -2; 4
>
> -32

15. $a_1 = 5$, $r = 3/4$; find a_6. (Set up only; do not calculate.)

            $a_6 = $ ___ (___) ⎯⎯

> 5; $3/4$; 5

16. $a_1 = -5$, $r = -1/3$; find a_8. (Set up only.)

            $a_8 = $ ___ (___) ⎯⎯

> -5; $-1/3$; 7

17. $a_4 = 12$, $a_5 = 9$; find a_{12}. (Set up only.)
    To find r we write

           $r = ($ ___ $) \div ($ ___ $) = $ ___ .

    To find a_1 **use the formula** $a_n = a_1 \cdot r^{n-1}$. **Here we can use** a_4 **as** a_n.

           ___ $= a_1($ ___ $)$ ⎯⎯

           $a_1 = $ _____

    Now set up a_{12}.

           $a_{12} = $ _____ (___) ⎯⎯

> 9; 12; $3/4$
>
> 12; $3/4$; 3
>
> $256/9$
>
> $256/9$; $3/4$; 11

18. $a_6 = 25$, $a_7 = 50$; find a_9.

           $a_9 = $ _____

> 200

19. $a_3 = 9$, $a_7 = 16/9$; find a_{11}.

           $a_{11} = $ _____

> $256/729$

13.3   Series and Applications

| 1 | Use summation notation. (Frames 1–5) |

| 2 | Find the sum of a specified number of terms of an arithmetic sequence. (Frames 6–16) |

| 3 | Find the sum of a specified number of terms of a geometric sequence. (Frames 17–25) |

| 4 | Solve applied problems involving sequences. (Frames 26–28) |

---

1. The sum of the terms of a sequence is called a

    _____. The Greek letter _____, denoted

    _____, is used to mean "_____." The letter

    _____ is commonly used as the _____ of sum-

    mation.

    series; sigma

    Σ; sum

    i; index

Evaluate the sums in Frames 2–5.

2. $\sum\limits_{i=1}^{4} 2 + i^2$

$$= (2 + \underline{}^2) + (2 + \underline{}^2) + (2 + \underline{}^2) + (2 + \underline{}^2)$$

$$= (2 + \underline{}) + (2 + \underline{}) + (2 + \underline{}) + (2 + \underline{})$$

$$= \underline{} + \underline{} + \underline{} + \underline{}$$

$$= \underline{}$$

    1; 2; 3; 4

    1; 4; 9; 16

    3; 6; 11; 18

    38

3. $\sum\limits_{i=4}^{6} 5i - 2$

$$= (5 \cdot \underline{} - 2) + (5 \cdot \underline{} - 2) + (5 \cdot \underline{} - 2)$$

$$= (\underline{} - 2) + (\underline{} - 2) + (\underline{} - 2)$$

$$= \underline{} + \underline{} + \underline{}$$

$$= \underline{}$$

    4; 5; 6

    20; 25; 30

    18; 23; 28

    69

4. $\sum\limits_{i=1}^{4} \dfrac{1}{i} = $ _____

    $\dfrac{25}{12}$

5. $\sum\limits_{i=2}^{5} \dfrac{i + 1}{12 - 2i} = $ _____

    $\dfrac{127}{24}$

6. If an arithmetic sequence has first term a_1 and common difference d, then the sum of the first _____ terms is given by the formulas

$$S_n = \text{_____}$$

or $$S_n = \text{_____}.$$

The first of these formulas is used if we know the first and _____ terms. The second formula is used if we know the first term and the

_____ _____ .

n

$\frac{n}{2}(a_1 + a_n)$

$\frac{n}{2}[2a_1 + (n - 1)d]$

last

common; difference

7. To find the sum of the first 10 terms of the arithmetic sequence having $a_1 = 8$ and $a_{10} = 30$, use the (*first/second*) formula from above.

$$S_{10} = \frac{\overline{}}{2}(\text{\_\_\_} + \text{\_\_\_})$$

$$= \text{\_\_\_\_}$$

first

10; 8; 30

190

8. To find the sum of the first 10 terms of the arithmetic sequence having $a_1 = 12$ and $d = -3$, use the (*first/second*) formula from above.

$$S_{10} = \frac{\overline{}}{2}[2(\text{\_\_\_}) + (\text{\_\_\_})(\text{\_\_\_})]$$

$$= \text{\_\_\_\_}(\text{\_\_\_} + \text{\_\_\_})$$

$$= \text{\_\_\_\_}$$

second

10; 12; 9; -3

5; 24; -27

-15

9. The sum $\sum_{i=1}^{6} (2i - 3)$ represents the series

$$\text{\_\_\_\_} + \text{\_\_\_\_} + \text{\_\_\_\_} + \text{\_\_\_\_} + \text{\_\_\_\_} + \text{\_\_\_\_}$$

This is an arithmetic sequence having $a_1 = $ _____ and $a_6 = $ _____ . The sum is thus given by

$$S_6 = \frac{\overline{}}{2}(\text{\_\_\_} + \text{\_\_\_}) = \text{\_\_\_\_}.$$

-1; 1; 3; 5; 7; 9

-1

9

6; -1; 9; 24

10. Any sum of the form

$$\sum_{i=1}^{n} (mi + p),$$

where m and p are real numbers, represents the

sum of the terms of an arithmetic sequence having

_____ _____ d = _____. Then,   common; difference; m

$$a_1 = m(\underline{\quad}) + \underline{\quad}$$ 1; p

$$= \underline{\quad} + \underline{\quad}.$$ m; p

To find, for example,

$$\sum_{i=1}^{8} (3i + 7),$$

m = _____, p = _____,   3; 7

$$a_1 = \underline{\quad} + \underline{\quad} = \underline{\quad}$$ 3; 7; 10

and d = _____. Hence,   3

$$\sum_{i=1}^{8} (3i + 7) = \frac{\overline{\quad}}{2}[2(\underline{\quad}) + \underline{\quad}(\underline{\quad})].$$ 8; 10; 7; 3

$$= \underline{\quad}(\underline{\quad} + \underline{\quad})$$ 4; 20; 21

$$= \underline{\quad}$$ 164

Find the sums indicated in Frames 11–14.

11. $\sum_{i=1}^{12} (3i - 11)$

$$a_1 = \underline{\quad}, \quad d = \underline{\quad}$$ −8; 3

The sum is _____.   102

12. $\sum_{i=1}^{10} (5 - 3i)$

$$a_1 = \underline{\quad}, \quad d = \underline{\quad}$$ 2; −3

The sum is _____.   −115

13. $\sum_{i=1}^{15} (2i + 5) = \underline{\quad}$ 315

14. $\sum_{i=1}^{10} \left(\frac{2}{3}i + 2\right) = \underline{\quad}$ $\frac{170}{3}$

15. To find the sum of the first 200 positive integers, write them as a series.

$$\underline{\quad} + \underline{\quad} + \underline{\quad} + \cdots + \underline{\quad}$$ 1; 2; 3; 200

Here we have an arithmetic sequence with
$a_1 =$ ____ and $a_{200} =$ ____. Thus the sum of
the first 200 terms is

$$\frac{\rule{1cm}{0pt}}{2}(\rule{1cm}{0pt} + \rule{1cm}{0pt}) = \rule{2cm}{0pt}.$$

1; 200

200; 1; 200; 20,100

16. Find the sum of the first 500 positive integers.

    _____

125,250

17. The sum of the first n terms of a geometric
sequence having first term a_1 and common ratio
r ($r \neq 1$) is given by the formula

$$S_n = \frac{\rule{1cm}{0pt}(1 - \rule{1cm}{0pt})}{\rule{1cm}{0pt}}.$$

a_1; r^n
$1 - r$

18. To find the sum of the first 7 terms of the
geometric sequence 3, 6, 12, 24, ..., we note
that $a_1 =$ ____ and $r =$ ____. Thus

$$S_7 = \frac{\rule{1cm}{0pt}(1 - \rule{1cm}{0pt})}{1 - \rule{1cm}{0pt}}$$
$$= 3(\rule{1cm}{0pt})$$
$$= \rule{1cm}{0pt}.$$

3; 2
3; 2^7
2
127
381

**Find the sum of the first six terms for each of the
geometric sequences of Frames 19 and 20. Set up
only; do not evaluate.**

19. $a_1 = -4$, $r = 3/2$

$$S_6 = \frac{\rule{1cm}{0pt}(1 - \rule{1cm}{0pt})}{1 - \rule{1cm}{0pt}}$$

-4; $(3/2)^6$
3/2

20. $a_2 = 10$, $a_3 = 5$

First find r: $r = (\rule{0.7cm}{0pt}) \div (\rule{0.7cm}{0pt}) = \rule{0.7cm}{0pt}$.
Then find a_1 by the formula $a_n = a_1 \cdot r^{n-1}$. If
we use a_2 as a_n, we find that $a_1 =$ ____. Thus

$$S_6 = \frac{\rule{1cm}{0pt}(1 - \rule{1cm}{0pt})}{1 - \rule{1cm}{0pt}}$$

5; 10; 1/2

20

20; $(1/2)^6$
1/2

21. A sum of the form $\sum\limits_{i=1}^{n} m \cdot p^i$ represents the sum
    of the first n terms of a geometric sequence
    having first term $a_1 =$ _____ and common ratio   mp
    $r =$ _____.   p

22. To find the sum $\sum\limits_{i=1}^{8} 3 \cdot 2^i$, we first find a_1:

    $a_1 = 3(2)^{\overline{}} =$ _____. Also, $r =$ _____. Hence   1; 6; 2

    $$S_8 = \frac{\underline{}(\underline{} - 1)}{\underline{}}$$ 6; 2^8
                                                               1

    $=$ _____.   1530

Find the sums indicated in Frames 23–25.

23. $\sum\limits_{i=1}^{4} 2 \cdot 3^i$

    $a_1 =$ _____, $r =$ _____   6; 3
    $S_4 =$ _____   240

24. $\sum\limits_{i=1}^{6} 3\left(\frac{2}{3}\right)^i$ (Set up only.)

    $a_1 =$ _____, $r =$ _____   2; 2/3

    $S_6 =$ _____ $\dfrac{2[(2/3)^6 - 1]}{2/3 - 1}$

25. $\sum\limits_{i=1}^{6} (-1)^i \cdot 2$

    $a_1 =$ _____, $r =$ _____   -2; -1

    Sum $=$ _____   0

Solve the applied problems in Frames 26–28.

26. A gardener wants to plant a triangular flower
    bed. He uses 17 plants for the bottom row.
    Each row has 4 fewer plants than the previous
    row. Suppose that there are 5 rows in the
    flower bed.

    (a) How many plants are planted in the top row?

This is a(n) (*arithmetic/geometric*) sequence

with a_1 = _____ and (d/r) = _____. Use the

formula a_n = _____ .

$$a_5 = \_\_\_\_$$

There (*is/are*) _____ plant(s) planted in the

top row.

| arithmetic |
| 17; d; −4 |
| $a_1 + (n - 1)d$ |
| 1 |
| is; 1 |

(b) **What is the total number of plants in the
flower bed?**

Find the sum of the terms of the sequence.

Use the formula S_n = _____ .

$$S_5 = \_\_\_\_$$

There are _____ plants altogether.

| $\frac{n}{2}(a_1 + a_n)$ |
| 45 |
| 45 |

27. **Health care workers find that each year .75 as
many cases of a certain disease are recorded
than the previous year. In 1991, 1600 cases
were recorded.**

(a) **How many cases will be recorded in 1993?**

This is a(n) (*arithmetic/geometric*) sequence

with a_1 = _____ and (d/r) = _____. Use the

formula a_n = _____ .

Here, find $a_{\_\_}$.

$$a_3 = \_\_\_\_\_(\_\_\_)^{\overline{}}$$

$$= _____$$

There will be _____ cases recorded in 1993.

| geometric |
| 1600; r; .75 or 3/4 |
| $a_1 r^{n-1}$ |
| 3 |
| 1600; 3/4; 2 |
| 900 |
| 900 |

(b) **How many cases will be recorded altogether
in the years 1991, 1992, and 1993?**

Find the sum of the first three terms.

$$S_n = _____$$

a_1 = _____, r = _____, n = _____

$$S_3 = _____$$

_____ cases will be recorded during the

three years.

| $\dfrac{a_1(1 - r^n)}{1 - r}$ |
| 1600; 3/4; 3 |
| 3700 |
| 3700 |

28. **Rona deposited $2000 at the end of each year for 5 years in an account paying 8% interest. What is the future value of this annuity?**

The first payment will earn interest for _____ | 4

years, the second for _____ years, the third for | 3

_____ years, the fourth for 1 year. The last | 2

payment earns no interest.

The total amount is

$$\underline{\hspace{1cm}}(\underline{\hspace{0.6cm}})^{\overline{\hspace{0.6cm}}} + \underline{\hspace{1cm}}(\underline{\hspace{0.6cm}})^{\overline{\hspace{0.6cm}}}$$
$$+ \underline{\hspace{1cm}}(\underline{\hspace{0.6cm}})^{\overline{\hspace{0.6cm}}} + \underline{\hspace{1cm}}(\underline{\hspace{0.6cm}}) + \underline{\hspace{1cm}}.$$

| 2000; .08; 4; 2000; .08; 3
| 2000; .08; 2; 2000; .08; 2000

This is a(n) _____ sequence. | geometric

The sum is given by

$$S_5 = \underline{\hspace{3cm}}$$ | $\dfrac{2000[1 - (1.08)^5]}{1 - 1.08}$

$$= \underline{\hspace{3cm}}.$$ | 11,733.20

The future value of the annuity is _____. | $11,733.20

13.4 Sums of Infinite Geometric Sequences

1 Find the sum of the terms of an infinite geometric sequence. (Frames 1-6)

2 Use the sum formula to write a repeating decimal as a quotient of integers. (Frames 7-10)

---

1. 3, 2, 4/3, 8/9, 16/27, ... is an example of an

_____ geometric sequence. If the constant | infinite

ratio r satisfies the condition _____, | $|r| < 1$

the sum of an infinite geometric sequence can

be found by the formula

$$\underline{\hspace{3cm}}.$$ | $S_\infty = \dfrac{a_1}{1 - r}$

The sequence at the beginning of this frame has

$a_1 = $ _____ and $r = $ _____. Since 2/3 < 1, the | 3; 2/3

sum is given by

$$\dfrac{\overline{\hspace{1cm}}}{1 - \underline{\hspace{1cm}}} = \underline{\hspace{1cm}}.$$ | 3;
| $\dfrac{2}{3}$ 9

Find each of the sums indicated in Frames 2–6.

2. $\frac{2}{3}, \frac{1}{3}, \frac{1}{6}, \frac{1}{12}, \cdots$

$a_1 = $ _____ , $r = $ _____

$S_\infty = $ _____

$\frac{2}{3}; \frac{1}{2}$

$\frac{4}{3}$

3. $12, -3, \frac{3}{4}, -\frac{3}{16}, \cdots$

$a_1 = $ _____ , $r = $ _____

$S_\infty = $ _____

$12; -\frac{1}{4}$

$\frac{48}{5}$

4. $\displaystyle\sum_{i=1}^{\infty} \left(\frac{3}{8}\right)^i = $ _____

$\frac{3}{5}$

5. $\displaystyle\sum_{i=1}^{\infty} 4\left(\frac{4}{3}\right)^i = $ _____

$|r| = 4/3 \nless 1$, so the sum does not exist

6. $\displaystyle\sum_{i=1}^{\infty} \frac{3}{4}\left(\frac{2}{3}\right)^i = $ _____

$\frac{3}{2}$

7. To write .343434... as p/q where p and q are integers, write .343434... as

.34 + .0034 + _____ + $\cdots$.

.000034

This is the sum of the terms of an infinite geometric sequence having $a_1 = $ _____ and $r = $ _____ .

.34; .01

$$S_\infty = \frac{\rule{2cm}{0.4pt}}{1 - \rule{1.5cm}{0.4pt}} = \frac{\rule{1.5cm}{0.4pt}}{\rule{1.5cm}{0.4pt}} = \frac{\rule{1.5cm}{0.4pt}}{\rule{1.5cm}{0.4pt}}$$

.34; .34; 34
.01; .99; 99

**Write each repeating decimal in Frames 8–10 as a
fraction of the form p/q where p and q are integers.**

8. .2525... = _____

$\frac{25}{99}$

9. .104104104... = _____

$\frac{104}{999}$

10. .636363... = _____

$\frac{21}{33}$

13.5 The Binomial Theorem

[1]   Learn how to construct Pascal's triangle. (Frames 1–7)

[2]   Evaluate factorials. (Frames 8–12)

[3]   Evaluate the binomial coefficient $\binom{n}{r}$ for specific values of n and r. (Frames 13–16)

[4]   Expand $(x + y)^n$ using the binomial theorem. (Frames 17–23)

[5]   Find the rth term in the binomial expansion of $(x + y)^n$. (Frames 24–28)

Expand the products in Frames 1–5.

1. $(x + y)^0 =$ _____

    1

2. $(x + y)^1 =$ _____

    $x + y$

3. $(x + y)^2 =$ _____

    $x^2 + 2xy + y^2$

4. $(x + y)^3 =$ _____

    $x^3 + 3x^2y + 3xy^2 + y^2$

5. $(x + y)^4 =$ _____

    $x^4 + 4x^3y + 6x^2y^2 + 4xy^3 + y^4$

6. The coefficients of the terms in the expansions above form 5 rows of a triangular array of numbers called _____ _____.

    Each number in the triangle is the _____ of the two numbers directly above it.

    Pascal's triangle
    sum

7. Complete the portion of Pascal's triangle given below.

```
                1
           ___      1
        1      ___      ___
           ___      ___      3      ___
        1      4      ___      ___      1
     1      ___      10      ___      5      ___
        ___      ___      15      ___      ___      6      1
```

    1
    2; 1
    1; 3; 1
    6; 4
    5; 10; 1
    1; 6; 20; 15

8. The number n! (read n-_____) is defined
   as follows for positive integers n.

   $$n! = \text{_____}$$

   $$0! = \text{\_\_\_\_}$$

	factorial
	$n(n-1)(n-2)\cdots$ $(3)(2)(1)$
	1

Evaluate the factorials in Frames 9–12.

9. $4! = \text{\_\_} \cdot \text{\_\_} \cdot \text{\_\_} \cdot \text{\_\_} = \text{\_\_\_}$ | 4; 3; 2; 1; 24

10. $6! = \text{\_\_} \cdot \text{\_\_} \cdot \text{\_\_} \cdot \text{\_\_} \cdot \text{\_\_} \cdot \text{\_\_} = \text{\_\_\_}$ | 6; 5; 4; 3; 2; 1; 720

11. $0! = \text{\_\_\_}$ | 1

12. $3! 5! = (\text{\_\_} \cdot \text{\_\_} \cdot \text{\_\_}) \cdot (\text{\_\_} \cdot \text{\_\_} \cdot \text{\_\_} \cdot \text{\_\_} \cdot \text{\_\_})$
    $= \text{\_\_} \cdot \text{\_\_\_\_}$
    $= \text{\_\_\_\_}$

3; 2; 1; 5; 4; 3; 2; 1
6; 120
720

13. The _____ coefficient, written $\binom{n}{r}$ is
    defined as

    $$\binom{n}{r} = \frac{\text{_____}}{\text{_____}}$$

binomial
$n!$
$r!(n-r)!$

Evaluate the binomial coefficients in Frames 14–16.

14. $\binom{5}{3} = \dfrac{\text{\_\_}!}{\text{\_\_}! \ \text{\_\_}!} = \dfrac{\text{\_\_} \cdot \text{\_\_} \cdot \text{\_\_} \cdot \text{\_\_} \cdot \text{\_\_}}{(\text{\_\_} \cdot \text{\_\_} \cdot \text{\_\_})(\text{\_\_} \cdot \text{\_\_})}$

    $= \dfrac{\text{\_\_\_\_}}{\text{\_\_\_\_}} = \text{\_\_\_}$

5; 5; 4; 3; 2; 1
3; 2; 3; 2; 1; 2; 1
120;
12   10

15. $\binom{7}{0} = \text{\_\_\_}$ | 1

16. $\binom{10}{8} = \text{\_\_\_}$ | 45

17. To expand $(x + y)^n$, use the _____ theorem. | binomial

18. For example, $(x + y)^3$ is

$$(x + y)^3 = \underline{\quad} + \left(\frac{}{1}\right)x^2y + \left(\frac{3}{}\right)xy^2 + \underline{\quad}$$

$$= x^3 + \underline{\quad} + \underline{\quad} + y^3$$

$x^3;\ 3;\ 2;\ y^3$

$3x^2y;\ 3xy^2$

Expand each binomial in Frames 19–23.

19. $(x + y)^7$

Use the binomial theorem with $n = \underline{\quad}$.

$$(x + y)^7 = x^7 + \binom{7}{1}\underline{\quad} + \binom{7}{2}\underline{\quad} + \binom{7}{3}x^4y^3$$

$$+ \binom{7}{4}x^3y^4 + \binom{7}{5}x^2y^5 + \binom{7}{6}\underline{\quad} + \underline{\quad}$$

$$= \underline{\hspace{5cm}}$$

7

$x^6y;\ x^5y^2$

$xy^6;\ y^7$

$x^7 + 7x^6y + 21x^5y^2$
$+ 35x^4y^3 + 35x^3y^4$
$+ 21x^2y^5 + 7xy^6 + y^7$

20. $(p - q)^5$

Use the binomial theorem.

$$(p - q)^5 = p^5 + \binom{5}{1}p^4(-q) + \binom{5}{2}\underline{\quad} + \binom{5}{3}p^2(-q)^3$$

$$+ \binom{5}{4}\underline{\quad} + (-q)^5$$

$p^3(-q)^2$

$p(-q)^4$

The second term is negative because $-q$ is raised to an $\underline{\quad}$ power. After that the signs of the terms will $\underline{\quad}$. Finally,

$$(p - q)^5 = \underline{\hspace{5cm}}.$$

odd

alternate

$p^5 - 5p^4q + 10p^3q^2$
$- 10p^2q^3 + 5pq^4 - q^5$

21. $(2x + y)^6 = (\underline{\quad})^6 + \binom{6}{1}(2x)^5y + \left(\frac{6}{}\right)(\underline{\quad})^4y^2$

$$+ \binom{6}{3}(2x)^3y^3 + \left(\frac{6}{}\right)(2x)^2y^4$$

$$+ \left(\frac{6}{}\right)(\underline{\quad}) + \underline{\quad}$$

$$= \underline{\hspace{5cm}}$$

$2x;\ 2;\ 2x$

4

$5;\ 2xy^5;\ y^6$

$64x^6 + 192x^5y$
$+ 240x^4y^2 + 160x^3y^3$
$+ 60x^2y^4 + 12xy^5 + y^6$

22. $(3r - s)^4 =$ _____ | $81r^4 - 108r^3s$
$+ 54r^2s^2 - 12rs^3 + s^4$

23. $\left(\dfrac{p}{2} - \dfrac{q}{3}\right)^3 =$ _____ | $\dfrac{p^3}{8} - \dfrac{p^2q}{4} + \dfrac{pq^2}{6} - \dfrac{q^3}{27}$

24. The rth term of the binomial expansion of
$(x + y)^n$ where $n \geq r - 1$ is given by
$$\left(\underline{\quad}\right)x^{\underline{\quad}} y^{\underline{\quad}}.$$
| n; $n - (r - 1)$; $r - 1$
$n - (r - 1)$

Find the indicated terms in Frames 25–28.

25. 4th term of $(x + y)^9$
    Write the 4th term of the binomial expansion
    $(x + y)^9$, as
    $$\dfrac{9!}{(9 - 3)!3!}\underline{\quad\quad} \text{ which is } \underline{\quad\quad}.$$
| $x^{9-3}y^3$; $84x^6y^3$

26. 8th term of $(x + y)^{13}$ _____ | $1716x^6y^7$

27. 7th term of $(2x - 3y)^{12}$
    First find $r - 1$ to avoid careless sign
    mistakes.
    $r - 1$, for $r = 7$, is 6. Then
    $$\dfrac{12!}{(12 - 6)!6!}(2x)^6(-3y)^6 = \underline{\quad\quad}$$
| $43,110,144x^6y^6$

28. 11th term of $(4x - 3y)^{12}$ _____ | $62,355,744x^2y^{10}$

13.6 Mathematical Induction

1. Learn the principle of mathematical induction. (Frames 1 and 2)

2. Use the principle of mathematical induction to prove a statement. (Frames 3 and 4)

---

1. Let S_n be a statement concerning the positive
   integer ___. Suppose that
| n

(a) S__ is true.

1

(b) For any positive integer k, k ≤ n,
     S__ implies S____.

k; k+1

Then _____ is true for every positive integer n.

S_n

2. In order to write a proof by mathematical in-
    duction, _____ steps are required.

two

Step 1 Prove that the statement is true
        for n = _____.

1

Step 2 Show that, for any positive integer
        k, k ≤ n, _____ implies _____.

S_k; S_{k+1}

Prove the theorems of Frames 3 and 4 by mathematical induction.

3. $1 + 3 + 5 + \cdots + (2n - 1) = n^2$

Let _____ represent this statement.

S_n

Write S_1: 1 = _____

1^2

Is this true? (*yes/no*)

yes

Assume the statement is true for some
natural number k so that

$1 + 3 + 5 + \cdots + (2k - 1) = $ _____.

k^2

The next number in the sequence after 2k - 1
is _____. Add this number to both sides.

2k + 1

$1 + 3 + 5 + \cdots + (2k - 1) + ($_____$)$

2k + 1

$= k^2 + ($_____$)$

2k + 1

On the right we have just _____, so that

$(k + 1)^2$

$1 + 3 + 5 + \cdots + (2k + 1) = $ _____. Thus,

$(k + 1)^2$

the truth of S_k implies the truth of _____,

S_{k+1}

and the theorem has been proven by mathematical

_____.

induction

4. $\dfrac{1}{1 \cdot 2} + \dfrac{1}{2 \cdot 3} + \dfrac{1}{3 \cdot 4} + \cdots + \dfrac{1}{n(n + 1)} = \dfrac{n}{n + 1}$

Let n = 1. Then

$$\frac{1}{1(1 + 1)} = \frac{1}{1 \cdot 2} = \frac{1}{1 + 1}.$$

Therefore, this statement (*is/is not*) true for
n = 1.

is

    Suppose the statement is true for _____.

k

Then

$$\frac{1}{1 \cdot 2} + \frac{1}{2 \cdot 3} + \frac{1}{3 \cdot 4} + \cdots + \frac{1}{k(k + 1)} = \underline{\hspace{1cm}}.$$

The next term on the left is _____.

$\dfrac{1}{(k + 1)(k + 2)}$

Add this to both sides of the equation.

$$\frac{1}{1 \cdot 2} + \frac{1}{2 \cdot 3} + \frac{1}{3 \cdot 4} + \cdots + \underline{\hspace{2cm}}$$

$\dfrac{1}{(k + 1)(k + 2)}$

$$= \frac{1}{k + 1} + \underline{\hspace{2cm}}$$

$\dfrac{1}{(k + 1)(k + 2)}$

Simplify on the right.

$$\frac{k}{k + 1} + \frac{1}{(k + 1)(k + 2)} = \frac{\underline{\hspace{1cm}} + 1}{(k + 1)(k + 2)}$$

$k(k + 2)$

$$= \frac{\underline{\hspace{1cm}}}{(k + 1)(k + 2)}$$

$k^2 + 2k + 1$

$$= \underline{\hspace{1cm}}$$

$\dfrac{k + 1}{k + 2}$

The truth of S_k implies the truth of _____,
proving the theorem.

S_{k+1}

13.7   Permutations

[1]   Use the fundamental principle of counting. (Frames 1–4)

[2]   Know the formula for P(n, r). (Frames 5–12)

[3]   Use permutations to solve counting problems. (Frames 13–16)

[4]   Use the fundamental counting principle with restrictions.
   (Frames 17 and 18)

---

1. The fundamental principle of counting states that
if one event can occur in m ways, and a second
event can occur in n ways, then both events can
occur in _____ ways, provided the outcome of

mn

the first does not influence the outcome of the
second. Such events are called _____

independent

events.

2. The fundamental principle of counting can be extended to _____ _____ of events.   any; number

Solve the problems of Frames 3-4.

3. **How many different mobile homes are available if the manufacturer offers a choice of 6 basic plans, 4 different exteriors, and 3 different woods for the interior paneling.**

   Use the _____ _____ of   fundamental; principle
   _____. We can break this down into   counting
   events. The first event is the selection of
   a basic plan, the second event is the selection of type of exterior, and the third event
   is the selection of interior paneling. None
   of these events depends on another. Therefore
   the number of different mobile homes is
   $6 \cdot 4 \cdot$ _____ = _____.   3; 72

4. **The Sharp family has decided to get a new dog, with all family members adding their preferences to the list. They have narrowed their choices down to 3 breeds and are still undecided as to the sex of the dog. They also have a list of 5 names. How many possibilities are there for the breed, sex, and name of the dog?** _____   30

5. For any counting number n, the product
   $$n(n-1)(n-2) \cdots (2)(1)$$
   is symbolized by _____, and this is read "____   n!; n
   _____."   factorial

6. 0! is defined to be equal to ____.   1

7. The number of _____ of n things taken   permutations
   ____ at a time is written _____.   r; P(n, r)

8. To find P(n, r), use the formula

$$P(n, r) = \underline{\hspace{3cm}}.$$

$$\dfrac{n!}{(n - r)!}$$

Find the permutations in Frames 9 and 10.

9. P(6, 4)

   Use the formula of Frame 2.

$$P(6, 4) = \dfrac{\overline{\hspace{1cm}}}{(\underline{\hspace{2cm}})!}$$

6

6 − 4

$$= \dfrac{6 \cdot 5 \cdot 4 \cdot 3 \cdot 2 \cdot 1}{\overline{\hspace{2cm}}}$$

2 · 1

$$= \underline{\hspace{2cm}}$$

360

10. P(7, 2) = \underline{\hspace{2cm}}

42

11. There is a special formula for P(n, n):

$$P(n, n) = \underline{\hspace{2cm}}.$$

n!

   For example,

$$P(5, 5) = \underline{\hspace{2cm}}$$

5!

$$= 5 \cdot 4 \cdot 3 \cdot 2 \cdot 1 = \underline{\hspace{2cm}}.$$

120

12. Find P(4, 4) = \underline{\hspace{2cm}}.

4! or 24

13. In how many ways can 7 children swing in 4 swings?

   This represents P(7, 4), which is \underline{\hspace{2cm}}.

840

14. In how many ways can 5 people ride 3 horses?

$$\underline{\hspace{2cm}}$$

60

15. **How many 4-letter radio-station call letters can be made if the first 3 letters are KEQ and no letter can be repeated? If repetitions are allowed?**

If we use KEQ as the first three letters
then there are _____ letters left to use that
are not KEQ. Therefore, there are _____ ways
if no letter can be repeated. If repetitions
are allowed, there are _____ ways.

23
23

26

16. The Child Care Co—op has a "dress up" box of
clothes. There are 7 hats, 4 long dresses,
and 6 pairs of shoes. How many different out-
fits, which include 1 hat, 1 long dress, and
1 pair of shoes, are possible? _____

$7 \cdot 4 \cdot 6 = 168$

17. How many ways can a three—digit area code be
formed if the middle digit must be 0 or 1 and
the first and third digits cannot be 0 or 1?
(Repetitions can occur in the first and third
digit.)

There are ____ choices for the first digit,
since 0 or 1 cannot be used.

8

There are ____ choices for the middle digit.

2

There are ____ choices for the third digit.

8

Thus, there are ____ • ____ • ____ = _____
ways to form the area code with the given
restrictions.

8; 2; 8; 128

18. How many ways can the seven days of the week
be arranged if Saturday and Sunday must be in
the last two positions and no repetitions are
allowed?

____ choices for the first day

5

____ choices for the second day

4

____ choices for the third day

3

____ choices for the fourth day

2

1 choice for the fifth day

____ choices for the sixth day

2

1 choice for the seventh day

There are ___ • ___ • ___ • ___ • 1 • ___ • 1 = _____
ways to arrange the days of the week given the
restrictions.

5; 4; 3; 2; 2; 240

13.8 Combinations

[1] Learn the formula $\binom{n}{r}$ for combinations. (Frames 1–5)

[2] Use the combinations formula to solve counting problems. (Frames 6–12)

[3] Use the combinations formula to find properties for special cases.
(Frames 13 and 14)

1. Permutations are used to find the number of ways
a group of items can be arranged in _____.
Combinations are used if order (*is/is not*) important.

order

is not

2. The number of combinations of n things, taken r
at a time is written

_____.

This number is found with the formula

$$\binom{n}{r} = \text{_____}.$$

$\binom{n}{r}$

$\dfrac{n!}{r!(n-r)!}$

Find the number of combinations in Frames 3–5.

3. $\binom{7}{3} = \dfrac{7!}{\text{\_\_\_}!(\text{\_\_\_\_\_})!} = \text{\_\_\_\_}$

3; 7 − 3; 35

4. $\binom{9}{5} = \text{\_\_\_\_}$

126

5. $\binom{20}{3} = \text{\_\_\_\_}$

1140

Solve the problems of Frames 6–12.

6. How many samples of 10 fish can be netted from an aquarium holding 20 fish of different varieties?

    We want

    $$\binom{20}{10} = \frac{20!}{(20-10)!10!} = \frac{20!}{10!10!} = \underline{\hspace{2cm}}.$$

    184,756

7. There are 25 students in Ms. Hugh's ninth-grade class and 3 are to be selected to be on the student council. How many ways can they be selected? $\underline{\hspace{3cm}}$

    $\binom{25}{3} = 2300$

8. Sundance Natural Food restaurant makes sandwiches with cheese, avocado, tomato, onion, alfalfa sprouts, and nuts. How many different sandwiches can they make with any 4 of the ingredients?

    The number of combinations is

    $$\binom{6}{4} = \frac{6!}{(6-4)!4!} = \underline{\hspace{1.5cm}}.$$

    15

9. If a seashell collection has 12 olives, 7 sand dollars, and 4 nautilus shells, how many samples of 3 can be drawn in which all 3 are olive shells?

    $$\binom{12}{3} = \underline{\hspace{1.5cm}}$$

    220

10. The English department faculty has decided to form a committee to mediate between administration and the students. There are 40 faculty members. The committee will consist of 7 people.

    (a) In how many ways can the committee be chosen?

    $$\binom{40}{7} = \underline{\hspace{2cm}}$$

    18,643,560

(b)   **If there are 15 women and 25 men on the faculty, how many committees could be chosen containing 4 women and 3 men?**

    First, find the combinations possible that have 4 women out of 15 women.

$$\left(\dfrac{}{4}\right) = \underline{\hspace{2cm}}$$ 15; 1365

Now, once the possible combinations of women are found, consider the possible combinations of men.

$$\left(\dfrac{}{3}\right) = \underline{\hspace{2cm}}$$ 25; 2300

Given any one of the _____ combinations of   1365
women, any one of the possible combinations
of men could be picked, so (1365)(2300) =
_____ possible combinations of 4 women   3,139,500
and 3 men that may be chosen from 15 women
and 25 men.

Solve the problems of Frames 11 and 12 by using combinations or permutations as required.

11.   How many ways can the letters of the word ZEBRA
     be rearranged?   _____   120

12.   How many different samples of 3 out of 5 books
     may be selected?   _____   10

13.   How many committees of 12 students can be formed
     from a class of 12 students?

$$\binom{12}{12} = \dfrac{\underline{}!}{\underline{}!\,\underline{}!} = \dfrac{\underline{}!}{\underline{}!} = \underline{}$$ 12; 12
                                           12; 0; 12; 1

In general, $\binom{n}{n} = \underline{}$, since there is only ____   1; 1
way to select a group of n from a set of n
objects when order is disregarded.

14. How many speakers can be chosen from an oratory group of 8 members?

$$\binom{8}{1} = \frac{\underline{\quad}!}{\underline{\quad}!\,\underline{\quad}!} = \underline{\quad}$$

8

1; 7; ^{8}

In general, $\binom{n}{1} = \underline{\quad}$, since there are $\underline{\quad}$ ways of choosing 1 object from a set of n objects.

n; n

13.9 Probability

1 Learn the terminology of probability theory. (Frames 1–5)

2 Find the probability of an event. (Frames 6–8)

3 Find the probability of the complement of E, given the probability of E. (Frames 9–12)

4 Find the odds in favor of an event. (Frames 13–18)

5 Find the probability of a compound event. (Frames 19–23)

1. In probability, each repetition of an experiment is called a _____. To define probability, start with an experiment having one or more _____, each of which is _____ likely to occur.

trial

outcomes; equally

2. The set of all possible _____ for an experiment makes up the _____ space for the experiment.

outcomes
sample

Write the sample space for each experiment in Frames 3 and 4.

3. Tossing a coin _____

{H, T}

4. Rolling a die _____

{1, 2, 3, 4, 5, 6}

5. The probability of an event E, written _____,
    is the ratio of the number of outcomes in the
    _____ space S that are favorable to E to the
    _____ number of outcomes in S.

$P(E)$

sample
total

Find the probability of each event.

6. Heads on a single toss of a coin.
        The sample space is _____, with ____
    elements. The event "toss of heads" is _____,
    with ____ element. The probability of heads is
            _____.

$\{H, T\}$; 2
$\{H\}$
1
1/2

7. A die is rolled and the result is at least 5.
        Sample space: _____
            Event: _____
        Probability: _____

$\{1, 2, 3, 4, 5, 6\}$
$\{5, 6\}$
2/6 = 1/3

8. A diamond is drawn from a deck of 52 cards.
    (There are 13 diamonds in the deck.)

                    _____

13/52 = 1/4

9. The complement of event E, written _____, is
    made up of all outcomes in the sample space
    that (*do/do not*) belong to E.

E'

do not

10. Since $E \cup E' = S$, then
            $P(E) + P(E') =$ _____.

1

11. Write this result in an alternate way:
            $P(E) =$ _____.

$1 - P(E')$

12. If $P(E) = 8/11$, find $P(E')$.

                    _____

3/11

13. The odds in favor of an event E is the ratio of
    $P(E)$ and _____.

$P(E')$

Find the odds in favor of each event in Frames 14–16.

14. $P(E) = \frac{2}{5}$

If $P(E) = 2/5$, then $P(E') =$ _____, and the
odds in favor of E are

$$\frac{\frac{2}{5}}{\rule{1cm}{0.4pt}} = \rule{1cm}{0.4pt} .$$

Odds of 2/3 are often written _____.

3/5
$\frac{3}{5}$; $\frac{2}{3}$
2 to 3

15. $P(E) = \frac{1}{7}$

Odds in favor of E: _____

1 to 6

16. $P(E') = \frac{9}{19}$

Odds in favor of E: _____

10 to 9

17. Suppose the odds in favor of E are 8 to 5.
Find $P(E)$.

Odds of 8 to 5 means 8 _____ outcomes
out of _____ total outcomes, so

$$P(E) = \rule{1.5cm}{0.4pt} .$$

favorable

13

$\frac{8}{13}$

18. Find $P(E)$ if the odds against E are 7 to 9.

_____

$\frac{9}{16}$

19. Two events that cannot occur at the same time
are _____ exclusive events.

mutually

20. For example, rolling a 5 and a 2 on a die at the
same time are mutually _____ events.

exclusive

21. Find the probability of alternative events with
the formula

$$P(E \text{ or } F) = P(\rule{1.5cm}{0.4pt})$$
$$= P(E) + \rule{1cm}{0.4pt} - \rule{1cm}{0.4pt} .$$

$E \cup F$

$P(F)$; $P(E \cap F)$

22. Find P(E or F) if P(E) = 1/2, P(F) = 1/3, and
    P(E ∩ F) = 1/12.

    The probability is

$$P(E \text{ or } F) = \frac{1}{2} + \underline{\hspace{1cm}} - \underline{\hspace{1cm}}$$

$$= \underline{\hspace{1cm}}.$$

$\dfrac{1}{3}; \dfrac{1}{12}$

$\dfrac{3}{4}$

23. Find P(M or N) if P(M) = 3/10, P(N) = 2/5, and
    P(M ∩ N) = 1/20.

$\underline{\hspace{3cm}}$

$\dfrac{13}{20}$

24. Complete the following properties of probability.

    For any events E and F,

        $0 \leq \underline{\hspace{1.5cm}} \leq 1$

        P(a certain event) = $\underline{\hspace{1cm}}$

        P(an impossible event) = $\underline{\hspace{1cm}}$

        P(E) = $\underline{\hspace{2cm}}$

        P(E or F) = P(E) + P(F) − $\underline{\hspace{2cm}}$

P(E)

1

0

1 − P(E′)

P(E ∩ F)

Chapter 13 Test

The answers for these questions are at the back of this Study Guide.

Write the first five terms for each of the following sequences.

1. $a_n = 3(n - 1)$ 1. _____

2. Arithmetic, $a_1 = 9$, $d = -2$ 2. _____

3. Geometric, $a_4 = 12$, $r = 1/3$ 3. _____

4. $a_1 = 4$, $a_n = 3a_{n-1}$ for $n \geq 2$ 4. _____

5. $a_n = n + 3$ if n is odd, and $a = 2a_{n-1}$
   if n is even.   5. _____

6. A certain arithmetic sequence has $a_6 = 23$
   and $a_8 = 31$. Find a_{20}.   6. _____

Find a_5 and S_5 for each of the following sequences.

7. Arithmetic, $a_3 = 2$, $d = -6$ 7. _____

8. Geometric, $a_7 = 16$, $a_8 = 32$ 8. _____

9. Find a_6 for the geometric sequence with
   $a_1 = 3x^2$ and $a_3 = 12x^8$.   9. _____

Find each of the following sums that exist.

10. $\sum_{i=1}^{5} (4i - 3)$ 10. _____

11. $\sum\limits_{i=1}^{4} \left(\frac{1}{3}\right)(9^i)$

11. _____

12. $\sum\limits_{i=1}^{\infty} -3\left(\frac{1}{2}\right)^i$

12. _____

13. $128 + 48 + 18 + \frac{27}{4} + \cdots$

13. _____

14. Monique harvested 16 tomatoes from her
     greenhouse one week. Each week there-
     after she harvested 10 more tomatoes
     than the previous week's harvest. Find
     the total number of tomatoes she harvested
     after 8 weeks.

14. _____

15. Suppose that a bacteria colony doubles in
     size each day. If there are 100 bacteria
     in the colony today, how many will there
     be exactly one week from today?

15. _____

16. Use the binomial theorem to expand
     $$(2k - 3)^4.$$

16. _____

17. Find the sixth term in the expansion of
     $$\left(x - \frac{1}{2}y\right)^7.$$

17. _____

18. Use mathematical induction to prove that
     the following statement is true for every
     positive integer n.
     $$2 + 4 + 8 + \cdots + 2^n = 2^{n+1} - 2$$

18. _____

Evaluate the following.

19. $P(9, 4)$

19. _____

20. $\binom{7}{3}$

20. _____

21. $\binom{9}{9}$

21. _____

22. A bakery sells pies in 4 different sizes. There are 6 fillings and 3 types of crust available. How many different pies can be made?

22. _____

23. In how many ways can 6 apprentices be assigned to 4 master workers?

23. _____

24. How many different 5 card "hands" can be dealt from a deck of 12 cards?

24. _____

A single die is tossed. Find the probability that the face that is up shows each of the following points.

25. A number less than 5

25. _____

26. 6 or an odd number

26. _____

27. A number not divisible by 2

27. _____

28. 10

28. _____

29. 3

29. _____

30. A sample of 5 light bulbs is chosen. The probability of exactly 0, 1, 2, 3, 4, or 5 bulbs having broken filaments is given in the following table.

Number broken	0	1	2	3	4	5
Probability	.15	.21	.33	.19	.08	.04

Find the probability that at most 3 filaments are broken.

30. _____

ANSWERS
TO
CHAPTER TESTS

ANSWERS TO CHAPTER TESTS

Chapter 1 Test Answers

1. 1   2. 11   3. −4   4. 8   5. 6   6. 3   7. 2

8. −3   9. 5/3   10. 2/3   11. 4   12. −11, −9, 4, 0 (or 0/5)

13. Integers and 3/8, −2/3, 1.4, $.\overline{7}$ 14. π, $-\sqrt{5}$, .121122...   15. All

16. True   17. False   18. True   19. True   20. 0   21. −9

22. −1/3   23. Commutative   24. Associative   25. Inverse

26. Identity   27. Inverse   28. Distributive   29. Symmetric

30. Transitive   31. Substitution   32. Commutative

33. Multiplication property of zero   34. Addition of equality

35. Trichotomy   36. Transitive   37. Reflexive

38. $[-7, \infty)$ 39. $(-\infty, -1)$ 40. $(0, 3]$

Chapter 2 Test Answers

1. $\{-9\}$ 2. $\{3\}$ 3. $\{-8/3\}$ 4. $\{3, -3\}$ 5. $\{4, -5\}$

6. $\{8/3, -8\}$ 7. $\{2, 0\}$ 8. $\{3, 1\}$ 9. $\{30\}$ 10. $\{-1\}$

11. $\{-3/4\}$ 12. $\varnothing$

13. $[3, \infty)$ 14. $[-1, 3]$ 15. $(-\infty, 5] \cup [7, \infty)$

16. $[5, \infty)$ 17. $[-2, 10]$ 18. $(-\infty, -9) \cup (-4, \infty)$

19. $(-\infty, \infty)$ 20. $\varnothing$ 21. $m = \dfrac{8 - 2n}{7}$ 22. $z = \dfrac{k - 3x + 5y}{2}$

23. Contradiction   24. ∅   25. {3, 4, 5, 6, 7, 8, 9, 12}   26. {3}

27. {1, 2, 3, 6, 9}   28. 51; 40   29. Length, 14; width, 8   30. $93.75

31. $11,000 at 8%, $7000 at 10%   32. 15 liters   33. 28 dimes, 9 quarters

34. 3 toppings   35. 6 miles per hour   36. 150°

Chapter 3 Test Answers

1. $\dfrac{16}{9}$ 2. $36x^3y^7$ 3. $\dfrac{p^{12}}{3q^2}$ 4. $\dfrac{4}{a^8b^5}$ 5. $\dfrac{1}{x^2}$

6. 9.04×10^{-4} 7. $12x^3y^3 - 9x^2y^4$ 8. $15y^2 - 14yz - 8z^2$

9. $16x^2 - 81y^2$ 10. $p^3 - 7p^2 + 18p - 18$ 11. $8a^3 - 20a^2 + 16a - 6$

12. $3x^2 - x - 5$ 13. $2y^2 - 6$ 14. $4x^2 + y^2 + 4xy - 12x - 6y + 9$

15. 3; 4   16. 4; -2   17. 3; -7   18. 1; none   19. Binomial

20. Monomial   21. $3(2a^2 + 4a + 3)$ 22. $7pq(p + 2)$ 23. $(x - 10)(x + 1)$

24. $(-p - 5)(p + 1)$ or $(p + 5)(-p - 1)$ 25. Cannot be factored

26. $(4a - 3k)(2a + 3k)$ 27. $(8m + 3n)(8m - 3n)$ 28. $(11b + 6)(11b - 6)$

29. $(x - 2)(x^2 + 2x + 4)$ 30. $(3p - 2)(9p^2 + 6p + 4)$

31. $(p + 2)(p^2 - 2p + 4)$ 32. $(2a^2 + 1)(4a^4 - 2a^2 + 1)$

33. $(a + 7)(-a - 7)$ 34. $(m - 4n)^2$ 35. $(8r + 3s)^2$

36. $(m + n)(k - r)$ 37. $(a - c)(3b + 2)$ 38. $(1 + x)(1 + y)$

39. $(2x + 1)(4y - 3)(4y + 3)$ 40. 1   41. {1/2, -4}

42. {5/3, -3/1}   43. {0, -2/3, 1/5}   44. 9, 11; or -11, -9

Chapter 4 Test Answers

1. $-\dfrac{9}{4}$ 2. $\dfrac{8p - 1}{p}$ 3. $\dfrac{3}{8r}$ 4. $(r + 1)(r - 4)$ 5. $\dfrac{2a - 3}{a + 4}$

6. $24y^3$ 7. $a(a - 3)(a - 1)$ 8. $\dfrac{2a^2}{(1 + a)(a - 1)}$ or $\dfrac{-2a^2}{(1 + a)(1 - a)}$

9. $\dfrac{(2r - 1)}{r(r - 1)}$ 10. $\dfrac{(m + 3)}{(m + 1)(m + 2)}$ 11. $\dfrac{8}{7}$ 12. $\dfrac{2}{4k + 2}$ or $\dfrac{1}{2k + 1}$

13. $p^2 - \dfrac{2}{3} + \dfrac{1}{3p}$ 14. $4x^2 - x + 4 + \dfrac{-40}{2x + 5}$ 15. $2a^3 - 4a^2 + 7a - 5$

16. $3x^3 - 2x^2 + 2x - 3 + \dfrac{-3}{x + 2}$ 17. $x^3 - x^2 - 4x + 8$

18. $x^3 + \dfrac{3}{5}x^2 - x + 2$ 19. Yes   20. $\{8/3\}$ 21. $\{0\}$ 22. $\{-1/4\}$

23. $\dfrac{xy^2 + x^2}{y^2 - x^2y}$ 24. $\sigma = 10$ 25. $\dfrac{2q}{q + 3}$ 26. $\dfrac{m + n}{m - n}$ 27. $-6, -5$

28. $12/5$ hours   29. 3 miles per hour

Chapter 5 Test Answers

1. 4 2. $\dfrac{1}{5}$ 3. 16 4. $\dfrac{1}{m}$ 5. $\dfrac{1}{p^{1/8}}$ 6. $\dfrac{1}{p^{3/2} \cdot q^{1/2}}$

7. $\dfrac{a^2b^3}{p^4}$ 8. 1 9. $6\sqrt{3}$ 10. $-5\sqrt{3}$ 11. $8\sqrt{2}$ 12. $\sqrt{m}$

13. $3m^2n^3\sqrt[3]{m}$ 14. 12 15. $28\sqrt{2}$ 16. $12\sqrt{7}$ 17. $\dfrac{3\sqrt{5}}{10}$

18. $7 - 2\sqrt{10}$ 19. 1 20. $-\dfrac{\sqrt{2} + 3}{7}$ 21. $\sqrt[4]{45}$ 22. $\{11\}$

23. $\{10\}$ 24. $\{-2\}$ 25. $\{7\}$ 26. $1 - 6i$ 27. $-16 + 0i$

28. $50 + 0i$ 29. $0 - i$ 30. $\dfrac{1}{2} - \dfrac{1}{2}i$

Chapter 6 Test Answers

1. $\{1/2, -4\}$ 2. $\{1/3\}$ 3. $\{6, -3/2\}$ 4. 4 miles per hour

5. $\{2 \pm \sqrt{6}\}$ 6. $\left\{\dfrac{9 \pm \sqrt{69}}{6}\right\}$ 7. $\{3/2\}$ 8. $\{1, 1/2\}$

9. $\{2, -2, 3i, -3i\}$ 10. $\{5\}$ 11. $\{1/3, -4\}$ 12. Two imaginary

13. Yes   14. $11/6; 2$ 15. $2x^2 + 5x - 3 = 0$ 16. $z = \pm\dfrac{\sqrt{2mk}}{2k}$

17. $t = \pm\dfrac{\sqrt{A^2 - r^2z^2y^2}}{zy}$ 18. 8 meters; 15 meters; 17 meters

19. 7 meters by 10 meters   20. 3 seconds   21. $[-1/2, 5/3]$

22. $(-\infty, -3/2) \cup (2, \infty)$ 23. $(-\infty, -3] \cup [-1, 4]$ 24. $[1, 3)$

Chapter 7 Test Answers

1. $-10/9$ 2. $-4/7$; $(2, 0)$; $(0, 8/7)$ 3. $2/5$; $(1/2, 0)$; $(0, -1/5)$

4. Undefined; $(5, 0)$; none   5. 0; none; $(0, -2)$ 6. $3x - 2y = -9$

7. $3x + y = 6$ 8. $x = 8$ 9. $3x - 2y = -7$ 10. $2x + 3y = 11$

11. $2x + y = -2$ 12. $2\sqrt{41}$ 13. $y = -\frac{1}{4}x + 5$, 9/4 hours or

2 hours, 15 minutes   14. Neither   15. Perpendicular   16. .4 inches

17.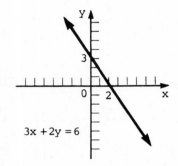

18.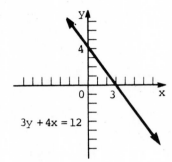

19.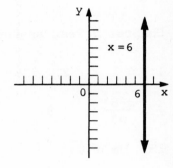

20.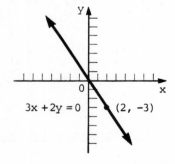

21.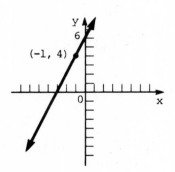

22.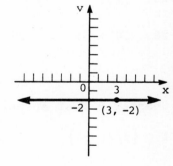

23.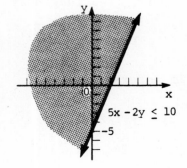

24.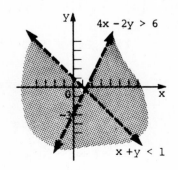

25.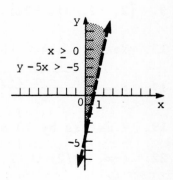

26. 5   **27.** 1600 feet   **28.** x = 80   **29.** Domain: $\{2, 3, 5, 9\}$;

range: $\{5, 11, 17\}$; function   **30.** Domain: $[-4, \infty)$; range: $(-\infty, \infty)$;

not a function   **31.** Both domain and range are $(-\infty, \infty)$; function

32. Domain: $(-\infty, \infty)$; range: $[-3, \infty)$; function

33. 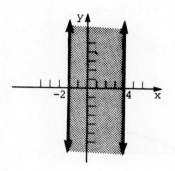   **34.**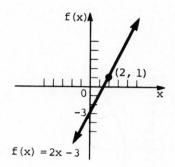

35. $f(x) = -\frac{2}{3}x + 2$ $f(2) = \frac{2}{3}$ $f(j) = -\frac{2}{3}j + 2$

Chapter 8 Test Answers

1. -13   **2.** -2   **3.** -1/7   **4.** 11

5.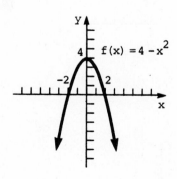

6. 128 feet; 4 seconds

7. Width: 37 inches; maximum area: 1369 square inches

8. Not symmetric

9. Symmetric to the x-axis, y-axis, and origin

10. Symmetric to the x-axis   **11.** Symmetric to the origin

12. Increasing on $(-\infty, -1)$ and $(1, \infty)$; decreasing on $(-1, 1)$; constant on no

interval   **13.** (4, -3); 3   **14.** (1/2, 0); 5/2

15.

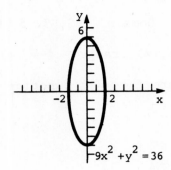

$-9x^2 + y^2 = 36$

16.

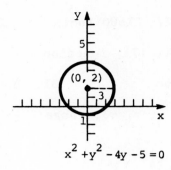

$x^2 + y^2 - 4y - 5 = 0$

17.

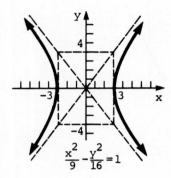

$\frac{x^2}{9} - \frac{y^2}{16} = 1$

18.

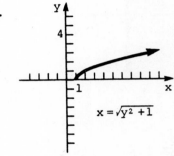

$x = \sqrt{y^2 + 1}$

19. Ellipse   **20.** Parabola   **21.** Hyperbola

22.

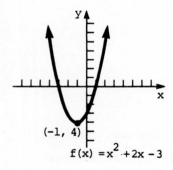

$(-1, 4)$

$f(x) = x^2 + 2x - 3$

23.

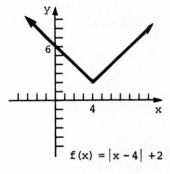

$f(x) = |x - 4| + 2$

24.

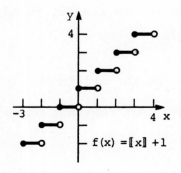

$f(x) = [\![x]\!] + 1$

25.

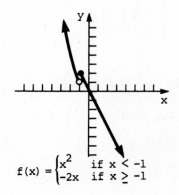

$f(x) = \begin{cases} x^2 & \text{if } x < -1 \\ -2x & \text{if } x \geq -1 \end{cases}$

Chapter 9 Test Answers

1.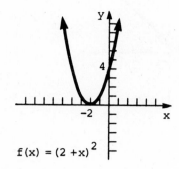

$$f(x) = (2 + x)^2$$

2.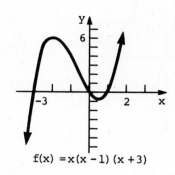

$$f(x) = x(x - 1)(x + 3)$$

3. $P(x) = x^3 - x^2 - 14x + 24$

4. No, since $P(4) = 152$.

5. Yes, since $P(-3) = 0$.

6. $P(x) = -\frac{2}{3}(x^3 - 4x^2 - 2x + 8)$

7. $1 + i$, $1 - i$, 2, -1

8. $P(x) = (x + 2)(x - 1)(x + 1)$

9. $\pm 1/2$, ± 1, ± 2; 1, -2, $-1/2$

10. ± 1, ± 2, ± 5, ± 10; no rational zeros

11. $P(0) = 2$ and $P(1) = -2$, so there must be a real zero between 0 and 1.

12.
$$
\begin{array}{r}
3\,)\ \ 1 \quad 2 \quad -5 \quad -6 \\
 3 \quad 15 \quad 30 \\
\hline
1 \quad 5 \quad 10 \quad 24
\end{array}
$$
← All are positive

$$
\begin{array}{r}
-4\,)\ \ 1 \quad 2 \quad -5 \quad -6 \\
 -4 \quad 8 \quad -12 \\
\hline
1 \quad -2 \quad 3 \quad -18
\end{array}
$$
← Signs alternate

13. Positive: 1 or 3; negative: 1

14. $P(-2) = -12$ and $P(-1) = 3$; -1.3

15.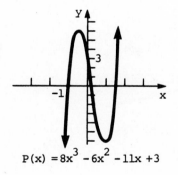

$$P(x) = 8x^3 - 6x^2 - 11x + 3$$

16.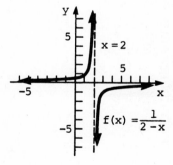

$$f(x) = \frac{1}{2 - x}$$

17.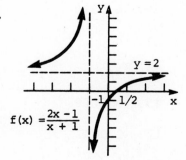

$y = 2$

$$f(x) = \frac{2x - 1}{x + 1}$$

18.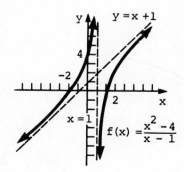

$y = x + 1$

$x = 1$

$$f(x) = \frac{x^2 - 4}{x - 1}$$

19.

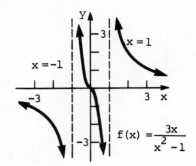

$$f(x) = \frac{3x}{x^2 - 1}$$

20.

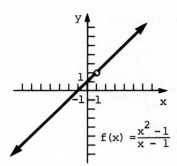

$$f(x) = \frac{x^2 - 1}{x - 1}$$

Chapter 10 Test Answers

1. True   **2.** $f^{-1}(x) = \dfrac{2 - x^5}{4}$

3.

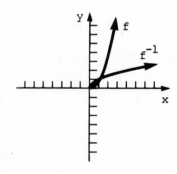

4.

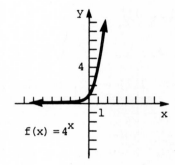

$$f(x) = 4^x$$

5.

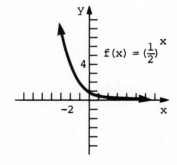

$$f(x) = (\tfrac{1}{2})^x$$

6.

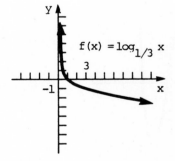

$$f(x) = \log_{1/3} x$$

7.

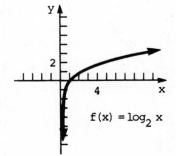

$$f(x) = \log_2 x$$

8. −2.2484   **9.** 4.0326

10. 2.6763   **11.** 4

12. −2   **13.** 3/2

14. −5   **15.** 3.3069

16. −.0605   **17.** 2.0437

18. 1.9208   **19.** {1/8}

20. $\{5\}$ **21.** $\{1.6397\}$ **22.** $\{8\}$ **23.** $\{8\}$

24. (a) 500 (b) 611 (c) $t = 17.3$ **25.** (a) $2617.29 (b) $2619.93

Chapter 11 Test Answers

1. $\{(-1, 4)\}$ **2.** $\{(2, 5)\}$ **3.** $\{(-1, 3)\}$ **4.** $\emptyset$ **5.** $\{(-8, 6)\}$

6. $\{(6, 5)\}$ **7.** $\{(1, -1)\}$ **8.** $\{(1, 1)\}$ **9.** $\{(-1, 2, 0)\}$

10. $\emptyset$ **11.** $\{(1, 3, 2)\}$ **12.** 6 liters   **13.** 40 miles per hour

14. 6 meters, 9 meters, and 15 meters   **15.** $\{(1, 1), (-1, -1)\}$

16. $\{(0, 3), (3, 0)\}$ **17.** $\{(3, 4), (-3, 4), (3, -4), (-3, -4)\}$

18.   **19.**

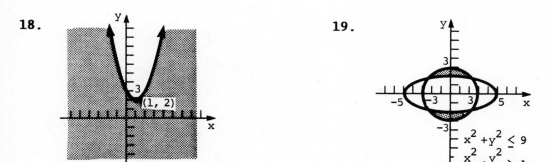

20. $\{(6/5, 6/5)\}$

Chapter 12 Test Answers

1. -48 **2.** -36 **3.** 37   **4.** -23 **5.** 0   **6.** -44

7. True; the corresponding rows and columns are interchanged.   **8.** False

9. False   **10.** True; each is equal to zero, since 2 rows in each are

identical.   **11.** -36 **12.** 56   **13.** $\{(3, 2)\}$ **14.** $\{(-1, 3)\}$

15. $\{(3, 2, -4)\}$ **16.** $\{(-1, 23, 16)\}$ **17.** $\{(2, -6)\}$

18. $\{(-3, 0)\}$ **19.** $\{(2, 1, -1)\}$ **20.** $\emptyset$

Chapter 13 Test Answers

1. 0, 3, 6, 9, 12   2. 9, 7, 5, 3, 1   3. 324, 108, 36, 12, 4

4. 4, 12, 36, 108, 324   5. 4, 8, 6, 12, 8   6. 79

7. $a_5 = -10$, $S_5 = 10$ 8. $a_5 = -4$, $S_5 = -2\ 3/4$ 9. $96x^{12}$ 10. 45

11. 2460   12. -3   13. 1024/5   14. 408   15. 25,600

16. $16k^4 - 96k^3 + 216k^2 - 216k + 81$ 17. $-\frac{21}{32}x^2y^5$

18. (i) $2^1 \overset{?}{=} 2^{1+1} - 2$

$2 \overset{?}{=} 4 - 2$

$2 = 2$

Therefore, S_1 is true.

(ii) Assume true for $n = k$:

$$2 + 4 + \cdots + 2^k = 2^{k+1} - 2$$
$$s + 4 + \cdots + 2k + 2^{k+1} = 2^{k+1} - 2 + 2^{k+1}$$
$$= 2 \cdot 2^{k+1} - 2$$
$$= 2^{k+2} - 2$$
$$= 2^{(k+1)+1} - 2$$

Therefore, if S_k is true, then S_{k+1} is true.

By mathematical induction, S_n is true for all positive integers.

19. 3024   20. 35   21. 1   22. 72   23. 360   24. 792

25. 2/3   26. 2/3   27. 1/2   28. 0   29. 1/6   30. .88